Metal Materials and Engineering

Metal Materials and Engineering

Edited by **Howard Currant**

CWILLFORD PRESS

New York

Published by Willford Press,
118-35 Queens Blvd., Suite 400,
Forest Hills, NY 11375, USA
www.willfordpress.com

Metal Materials and Engineering
Edited by Howard Currant

International Standard Book Number: 978-1-68285-035-0 (Hardback)

Contents

Preface

The prime focus of materials engineering is the discovery and applications of materials. Properties of metals such as malleability and ductility have rendered them useful in diverse areas and there has been tremendous progress in their study. Latest researches related to topics like structure, crystallography, bonding, synthesis and processing of gold and other metals have been discussed thoroughly in this book. It is highly recommended for students of diverse branches of engineering, materials science and associated disciplines. Also included within this book are various innovative topics which enthusiastic students and scholars can take up for research.

This book is the end result of constructive efforts and intensive research done by experts in this field. The aim of this book is to enlighten the readers with recent information in this area of research. The information provided in this profound book would serve as a valuable reference to students and researchers in this field.

At the end, I would like to thank all the authors for devoting their precious time and providing their valuable contributions to this book. I would also like to express my gratitude to my fellow colleagues who encouraged me throughout the process.

Editor

Conference report: gold highlights at the International Conference on Nanomaterials and Nanotechnology 2011 (ICNANO-2011) in Delhi, India, 18–21 December 2011

Sónia A. C. Carabineiro

Nano-biomaterials and biomedicals

The conference started with a plenary lecture from one of the conference chairs, *Anthony Turner* (Linköping University, Sweden), who spoke about 'Nanomaterials for biosensors and bioelectronics' and showed that several sensors contain gold films or gold surfaces. In fact, in surface plasmon resonance imaging, the sensor surface is almost always a thin layer of gold. Most immobilisation techniques involve the first layer of a chemical linker directly bound to the gold surface, allowing subsequent anchoring of molecules of interest [1]. Gold on glass electrodes are also starting points for the fabrication of artificial enzyme electrodes [2] and Au sputtered layers in the template synthesis of polyaniline nanostructures (Fig. 1) [3]. Turner was the winner of the Acharya Vinoca International award 2011 for his work on biosensors and bioelectronics (Fig. 2). This award is named after Vinayak Narahari Bhave, an Indian advocate of non-violence and human rights, and is given annually by the Vinova Research Institute of India, for notable and outstanding research in *Materials Science and Technology.*

Ashutosh Tiwari (from the same university) was the winner of the Nano Award 2011 (Fig. 2), for his work on gold nanobioelectronics. This award is honoured annually by the Vinova Bhave Research Institute for notable and outstanding research in the field of *Nanoscience and Nanotechnology.* Tiwari's plenary lecture dealt with the fabrication on an amperometric biosensor for the quantitative determination of urea in aqueous medium using hematein, a pH-

S. A. C. Carabineiro (✉)
Laboratory of Catalysis and Materials, Department of Chemical Engineering, Faculty of Engineering, University of Porto, Rua Dr. Roberto Frias, s/n, 4200-465 Porto, Portugal
e-mail: scarabin@fe.up.pt

sensitive natural dye [4]. The urease (Urs) was covalently immobilised onto an electrode made of gold nanoparticles functionalized with hyperbranched polyester-Boltron® H40 (H40–Au) coated onto an indium–tin oxide (ITO)-covered glass substrate (Fig. 3). Chitosan/gold–MPA nanocomposites for sequence-specific oligonucleotide detection were also mentioned. A 20-mer single-stranded oligodeoxyribonucleotide was covalently probed onto the nanocomposite electrode, made up of an ITO glass surface coated with chitosan, which is bonded with carboxyl functionalized thiol capped gold nanoparticles [5]. TEM micrograph of the composite electrode showed that gold nanoparticles had the diameter ranging from 4 to 16 nm.

Robert Gengan (Durban University of Technology, South Africa) spoke about the preparation of novel 3-amino-9-ethyl carboazole functionalized gold nanoparticle synthesis. The cytotoxicity of aflatoxin B_1, 3-amino-9-ethylcarboazole and gold nanoparticles on A549 cell lines was determined using a bioassay.

Soma Chattopadhyay (Illinois Institute of Technology, USA) spoke about the synthesis and EXAFS study of Au core–Ag shell nanoparticles inside unmodified horse spleen apoferritin protein, with Au core diameter smaller than 2 nm.

Alok Pandya (Gujarat University, India) talked about a water-soluble glucose biosensor based on an ultrasensitive boronic acid-calix[4]arene functionalized gold nanoprobe. This molecular receptor shows naked eye colour change from pink to blue due to aggregation, as confirmed by other techniques.

Neelam Verma (Punjab University, India) spoke about gold and silver nanoparticles used as sensors for electrochemical detection of organophosphorous pesticides. The lowest detection limit achieved for methyl parathion detection was 0.25 ppb, and for chlopyriphos, it was 1 ppt.

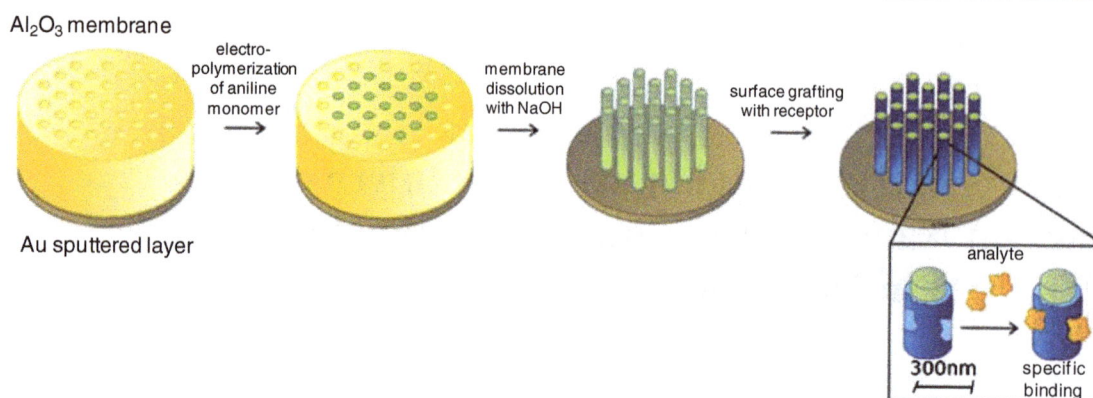

Fig. 1 Scheme of the template synthesis of polyaniline nanostructures (adapted from [3])

Minni Singh's oral presentation (from the same university) was about the enhanced response in electrochemical detection of glucose using model enzyme glucose oxidase by incorporating bimetallic Au–Pt nanoparticles in the conducting immobilisation matrix polypyrrole. The improved sensitivity of detection is particularly important for clinical diagnostics, food safety and environmental applications.

Dinesh Kumar (Banasthali University, India) spoke about core–shell (SiO_2@Ag/Au), metallic (Ag/Au) and capped metallic (Au/Ag@citrate) nanomaterials used as sensors for heavy metal ions (Cd(II), Zn(II), Fe(III) and Pb(II)) detection in water. These materials showed good performance and were simple, fast and possible to use by non-experts.

C.S. Pundir (Maharshi Dayanand University, India) talked about the immobilisation of oxalate oxidase onto gold nanoparticle–porous $CaCO_3$ microsphere hybrid encapsulated in silica sol and deposited on a Au electrode, for amperometric determination of oxalate in biological fluids (urine, plasma, fruits and vegetables). The biosensor showed to be reliable, have high sensitivity and was able to measure concentrations of oxalate as low as 1 µM.

Priyanka Sharma (Institute of Microbial Technology, India) presented a poster on the synthesis and characterisation of gold-coated iron oxide core–shell nanoparticles functionalised with receptor molecules (antibodies or proteins) to be used as functional biomaterials. They proved to be viable for exploiting the gold surface protein-binding reactivity for bioassay and the iron oxide core magnetism for magnetic bioseparation.

A. Chopra (from the same institute) presented a poster on a novel low-cost laser-ablated gold sensor modified with multiwalled carbon nanotubes using cysteamine. The presence of nanotubes on the surface of the Au electrode improved the direct transfer of glycated haemoglobin due to its intrinsic activity and a linear reproducible response was obtained. The method presented proved to be selective, sensitive, inexpensive and promising since the presently available tests for this haemoglobin are quite complicated and costly.

Deepika Bhatnaga's poster (Central Scientific Instruments Organization, India) and *N. Chauhan's* poster (Kurukshetra University, India) also reported on the use of multiwalled carbon nanotubes to modify a gold screen-printed electrode for direct electrochemistry and label-free detection of haemoglobin, and for a nanocomposite of Au nanoparticles and polyaniline for improved amperometric determination of lysine, respectively.

Prem Pandey's poster (Banaras Hindu University, India) reported on a prussian blue/gold nanoparticle nanocomposite to be used in the electrocatalytic determination of glutathione, with promising results.

Bindu Sharma's poster (Karnatak University Dharwad, India) dealt with the synthesis and optical characterization of biocompatible fluorescent gold nanoparticles, while *Krishnan Anand* (Durban University of Technology, South Africa) presented a poster on thiol-dual ligand gold nanoparticles (synthesis and bioassay).

Shobhana K. Menon (Gujarat University, India) presented a poster on a rapid detection of codeine using gold

Fig. 2 Anthony Turner (*left*) and Ashutosh Tiwari (*right*) from Linköping University, Sweden, receiving the Acharya Vinoca International Award 2011 and the Nano Award 2011, respectively, for their work on gold biosensors and bioelectronics (see text for details)

Fig. 3 a Preparation of hyperbranched gold (H40–Au) nanoparticles and **b** fabrication of H40–Au/ITO and Urs/H40–Au/ITO electrodes (adapted from [4])

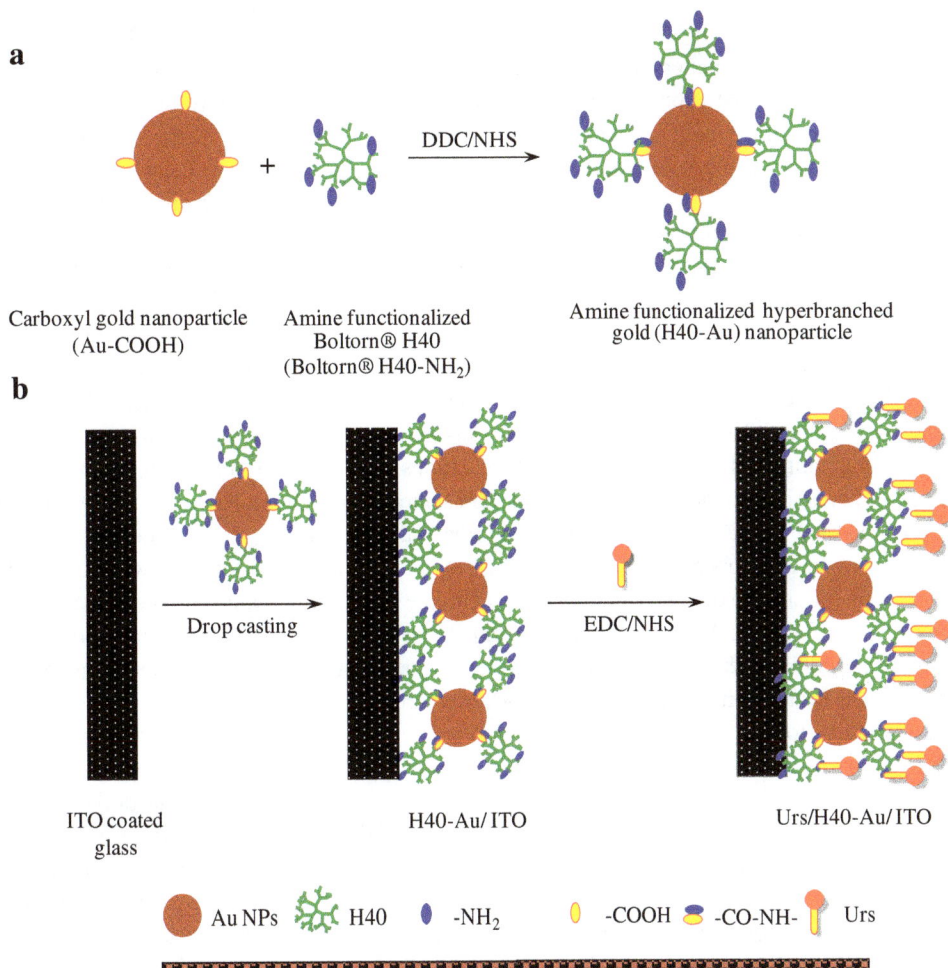

method using chitosan as a reducing/capping agent. These Au nanoparticles of ~10 nm were then used as carriers for the controlled delivery of the antituberculosis drug rifampicin and its catalytic activity was examined for gram-positive and gram-negative bacteria with promising results.

Balu A. Chopade's group (University of Pune, India) presented a poster on the use of *Dioscorea bulbifera* tuber extract (also called 'air potato', a plant with therapeutic use in Indian and Chinese traditional medicine) mediated synthesis of gold nanoparticles (gold nanotriangles, nanoprisms, nanotrapezoids and nanospheres). These bioreducted gold nanoparticles showed potent anticancer activity against three cancer cells lines, namely HL60, MCF7 and HeLa.

S.K. Pandey (Punjab University, India) presented a poster on gold nanoparticles induced chemiluminescence for VI antigen detection (a virulence factor of typhoid fever caused

nanoparticles as a probe. In this work, nanoparticles at nanomolar concentration can be clearly observed by the naked eye and allow sensitive detection with minimal consumption of drug samples from a crime scene. This can have interesting applications in forensic science since codeine and heroin drugs (Fig. 4, left and middle) are rapidly metabolised to morphine (Fig. 4, right) and the detection of the latter alone does not provide complete information.

Arghya Bandyopadhyay (University of Kalyani, India) presented a poster on the synthesis of nanofilm assemblies of gold protected by monodispersed iron oxide core metal shell magnetic nanoparticles. This novel nanocomposite may be used to develop a biosensor for detecting biomolecules quantitatively and qualitatively.

S. Malathi (University of Madras, India) presented a poster on the synthesis of gold nanoparticles by a green

Fig. 4 Molecular structures of codeine (*left*), heroin (*middle*) and morphine (*right*) drugs

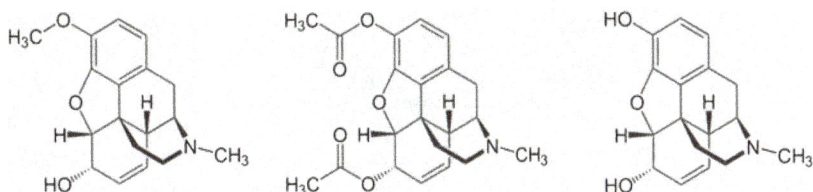

by *Salmonella enterica* serovar Typhi that is a life-threatening systemic infection and a major public health problem in developing countries). The proposed method is promising for the determination of clinically important bioactive analytes and may prove to be valuable in detecting important virulence strains of pathogens.

Saptarshi Chatterjee (University of Kalyani, India) presented a poster on the successful preparation of glutathione-functionalised gold nanoparticles used on a novel method of transformation of plasmid DNA in *Escherichia coli*. The process is less time consuming and increased the transformation efficient when compared to conventional methods. Plus, the Au nanoparticles are non-toxic making this gene delivery method suitable for biotechnological applications.

S. Lata (Maharshi Dayanand University, India) presented a poster dealing with nickel oxide nanoparticles/carboxylated multiwalled carbon nanotubes/polyaniline hybrid films electrodeposited on the surface of a gold electrode. Cytochrome c oxidised obtained from goat heart was covalently immobilised onto the hybrid film in order to construct an enzyme electrode. A biosensor was further constructed, which gave accurate and satisfactory results for the determination of cytochrome c on different serum samples.

Nano-fabrication, characterization and properties

V. N. Bhoraskar's plenary lecture (University of Pune, India) was about the synthesis and modification of Au, Ag, Cu, CdS and Ni nanoparticles by electron irradiating the respective chemical solutions with 6.5 MeV electrons and also the thin films of the respective elements by 15 keV electrons [6]. The average size of the particles could be tailored in the range of 50–130 nm for gold (Fig. 5). It was shown that Au nanoparticles could diffuse in polymers up to a depth of ~2 μm, using this method.

Habib Ullah's oral presentation (Pusan National University, Korea) concerned a work carried out in collaboration

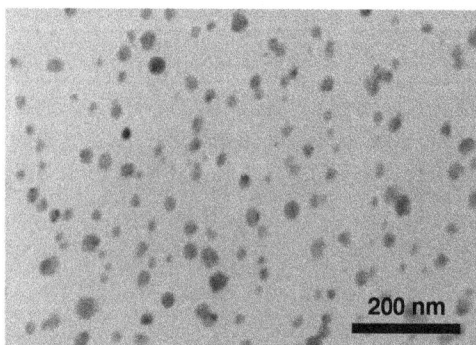

Fig. 5 TEM image of a 15 keV electron-irradiated Au coating (adapted from [6])

with the American International University of Bangladesh, dealing with water-soluble gold nanoparticles coordinated to poly(vinylpyrrolidone) (PVP). The final gold colloids were very stable as the PVP molecules coordinated through the C–N and C=O sites, instead of the C=O site alone.

Ali Ayati's oral presentation (Ferdowsi University of Mashhad, Iran) concerned the synthesis of stabilised 45 nm gold nanorods generated by a green chemistry type process using $H_6[PMo_9V_3O_{40}]$, a harmless polyoxometalate (POM), under UV irradiation and in the absence of any surfactant or seed. This POM plays the role of a photocatalyst, reducing agent and efficient stabiliser.

Biswajit Chowdhury (Indian School of Mines, India) described the preparation of ceria-based mixed oxide solid solutions having nanocrystallinity, where not only the pore size but also the defect sites can be created in the mixed oxide by doping lower valent cations. The gold nanoclusters on the oxide surface can dissociate the chemisorbed oxygen which diffuses through the defect oxides, providing excellent activity towards benzyl alcohol oxidation using molecular oxygen.

Sajid Ali Ansari (Aligarh Muslim University, India) presented a poster on polyaniline silver and gold composites prepared by in situ polymerization of aniline and ammonium peroxydisulphate as oxidising agent and Ag and Au colloidal nanoparticle solutions (reduced with sodium borohydride). These composites showed to have good optical and dielectric properties.

S. Tripathi (UGC-DAE Consortium for Scientific Research, India) in a collaborative work with the Karlsruher Institut für Technologie in Germany and the Catholic University of Leuven, Belgium, presented a poster with results of structural and magnetic characterisation of Co/Au system with two different compositions where Co was grown as a discontinuous layer sandwiched between a pair of Au layers. Interesting magnetic and structural transformations of the layers were observed due to the formation of a non-magnetic phase at the interface with temperature increase.

Parul Khurana (Banasthali University, India) had a poster on the synthesis and plasmonic properties of Au–Ag core–shell nanostructures. Ag and Au nanoshells were grown on silica microspheres with diameters ranging from 50 to 60 nm. Reduction of Au and Ag precursors onto gold decorated microspheres resulted in increasing gold coverage on the silica core. The product may also have many potential applications in optical, magnetic, biochemical and biomedical fields.

Sheenam Thatai (from the same group) had a poster dealing with the detection of Cd^{2+}, Zn^{2+} and Pb^{2+} ions in water using Au@citrate and SiO_2@Au nanoparticles as highly selective and sensitive nanosensors. The presence of metal ions changes the colour of the solution, visible to the naked eye, and causes aggregation of nanoparticles.

F. Hashemi (Khayyam University, Iran) presented a poster on theoretical investigations of the substituent effect on the molecular wire, Au/guanine/Au, performed using density functional calculations by considering the influence of an external electric field. The obtained results showed that an organic molecule can be functionalised as a molecular diode.

Nano-advanced materials

Palani Barathi (Velore Institute of Technology University, India) spoke about a nickel hexcyanoferrate-modified disposable gold nail electrode for electrocatalytic response of hydrazine in neutral pH. The sodium sulphate–sodium acetate pH 7 solution showed a well-defined, reversible and stable voltammetric responses and enhanced catalytic activity towards hydrazine without any excess alkali metal cation added solution. Under optimal conditions, the modified electrode showed a linear response for hydrazine in the concentration range from 0.5 to 10 mM.

S.K. Boruah (Gauhati University, India) presented a poster dealing with green synthesis of gold nanoparticles using a simple, fast, low-cost and economical technique, using fresh leaves and buds of *Camellia sinensis* tea (Fig. 6). The polyphenols present in the young leaves of tea extract reduce $HAuCl_4$ at room temperature. The core size of gold nanoparticles decreases as the amount of tea extract increases. Particles ranging from 9 to 12 nm were obtained.

The poster of *M. Yazdani* (Shiraz University, Iran) was about the synthesis of gold nanoparticles with imidazolium-based ionic liquids, which show preferential binding affinity towards gold crystal surfaces. 1-Dodecyl-3-methyl imidazolium tryptophan (long chain) and 1-ethyl-3-mehyl imidazolium tryptophan (short chain) were used, in order to test the effect of steric repulsion on the gold nanoparticles stability, being shown that the longer chain ionic liquids produced better results. Those nanoparticles showed to have a hydrophilic nature.

B. Batra (Maharshi Dayanand University, India) presented a poster on immobilisation of tyramine oxidase (purified from black gram seeds) on citrate-coated silver nano particles, deposited on a Au electrode, through cysteine layer, to obtain a tyramide biosensor that showed good performance in the detection of tyramide in beer and sauce, with a detection limit of 0.01 mM. Its advantage is that it does not suffer from leaching of the enzyme and measures this compound specifically.

A. Kedia (University of Delhi, India) presented a poster describing a simple, versatile and environmentally friendly one-step room temperature chemical synthesis route for the preparation of polyvinyl pyrrolidone functionalised size/shape-controlled gold nanoparticles by in situ polymerisation of the N-vinyl pyrrolidone monomer using $HAuCl_4$ as the oxidant. Water molecules act as a nucleophile that attacks the gold–vinyl complexes.

Ida Tiwari (Banaras Hindu University, India) presented a poster dealing with the addition of gold nanoparticles to multiwalled carbon nanotubes which were used to modify a glassy carbon electrode that showed good electrocatalytic activity towards the oxidation of NADH enzyme, better adhesion properties, good stability and no leaching of nanocomposite, thus being a promising electrochemical sensor for this enzyme in the future.

A collaborative work of *Naheed Ahmad* et al. (Patna University, India) and the Universities of Magadh, India, and of Aveiro, Portugal, dealt with the rapid green synthesis of silver and gold nanoparticles from the biowaste of pomegranate fruit (*Punica granatum*) at room temperature. The extracts of all parts of the fruit have therapeutic properties and the biosynthetic products along with reduced cofactors play an important role in the reduction of the Au salt to nanoparticles of ~10 nm, which showed to be quite stable.

Fig. 6 *C. sinensis* leaves and flower used in tea

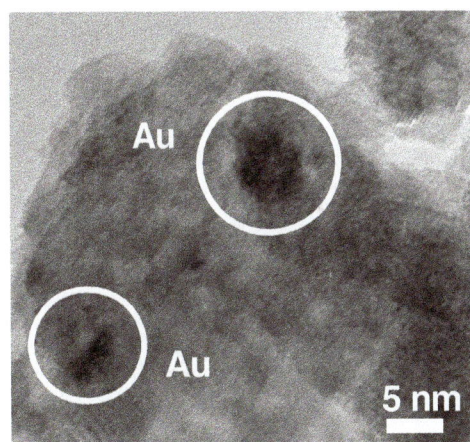

Fig. 7 HRTEM image of a Ce–Mn–O support (with a molar percentage of 30 % Mn and 70 % Ce) with gold nanoparticles (adapted from [7])

Sónia Carabineiro (University of Porto, Portugal) presented two posters dealing with gold nanoparticles supported on oxides, used for CO oxidation, namely, Ce–Mn–O composite materials. It was shown that the addition of Ce to Mn_xO_y produced supports with larger surface area, which led to smaller gold nanoparticles (Fig. 7) and consequentially improved catalytic activity. Another poster dealt with the heterogenation of organometallic complexes of several metals on carbon materials, to be used as catalysts for alkane oxidation. Heterogenised gold complexes showed promising results.

Deepshikha (Amity University, India) had a poster on the synthesis of polymer nanocomposites based on nanostructured polyaniline, gold nanoparticles and graphene nanosheets, by in situ polymerization. The nanodispersion could also be electrodeposited to produce a uniform nanofilm on an indium tin oxide surface. These composites showed good electrochemical properties and conductivity.

M. Misra's poster (Central Scientific Instruments Organization, India) also dealt with indium tin oxide, used as substrate for gold nanoparticles caped TiO_2 nanorod arrays for use in dye-sensitised solar cell technology. The Au nanoparticles on TiO_2 strongly absorb light, due to the localised surface plasmon resonance, and thereby promote light absorption of the dye.

J.P. Singha (King Saud University, Saudi Arabia) presented a poster dealing with the immobilisation of Pt, Au and Pt–Au nanoparticles on highly ordered TiO_2 nanotubes by pulse electro-deposition. The electrochemical behaviour of these composites was examined through the study of oxygen reduction.

Isha Mudahar (Khalsa College, India) presented a poster dealing with density functional calculations on pure dimers (C_{60}–C_{60}) and dimers doped with N and B ($C_{60-x}N_x$–$C_{60-x}B_x$) with gold contacts, in order to investigate the charge transfer in molecular nano junctions across C_{60} dimers. The results obtained suggest that there is a charge transfer from C to Au atoms.

Acknowledgments I am grateful to Prof. Anthony Turner for providing me the slides of his plenary lecture. Fundação para a Ciência e Tecnologia (FCT) is acknowledged for funding (CIENCIA 2007 programme and project PTDC/QUI-QUI/100682/2008, financed by FCT and FEDER in the context of Programme COMPETE).

References

1. Scarano S, Mascini M, Turner APF, Minunni M (2010) Review: surface plasmon resonance imaging for affinity-based biosensors. Biosens Bioelectron 25:957–966
2. Lakshmi D, Withcombe MJ, Davis F, Chianella I, Piletska EV, Guerreiro A, Subrahmanyam S, Brito PS, Fowler SA, Piletsky SA (2009) Chimeric polymers formed from a monomer capable of free radical, oxidative and electrochemical polymerisation. Chem Commun 9:2759–2761
3. Berti F, Todros S, Lakshmi D, Chianella I, Ferroni M, Piletsky SA, Turner APF, Marrazza G (2010) Quasi-monodimensional polyaniline nanostructures for enhanced molecularly imprinted polymer-based sensing. Biosens Bioelectron 26:497–503
4. Tiwari A, Aryal S, Pilla S, Gonga S (2009) An amperometric urea biosensor based on covalently immobilized urease on an electrode made of hyperbranched polyester functionalized gold nanoparticles. Talanta 78:1401–1407
5. Cao S, Mishra R, Pilla S, Tripathi S, Pandey MK, Shah G, Mishra AK, Prabaharan M, Mishra SB, Xin J, Pandey RR, Wu W, Pandey AC, Tiwari A (2010) Novel chitosan/gold-MPA nanocomposite for sequence-specific oligonucleotide detection. Carbohydr Polym 82:189–194
6. Mahapatra SK, Bogle KA, Dhole SD, Bhoraskar VN (2007) Synthesis of gold and silver nanoparticles by electron irradiation at 5–15 keV energy. Nanotechnology 18(13):135602
7. Carabineiro SAC, Silva AMT, Dražić G, Tavares PB, Figueiredo JL (2012) CO oxidation using gold supported on Ce-Mn-O composite materials. In: DiLoreto D, Corcoran I (eds) Carbon monoxide: sources, uses and hazards. Nova Publishers, Nova Science Pub Inc., New York

2

Alternative preparation of size-controlled thiol-capped gold colloids

**Martin Makosch · Václav Bumbálek · Jacinto Sá ·
Jeroen A. van Bokhoven**

Abstract Colloidal nanoparticles find application in chemistry, biology, and life science. We report an alternative preparation method for thiol-capped gold colloids by leaching of premade particles on a support. Via this method, monodispersed particles in the size of 2 to 2.5 nm can be obtained whereas the occurrence of bigger particles is restricted.

Keywords Gold · Colloid · Nanoparticle · Thiol-capped

An increasing interest in colloidal nanoparticles developed over the past years as an ever-growing number of applications in chemistry, biology, and life science emerged [1]. In biology and life science, these colloids are used for labeling [2], (drug) delivery [3], heating [4], and sensing [5] whereas supported [6] or unsupported [7] nanoparticles in chemistry are used especially in the field of catalysis. Although numerous preparation methods for colloidal nanoparticles exist [8–10], scientists still look for alternative routes as the demand for small monodispersed nanoparticles in solution is high. The challenge in the preparation of colloid gold nanoparticles is the size control especially in the regime below 5 nm as gold tends to sinter easily. Gold colloids are typically synthesized through reduction of a gold precursor in solution. We propose an alternative preparation method which yields monodispersed organic thiol-capped gold colloids in the size of 2 to 2.5 nm and prevents the occurrence of nanoparticles bigger than 5 nm.

M. Makosch (✉) · V. Bumbálek · J. A. van Bokhoven
Institute for Chemical and Bioengineering, ETH Zurich,
Wolfgang-Pauli Strasse,
8093 Zurich, Switzerland
e-mail: m.mkosch@gmx.net

J. Sá · J. A. van Bokhoven
Paul Scherrer Institute (PSI), Villigen, Switzerland

The method is based on the leaching of premade particles that are attached to a support. Because of the ability to control the gold particle size in supported gold [10], the method provides new opportunities to synthesize gold colloids of well-defined size.

The gold colloids were obtained from Au/Al_2O_3 colloid precursors which were prepared by a deposition precipitation with urea method according to [11]. The Au/Al_2O_3 colloid precursor and the colloid were characterized by transmission electron microscopy (TEM), atomic absorption spectroscopy, and high-energy resolution fluorescence-detected X-ray absorption spectroscopy (HERFD XANES). All experimental details of Au/Al_2O_3 colloid precursor synthesis, colloid preparation, and characterization can be found in the supplementary information. In short, a suspension of the Al_2O_3 support, urea, and an appropriate amount of $HAuCl_4 \cdot 3H_2O$ in water was prepared and stirred at 80 °C for 16 h under the exclusion of light. The powder was filtered, washed three times with water, and dried in vacuum overnight. After pretreatment in a flow of 5 % H_2/He at 300 °C, the Au/Al_2O_3 precursor was mixed with the corresponding organic thiol in degassed (with N_2) ethanol at a molar Au/thiol ratio of 1/20 for 16 h. After that, the catalyst was filtered, washed three times with EtOH, and dried in vacuum overnight. We refer to the 1-dodecanethiol-capped Au/Al_2O_3 colloid precursor as "Au_{C12}/Al_2O_3" in the following. The gold colloid was obtained by heating Au_{C12}/Al_2O_3 in toluene at the corresponding temperature. We refer to the 1-dodecanethiol-capped gold colloid as "Au_{C12} colloid" in the following.

The transition of the gold nanoparticles from the Al_2O_3 support to the liquid phase is visualized in Fig. 1. Figure 1 shows a characteristic TEM picture of Au_{C12}/Al_2O_3 before and the Au_{C12}/Al_2O_3 and the liquid toluene phase after heating in toluene at 80 °C.

Before heating in toluene, the thiol-capped Au nanoparticles were monodispersed and widely distributed on the Al_2O_3 support with a mean particle size of 2.8 nm.

Fig. 1 Characteristic TEM picture of Au_{C12}/Al_2O_3 before (**a**) and after (**b**) heating at 80 °C in toluene and the liquid phase after heating at 80 °C

After heating, the Al_2O_3 support showed only residual big nanoparticles (>5 nm) on the surface of the Al_2O_3 support whereas all smaller sized nanoparticles had disappeared. These particles were found in the toluene phase which turned from colorless to dark red, because of the gold colloid. The mean particles size was 2.7 nm suggesting, combined with the HERFD XANES data (supplementary information), that the gold nanoparticles leached into the liquid toluene phase as a whole. We investigated the effect of temperature on the colloid. Figure 2 shows the particle size distribution (left) and a characteristic TEM picture (right) of the Au_{C12} colloid after heating at 80 °C (a) and 100 °C (b) in toluene.

The particle size distribution obtained by heating the Au_{C12}/Al_2O_3 catalyst in toluene at 80 °C (a) resulted in a Gaussian shape with a maximum at 2.2 nm. The characteristic TEM picture shows spherical monodispersed gold nanoparticles of the same size. Heating the Au_{C12}/Al_2O_3 at 100 °C (b) in toluene resulted also in spherical Au_{C12} nanoparticles with a Gaussian particle size distribution, but much narrower compared to (a). The maximum was located at 1.8 nm. Additionally to the temperature effect, we investigated the effect of the molar Au/S ratio during the Au_{C12}/Al_2O_3 treatment and subsequent heating at 80 °C on the resulting Au_{C12} colloid. Table 1 summarizes the effect

Fig. 2 Particle size distribution (*left*) and characteristic TEM image (*right*) of the Au_{C12} colloid after heating at **a** 80 °C and **b** 100 °C

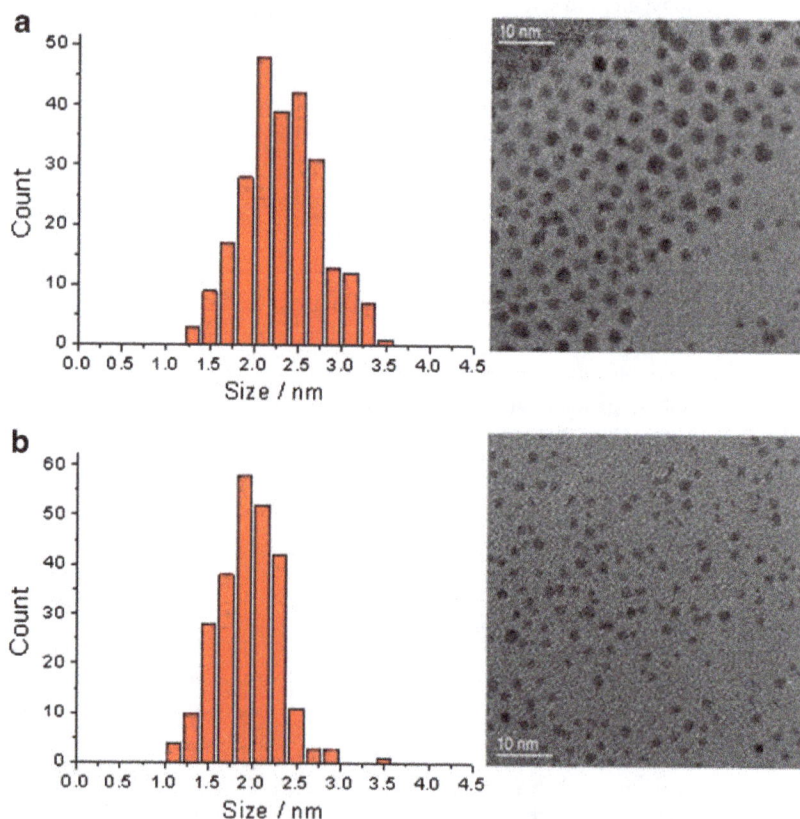

Table 1 Dependence of Au/S ratio on Au colloid yield for the Au_{C12} colloid after heating the Au_{C12}/Al_2O_3 colloid precursor in toluene at 80 °C

Molar Au/S ratio	Au colloid yield (%)
1:20	66.9
10:1	33.2
25:1	6.2
50:1	–
75:1	–

Table 2 Dependence of organic thiol on Au colloid yield after heating in toluene at 80 °C

Thiol	Au colloid yield (%)
1-Octadecanethiol	78.5
1-Dodecanethiol	66.9
1-Propanethiol	62.5
1,6-Hexanedithiol	11.5
1-Thioglycerol	7.3

Table 3 Dependence of solvent on Au colloid for the Au_{C12} colloid yield after heating in toluene at 80 °C

Solvent	Au colloid yield (%)
n-Hexane	74.6
Toluene	66.9
Ethanol	24.5
Water	23.4

on the yield of Au_{C12} colloid for different molar Au/S ratios of the Au_{C12}/Al_2O_3 colloid precursor.

Whereas no effect on the particle size was observed, we found that the gold colloid yield in toluene strongly depended on the molar Au/S ratio during the preparation of the Au_{C12}/Al_2O_3 colloid precursors. Whereas no leaching was observed for Au/S ratios of 75:1 and 50:1, gold nanoparticles were transferred to the liquid phase starting at ratios of 25:1. Higher gold colloid yields could be achieved via the application of different organic thiols, without affecting the size (Table 2). The maximum amount of gold that could be brought into solution was almost 80 %, which is likely because the bigger nanoparticles remain on the Al_2O_3 support as observed in Fig. 1b.

In general, thiols with a similar polarity as the solvent toluene (1-propanthiol, 1-dodecanethiol, 1-octadecanethiol) yielded higher gold colloid concentration, whereas thiols with an adjacent polar group resulted in lower concentrations. Within a row of the unpolar thiols, a longer $CH_3(C_nH_{2n})$ chain yielded a higher gold colloid concentration with a maximum of 78.5 % observed for 1-octadecanethiol in toluene. The same trend was observed within the polar thiols. A similar tendency was observed for the preparation of the Au_{C12} colloid in different solvents (Table 3).

The gold colloid yield strongly depended on the solvent used during preparation: the more polar the solvent and thus the more interaction of the $CH_3(CH_2)_{12}$ chain with the solvent, the higher the gold colloid yield. The maximum Au_{C12} colloid yield was obtained in the most unpolar solvent, n-hexane, whereas as much as 50 % less was observed in the most polar solvent, water. According to these observations, we propose a model for the fabrication of the size-controlled thiol-capped gold nanoparticles as depicted in Scheme 1.

Scheme 1 Schematic representation for the formation of thiol-capped gold nanoparticles

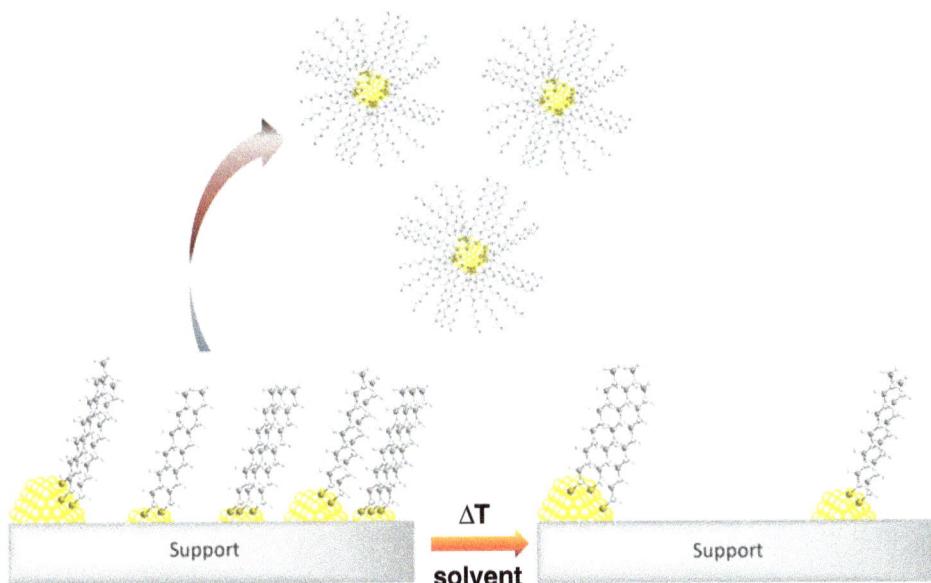

Due to the interaction of the long $CH_3(C_nH_{2n})$ chains with the solvent, the organic thiol-capped nanoparticles are transferred to the liquid phase in the same structure as they had on the support. The better the interaction between the organic thiols and the solvent, the higher the Au colloid yield, whereas this interaction is not strong enough to rip of particles bigger than 5 nm.

To summarize, we showed that monodispersed thiol-capped gold colloids in the range of 2 to 2.5 nm can be synthesized by initial treatment of Au/Al_2O_3 colloid precursors with organic thiols and subsequent heating in a solvent. Size control during preparation is given as particles bigger than 5 nm remain on the Al_2O_3 surface whereas all smaller sizes are transferred to the liquid phase. Variation of colloid precursor treatment and extraction temperature enables fine tuning of particle size whereas the yield of the gold colloid can be maximized via the right combination of organic thiol and solvent. This new route opens a new horizon for the production of colloidal particles in the size range below 5 nm.

References

1. Sperling RA, Rivera Gil P, Zhang F, Zanella M, Parak WJ (2008) Chem Soc Rev 37:1896
2. Agasti SS, Rana S, Park M-H, Kim CK, You C-C, Rotello VM (2010) Adv Drug Deliver Rev 62:316
3. Duncan B, Kim C, Rotello VMJ (2010) Control Release 148:122
4. Wilson BC (2010) Handbook of Photonics for Biomedical Science. CRC Press, Boca Raton
5. Dykman L, Khlebtsov N (2012) Chem Soc Rev 41:2256
6. Jia C-J, Liu Y, Bongard H, Schüth FJ (2010) Am Chem Soc 132:1520
7. Liang X, Wang Z-j, Liu C-j (2010) Nanoscale Res. Let 5:124
8. Xu B, Song RG, Wang C (2011) Adv Mater Res 415–417:648
9. Liu S, Chen G, Prasad PN, Swihart MT (2011) Chem Mater 23:4098
10. Bond GC, Louis C, Thompson DT (2006) Catalysis by gold. Imperial College Press, London
11. Hugon A, Delannoy L, Louis C (2008) Gold Bull 41:127

3

The fire assay reloaded

Paolo Battaini · Edoardo Bemporad · Daniele De Felicis

Abstract The fire assay process is still the most accurate and precise method for measuring the gold content in gold alloys. Scanning electron microscopy and transmission electron microscopy have been applied to observe the change in microstructure of the samples undergoing the fire assay process. The performed observations reveal that the microstructure of the specimen is more complex than expected. Before the parting stage, the specimen is not a perfect gold–silver binary alloy but contains also copper–silver oxides and other residual compounds. The parting stage appears to be a dealloying process leading to a nanoporous gold nanostructure. What observed after partition explains the evolution of the shape and colour of the specimen and may allow for a better comprehension of the procedure and an improvement in the method.

Keywords Fire assay · Nanoporous gold · Dealloying · Parting stage · Inquartation

Introduction

The fire assay process has been well known since ancient times as a method for measuring the gold content of precious alloys. In fact, it was first mentioned by the Egyptians on a cuneiform tablet that dates back to about 1360 B.C. One of the first written proofs of the fire assay process lies in the *De la Pirotechnia*, by the Italian Vannoccio Biringuccio [1], born in Siena in 1480. Even now, it is still the most accurate and precise method, besides being the most popular in the goldsmith's field. This notwithstanding, very little information is available about the microstructural evolution of the specimen undergoing the different steps of the process. The present work is aimed at describing this evolution by means of scanning electron microscopy, transmission electron microscopy, as well as other analytical techniques. The performed observations reveal some unexpected microstructural changes, which cast new light on the whole process and allow for understanding the remarkable effectiveness of the method. Knowing why each step of a procedure is performed in turn helps the operator to avoid making mistakes.

Materials and methods

Different methods have been applied to study the microstructural evolution of materials undergoing the fire assay process. The surface of the specimen has been investigated by scanning electron microscopy (SEM) equipped with electron microanalysis (energy dispersive spectrometry, EDS), to determine its chemical composition. The same techniques have been applied to study the related metallographic sections. The nanostructures detected during the analyses and the microstructural features have been explored by focused ion beam (FIB) coupled with electron beam for imaging, ion milling and deposition [2]. Some analyses on thin foils extracted by FIB have been performed by transmission electron microscopy (TEM). The crystal structure of the thin foils has been studied by selected area electron diffraction inside the TEM. A calorimeter has been adopted to study the transition which occurs during the last annealing stage of the sample.

Stages of the procedure, results and discussion

The results of the microstructural analysis of the samples at each stage of the fire assay process will be given. The fire assay method is described in the International Standard ISO 11426:1997. The present work is not aimed at giving the details of the process and the operating procedure which are well known [3].

P. Battaini (✉)
8853 Spa, Pero, Milan, Italy
e-mail: battaini@esemir.it

E. Bemporad · D. De Felicis
Mechanical and Industrial Engineering Department,
University of Rome "Roma Tre", Rome, Italy

Fig. 1 Lead pouch containing the specimen and Ag, closed and placed on a cupel made of magnesium oxide, ready to undergo the in-furnace cupellation process

Sampling and cupellation stages

The first stage of the fire assay process is sampling. Sampling is a crucial stage, which is carried out according to the kind of product to control—ingot, jewel, coin, etc. It will generally be a chip or a small fragment of a material. The sample will have to weigh about 250 mg to the maximum and to be weighed by means of precision balance sensitive to the hundredth of milligram.

Silver and lead are added to the specimen as described in the following. A sheet of pure lead of about 3 g is prepared for the assay of alloys whose titer is approximately 18 carats. The analyst makes an assumption about the kind of alloy the specimen has been taken from, then adjusts the weight of the lead sheet by increasing it by about 4 g, when dealing with white or red gold. The weight of the lead sheet may also be about 7 g for a 14-carat alloy. An amount of pure silver—at least 999.5 ‰ pure Ag, is prepared as well, according to the

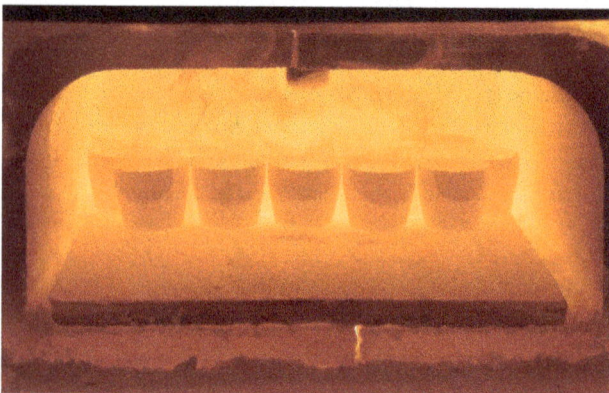

Fig. 3 At the end of the cupellation process, the beads remain on the cupels, gleaming and still

presumed gold content of the sample. This process is also known as inquartation and is aimed to dilute the gold in the sample. The small lead sheet, pouch-like folded is used to wrap up the sample and the silver. The pouch is then placed on a previously heated porous cupel (Fig. 1) made of magnesium oxides, already placed into the furnace at 1,150 °C.

After that, the door of the furnace is kept closed for a short period of time in order to allow the temperature to increase to 1,150 °C, then is left slightly opened to allow the oxygen of the air to enter. The cupellation process starts when the first oily droplets of litharge appear, floating on the cupel which absorbs them later. A small amount of litharge volatilizes releasing fumes (Fig. 2).

By the end of the process, an iridescent-banded film produced by extremely thin films of fluid litharge is observed on the surface of the residual bead. In the end, this thin film also

Fig. 2 Cupels containing the specimens in the furnace chamber at 1,150 °C. The fumes are due to litharge (lead oxides, which takes place because of the presence of air). On the front of the cupels, a dark stain is visible, caused by lead oxides and other non-noble metals (base metals) oxides, which are about to be absorbed by the cupel

Fig. 4 At the end of the cupellation process, the upper surface of the beads is bright and smooth. Very often, as shown in the figure, very different solidification structures are observed (fine grain, on the *left bead*, coarse grain on the *right bead*). The *lower part* of the figure shows the surfaces of the two beads previously in physical contact with the cupels. They are rough and matt

Fig. 5 Aspect of the cupel after the cupellation process. The greenish stain is due to the absorption of Pb and base metals oxides

absorption of oxides occurs among most of the section, as shown in Fig. 6. In the same figure, the acquisition positions of EDS spectra and the semi-quantitative results are also reported. Interestingly enough, a high Pb concentration and a significant amount of Cu are detected even into the cupel, which gives evidence of the efficacy of the absorption process. Silver is identified in the region where the bead has touched the cupel. In this area, the EDS microanalysis cannot reveal Au concentrations lower than 1 % in weight, also because of the concurrent presence of Pb. As a consequence, the presence of Au is not to be excluded—at least in this area, where it could be associated with Ag. In this case, a common error regarding the final measurement of the gold titer would occur, due to the absorption of gold by the cupel.

The final aim of the cupellation process is to obtain a binary gold–silver alloy, and it is achieved only if the requirements about the weight of inquarted lead and silver are satisfied and the oxide formation occurs correctly, with the appropriate oxygen contribution. At this stage, the operator's skill is fundamental.

However, it must be pointed out that if the alloy contains also Pt, Pd, Ir, Rh, Ru and Os, these elements remain in the bead together with Au and Ag. Ir, Rh, Ru and Os concentrate mainly at the bottom of the bead. In this work, this possibility is not considered and the details about how to deal with this case are not given [4].

In order to examine the microstructure of the bead, some SEM observations have been performed on its surface and metallographic sections.

disappears and the melted mass remains, gleaming and still (Fig. 3). Figure 4 shows the typical aspect of the beads obtained in this way.

At the end of the process, the cupel turns greenish due to the absorption of Pb and base metal oxides (Fig. 5). The

Fig. 6 The cupel section is mainly saturated with oxides. The chemical composition in the four positions marked by the *arrows* has been determined by EDS. This technique provides for semi-quantitative results. The element concentration changes from cupel to cupel. Ni and Zn are other frequently detected elements. Of course, the results can change from bead to bead

		Weight %		
	Point A	Point B	Point C	Point D
O	35.62	40.87	45.79	36.05
Na	0.17	-	1.22	-
Mg	36.08	52.14	45.43	31.11
Si	1.20	-	6.18	2.80
Fe	0.30	0.37	-	0.18
Cu	0.82	0.62	-	0.83
Ag	0.40	-	-	-
Pb	25.41	6.00	-	28.62
Ca	-	-	1.37	0.42

The upper part of the bead shows grains whose size changes significantly from specimen to specimen (Fig. 7). This behaviour could be reasonably ascribable to different causes, such as the dissimilar content of Ag and Au of the beads, the different solidification rates and the presence of trace elements which act as solidification nuclei. Carbon-rich residues are also detected on the surface.

The bottom of the bead, which touches the cupel, appears rough (Fig. 8), due to the imprint left by the granules of magnesium oxides of the cupel. Lead-base residues and other kind of oxides are found on this side of the bead, as detected by EDS, despite the brushing performed by the operator according to the procedure. This means that the surface cleaning is never accomplished. However, these residues are thought to be removed almost completely by the subsequent partition stage, as they are very superficial and the acid sweeps them away very easily. Recent studies support this idea,

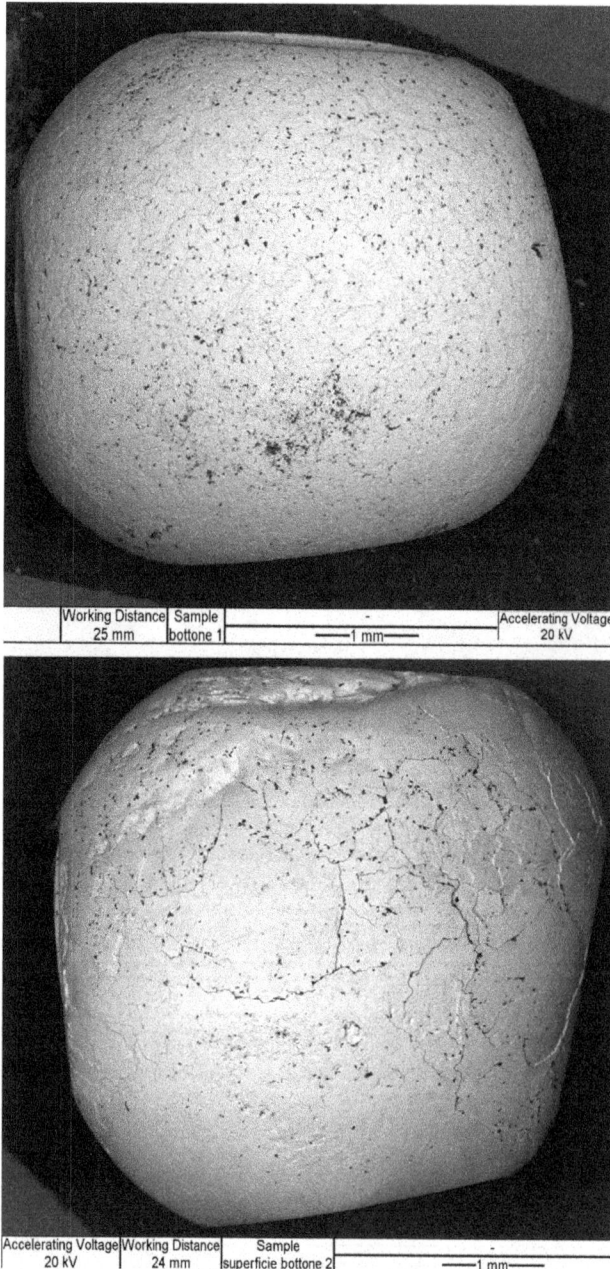

Fig. 7 SEM, backscattered electron image (BEI). Surface of the two beads. Different grain sizes are visible. The *small dark particles* are carbon-rich residues

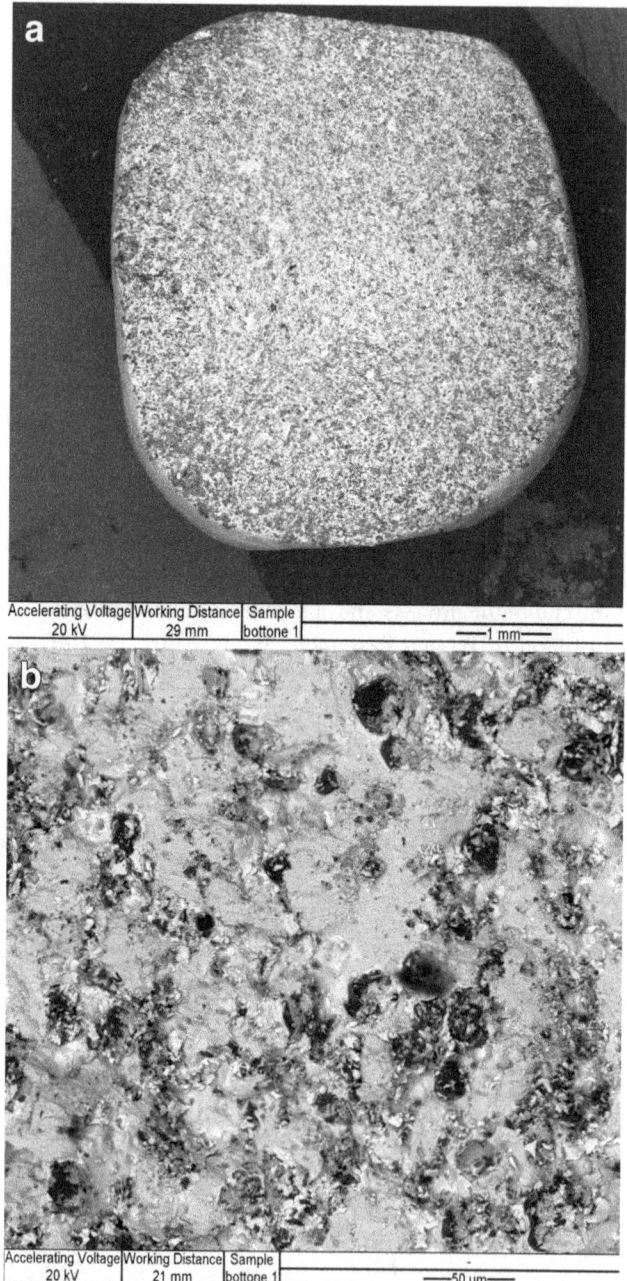

Fig. 8 SEM, backscattered electron image (BEI). **a** Lower surface of the bead. The surface is rough due to its physical contact with the cupel. Many oxides are steeped in it, even though the bead has been brushed, as per procedure. **b** Detail of the bottom of the bead. Oxides, residues of the cupel, scales of lead oxides (white areas) and carbon-rich residues are left despite brushing

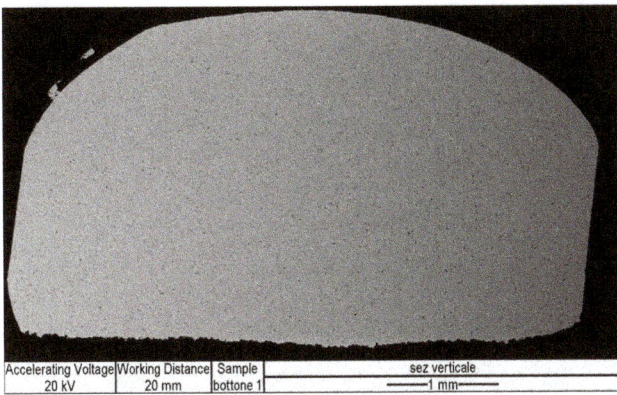

Fig. 9 SEM backscattered electrons image (BEI). Polished vertical metallographic section of the bead seen in the *upper* part of Fig. 7. Its microstructure is homogeneous, and shows a fine-net silver micro-segregation

showing that the cleaning of this side of the bead can cause only minor measurement errors [5]. However, the operator must be aware of that.

Figure 9 shows the aspect of a vertical section of the bead, perpendicular to the supporting plane of the cupel.

When using the backscattered electron signal of the SEM, a micro-segregation of Ag appears (Figs. 9, 10 and 11). The Ag weight concentration measured by EDS microanalysis along the segregation bands ranges between 74 and 76 % and decreases to about 70 % outside the segregation bands. Obviously, the concentrations reported here concern the bead considered as an example and are a function of the weight of the inquartation Ag used. As a consequence, they change from specimen to specimen.

Small isles mainly made of copper and silver oxides are also observed (Figs. 12 and 13). The horizontal section of the bead shows the same microstructural characteristics as the vertical one.

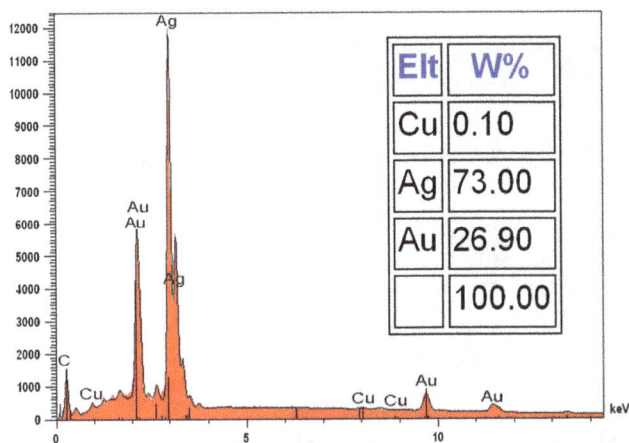

Fig. 11 SEM backscattered electrons image (BEI). Detail of Fig. 9. The fine network due to the micro-segregation of silver is visible. The weight concentration of silver on the boundaries of the net is about 75 %. The small dark isles are composed of Cu and Ag oxides. See their detail in Fig. 12. The morphology of segregation changes from bead to bead

At the end of this section, it is important to notice that the bead is not a homogeneous and pure gold–silver alloy, as expected, but contains some oxides on the bottom surface and inside its section.

Cornet stage

The bead is hammered, annealed and rolled to a thin strip. Then, the rolled section is flame annealed and

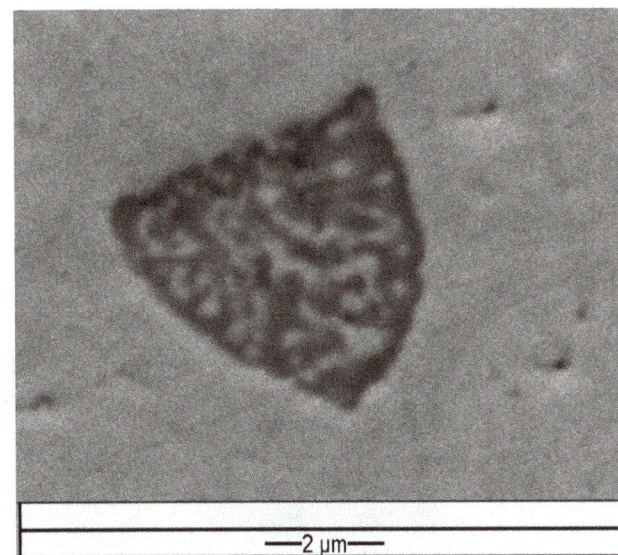

Fig. 10 EDS spectrum acquired on the metallographic section of Fig. 9. The corresponding semi-quantitative chemical analysis is given in the table. The bead contains Cu. The weight ratio Ag/Au is about 2.7, which is appropriate, as it should range between 2 and 3. Results change from bead to bead. The presence of other trace elements cannot be excluded, due to the relatively low sensitivity of the EDS technique

Elt	W%
Cu	0.10
Ag	73.00
Au	26.90
	100.00

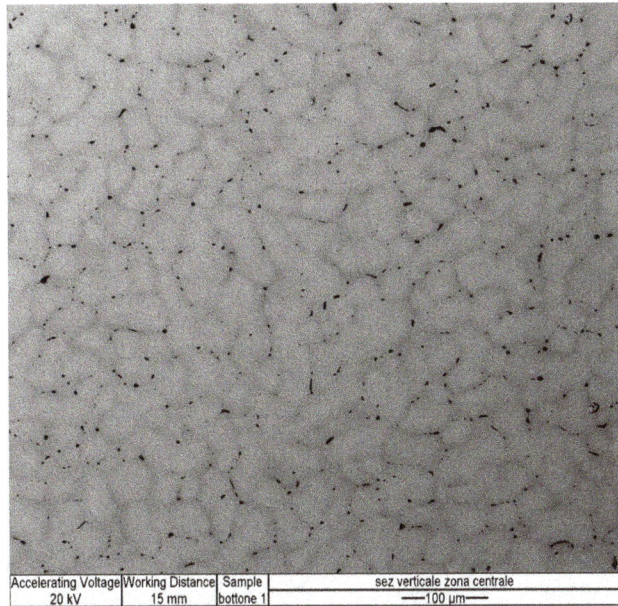

Fig. 12 Detail of one of the dark isles (*spots*) of Fig. 11, made of Cu and Ag oxides (see the correspondent EDS spectrum in Fig. 13)

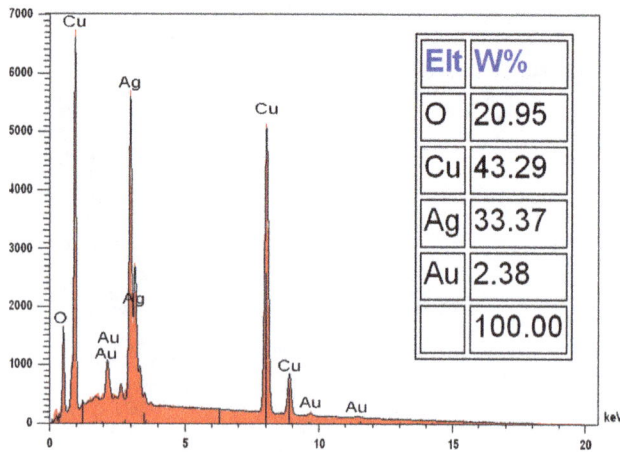

Fig. 13 Typical EDS spectrum acquired on the dark isles visible in Figs. 11 and 12. They are mainly made of Cu and Ag oxides

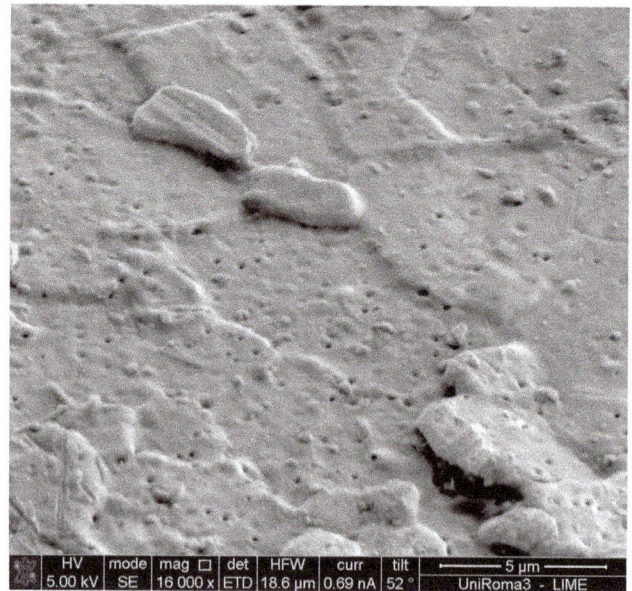

Fig. 15 SEM image of the surface of the work-hardened strip after flame annealing. The re-crystallization reveals the grains of the material. In addition, a dispersion of small voids is observed

subsequently rolled up to a cornet-like shape (Fig. 14). The rolling must not be too tight, so as to allow the wettability of the whole surface. Furthermore, particular attention must be paid in order to keep the previous bottom of the bead—the one contacting the cupel, to the outside. By doing this, the oxides seen on this side of the bead will be on the outer surface of the cornet. Now, the cornet is ready to be put into an acid solution in the following partition process.

An analysis by TEM has been performed on thin foils obtained from the surface of the annealed strip. The TEM thin foils are prepared by thinning the material by means of the so-called dual beam system (FEG-FIB) available in a focused ion beam apparatus [2]. Figure 15 shows the surface of the strip where the thin foils have been obtained. The presence of some grains in relief and the dispersion of fine round pores give evidence that annealing has taken place.

The effect of annealing on re-crystallization is clearly visible. The nano-voids created during dislocation coalescence characterizing the surface also affect the bulk, down to about 3 μm (Figs. 16 and 17). They are present in this zone because the plastic deformation of the metal is higher in this thin layer of the section, due to the rolling process.

The dislocation density decreases dramatically after flame annealing (Fig. 18), which is also supported by electron diffraction, as well as the twin crystals presence (Fig. 19). These observations show that the cause of the surface nano-voids is due to the re-absorption of dislocations.

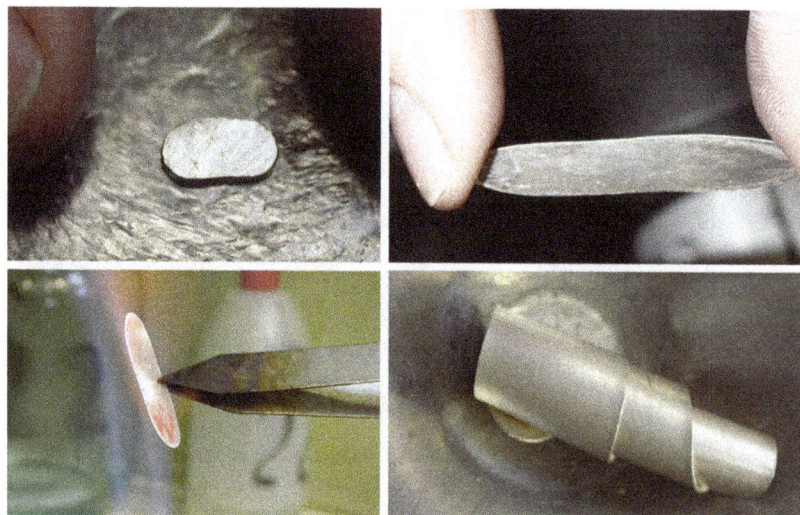

Fig. 14 The four stages of the cornet preparation are here represented. First the bead is hammered, flattened and rolled to a thin strip (top right), then flame annealed by means of a Bunsen apparatus and cornet shaped by hand

Fig. 16 At the end of the thinning procedure, the thin foil is pulled out of the specimen, ready to be analysed by TEM. The surface of the strip is the upper part of the thin foil

Parting stage

The parting stage allows the separation of gold from silver, providing a cornet made of gold only. The process occurs in two stages, by keeping the cornet in boiling nitric acid. The first stage makes use of nitric acid in 22° Bè (bè degrees) aqueous solution, the second one of nitric acid in 32° Bè aqueous solution (Fig. 20).

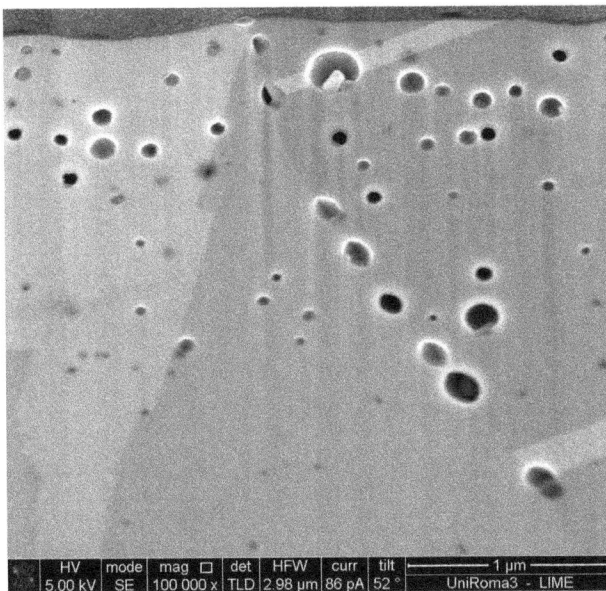

Fig. 17 SEM image of the thin foil obtained from the strip after flame annealing. See the grain re-crystallization and the presence of round nanovoids down to 3 μm in depth from the surface

At the end of the process, a brownish cornet—very difficult to handle, is obtained (Fig. 21). This cornet contains only gold. If too much silver, with respect to the initial specimen gold content, is added in the cupellation process, the cornet breaks into small pieces or even reduces to powder during partition. The microstructural evolution of the material during partition has long been unexplained. In fact, partition is simplistically referred to as a process in which silver is removed by means of nitric acid, which leaves only gold atoms inside the cornet. This description is not enough to describe thoroughly what happens. The reason why the cornet preserves its initial geometric shape (Fig. 21) is not clear. Since it is made of a quite homogeneous binary gold–silver alloy, a selective dissolution of silver should lead to a powder-like gold precipitation at most. Therefore, the reason why the cornet keeps its geometrical shape has been investigated by performing microstructural studies after partition.

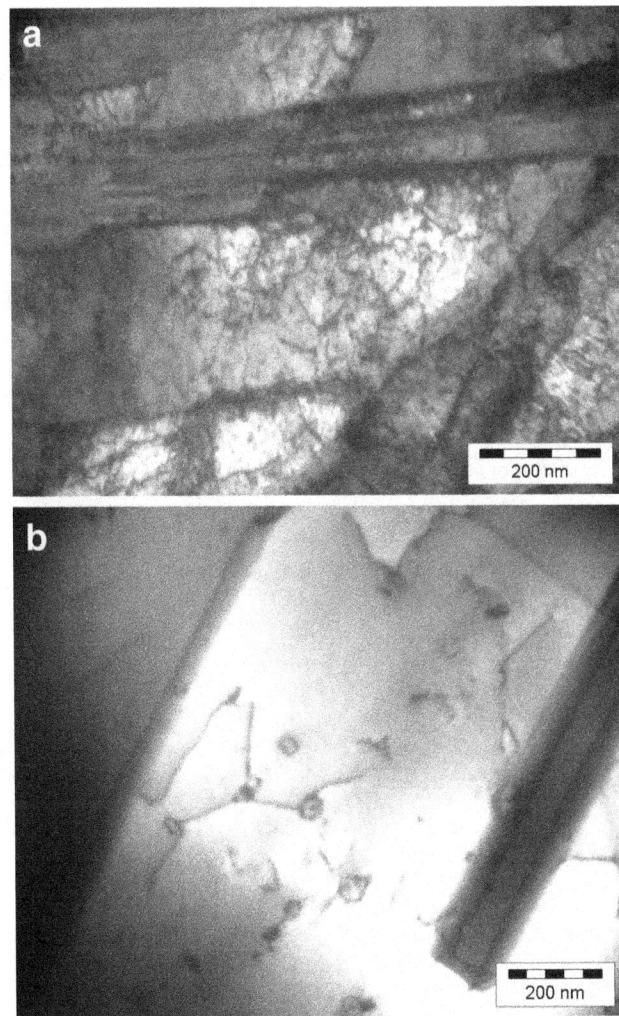

Fig. 18 TEM images. **a** Detail of the microstructure of the work-hardened strip. **b** Detail of the strip after flame annealing. The material re-crystallization causes a significant decrease in dislocation density. Very few dislocations and twin crystals appear in this image

Fig. 19 The electron diffraction pattern corresponding to the area in Fig. 18b confirms the considerable decrease in the dislocation density (the diffraction spots are dot-like and not elongated) and the presence of twin crystal, proved by twin spots next to the main ones

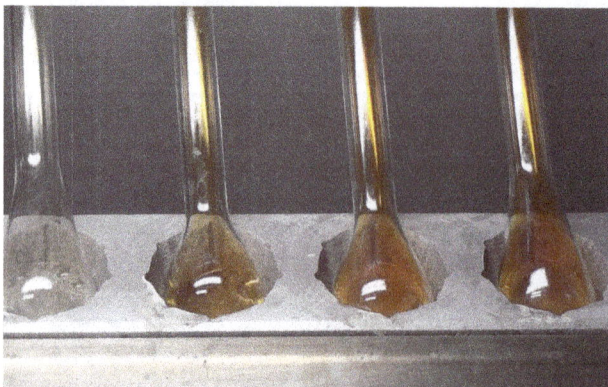

Fig. 21 The same cornet before and after partition is shown in the *upper and lower part* of the figure, respectively. See the size reduction that the cornet undergoes during the process. After partition, the cornet becomes typically brown, but preserves its initial shape

Figure 22 shows a detail of the cornet surface after partition (dark brown cornet in Fig. 21). The surface has the typical appearance of the nanoporosity obtained after dealloying of a Ag-Au alloy [6–8]. This microstructure is characterized by a considerable isotropy, not only at the surface of the cornet but also inside it (Figs. 23 and 24). The dealloying process and the microstructure of nanoporous gold have been studied and a mechanism of nanoporosity development has been proposed [9–17].

Another microstructural feature which turns out after the partition process is highlighted in Figs. 25 and 26. Along the cracks which reveal the inner part of the cornet, intergranular surfaces or better still the surfaces corresponding to the previous grain boundaries of the annealed strip are visible. This means that the transformation that occurred during partition

has taken place on a nanometric scale, preserving memory of larger sized structures such as the initial crystal grains. The gold atom movement giving rise to the nanoporous gold has occurred 'locally', without involving wider scale diffusion phenomena. This aspect is perfectly in agreement with the model describing the nanoporous gold genesis. Furthermore, the shrinkage of the cornet which preserves its morphology (Fig. 21) is a typical feature of the dealloying process [7].

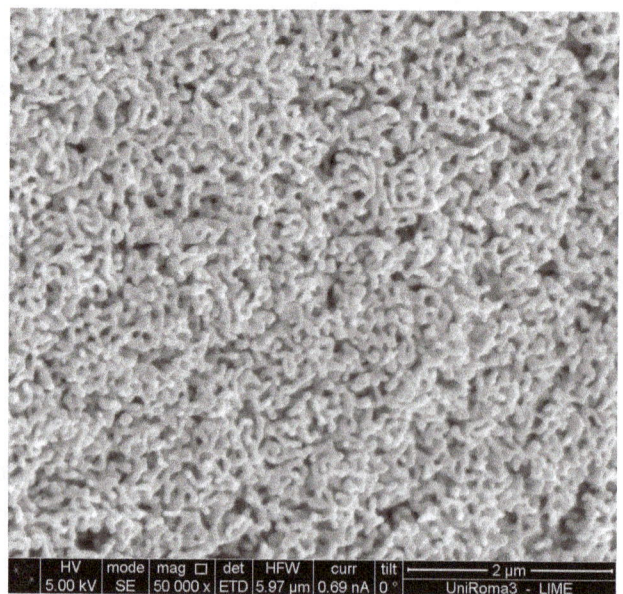

Fig. 20 Dissolution of silver during the first stage of partition, in 22° Bè boiling nitric acid. The amber fumes are produced by the dissolution of silver. The boiling acid is visible at the bottom of the flask

Fig. 22 Cornet surface after partition (brown cornet in the lower part of Fig. 21). A granular structure is observed, whose granule size is uniform

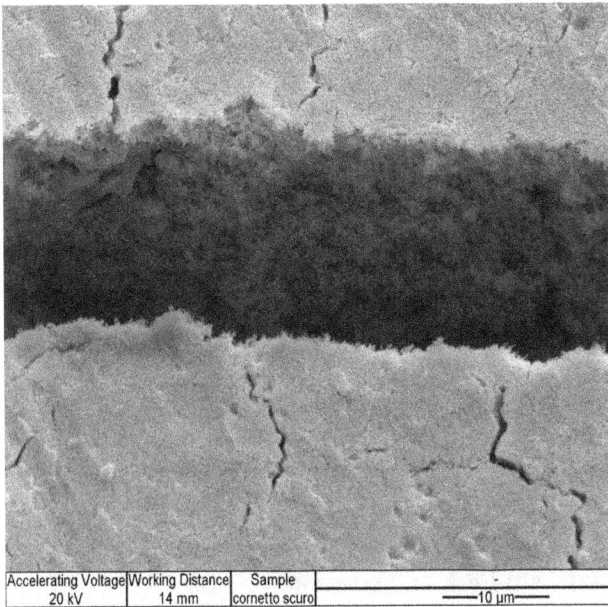

Fig. 23 SEM image. Fractures on the cornet surface after partition. The isotropy of nanoporosity is seen also inside the cornet

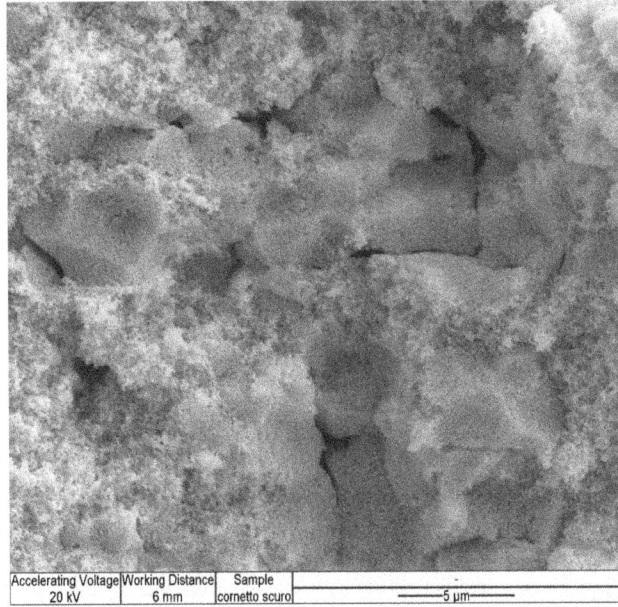

Fig. 25 SEM image. By breaking the cornet after partition, it is possible to detect the previous grain structure. The transformation of the material into a sponge structure has occurred on a nanometric scale, preserving the memory of the previous crystal grains. The grains appear detached from one another

In the light of the dealloying process, the geometry of the nanoporous gold (pore size, ligament size, gold cluster size) depends on the process of parting and on the initial silver–gold ratio. This means that good results are possible only if all the process parameters are respected. Furthermore, the final microstructure of the cornet after partition may change from a laboratory to another, due to slightly different equipments and procedures. It would be interesting to know if suggestions can be given in order to get the best results, for example reducing cracks in the nanoporous structure of the cornet.

In order to better understand the microstructure of the nanoporous cornet, a TEM thin foil has been prepared.

Figures 27 and 28 are TEM images of the nanoporous gold at different magnification. The nanoporous gold shows an inner fragmentation into gold nanocrystals, just a few nanometers sized. The nanocrystals are separated by twin crystals boundaries, as shown by selected area electron diffraction

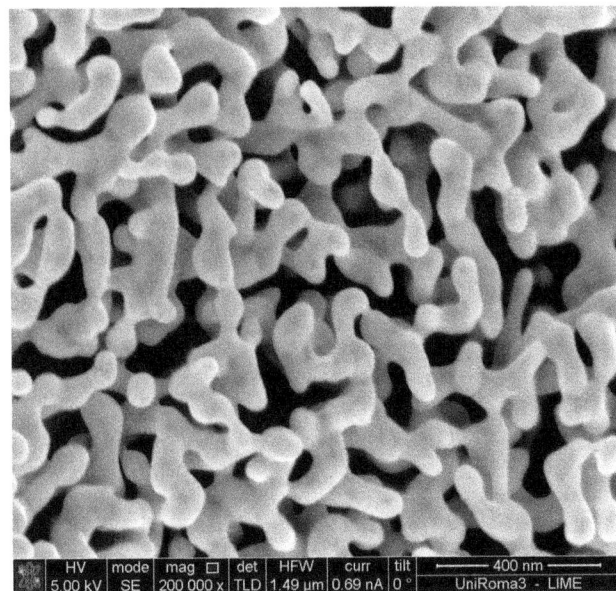

Fig. 24 Detail of the cornet microstructure after the partition process. A nanoporous morphology characterizes the whole specimen

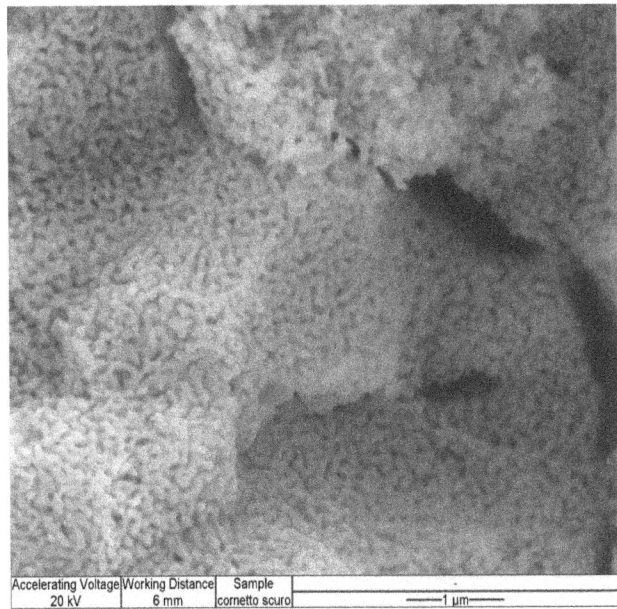

Fig. 26 SEM image. Detail of Fig. 25 showing an ex grain boundary region

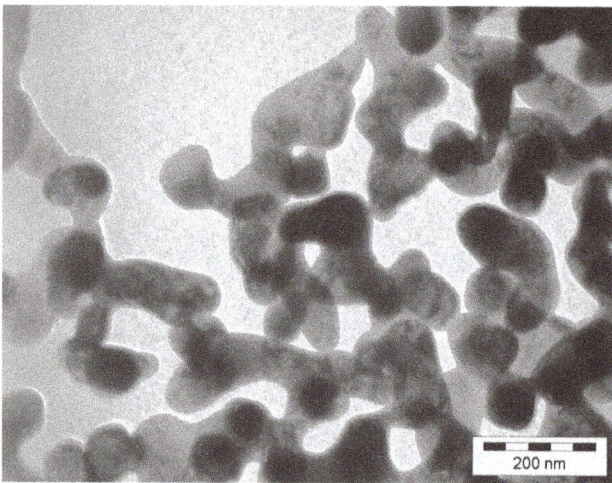

Fig. 27 TEM image of the nanoporous microstructure

Fig. 29 Selected area diffraction shows that the crystal structure of the sponge is pure gold and is characterized by twin crystals, as proved by the twin spots (see the circle in the SAD image)

(Fig. 29), which means that the nanoporous gold is composed of a pure gold nanocrystals cluster.

The fact that the cornet (Fig. 21) has a nanoporous microstructure explains the reason for its brown colour. This optical characteristic originates from the nanoscale morphology. In general, it can be said that, due to the nanoporosity, new modes of interaction of the light with the gold became important. In particular, interactions involving electronic oscillations called surface plasmons are possible. Furthermore, the gold behaves optically as a metal towards higher wavelengths (in the near-IR region) in comparison with bulk gold. For this reasons, the cornet appears brown in colour [18–22]. This very interesting behaviour of nanoporous gold and gold nanoparticles has been studied also for many different applications, such as monitoring body fluids, new electronic devices and decorative effects [23–31].

The annealing process of the cornet

After the partition process, the cornet is fragile due to its microstructure. At this point, the fire assay procedure entails annealing the cornet to make it easy to handle again. Annealing involves placing the cornet on a cupel which is put into the furnace at 900–950 °C. At the end of the process, the cornet has the typical gold colour and can be easily handled (Fig. 30). Obviously, this final annealing causes a decrease in its size, which adds to the one already occurred during partition. The nanoporous microstructure disappears in the final cornet.

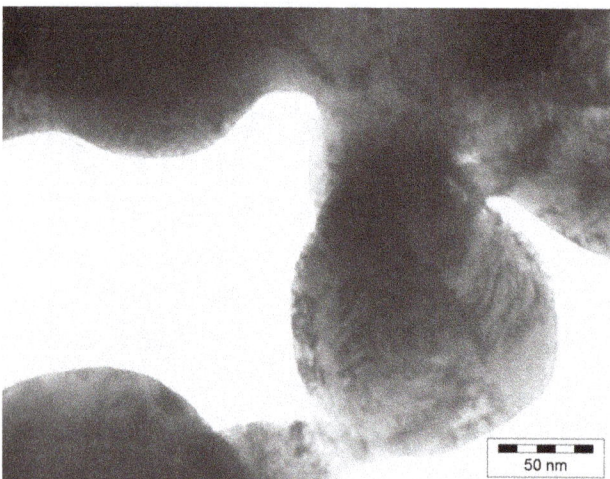

Fig. 28 Detail of the nanoporous microstructure. Electron diffraction can be performed on a selected area (SAD) whose diameter can be reduced down to 20 nm, and whose position can change

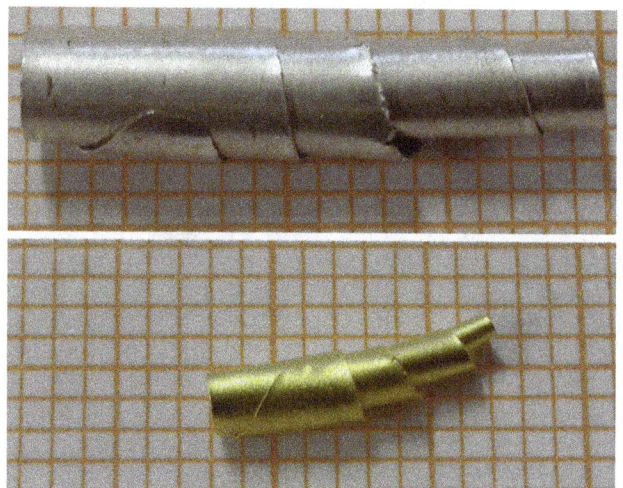

Fig. 30 The same cornet before partition and after final annealing is shown in the *lower and upper parts* of the figure, respectively

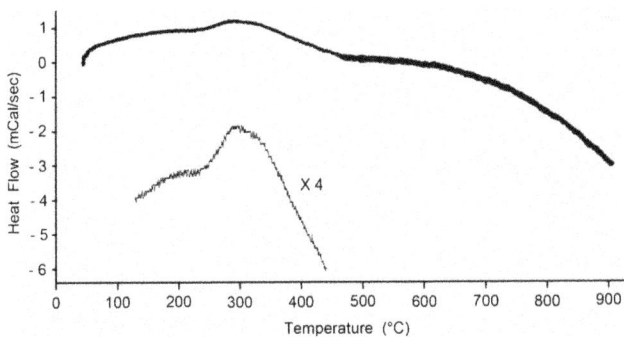

Fig. 31 The calorimetric curve shows only an exothermic reaction between 220 and 350 °C (see the *lower curve*, which is a magnification of the upper one). The peak of the exothermic reaction is at about 300 °C

A calorimetric analysis has been performed to define the temperature range in which the process of the nanoporous gold evolution into a bulk structure takes place.

The temperature inside the calorimeter has been increased at a speed of 7 °C/min and the specimen has been kept in nitrogen flow. The obtained graph is shown in Fig. 31. This measurement, repeated on other cornets, makes it clear that the nanoporous gold changes its structure at temperatures which are much lower than the gold melting temperature. At a temperature of 500 °C, the cornet already shows the characteristic colour of gold. SEM analyses reveal that at 500 °C the nanoporous structure is almost completely destroyed for the most part, at least on the cornet surface (Fig. 32).

Even this behaviour of the nanoporous gold is well known. In fact, the most frequently used method to modify the morphology of nanoporous gold has been the thermal treatment

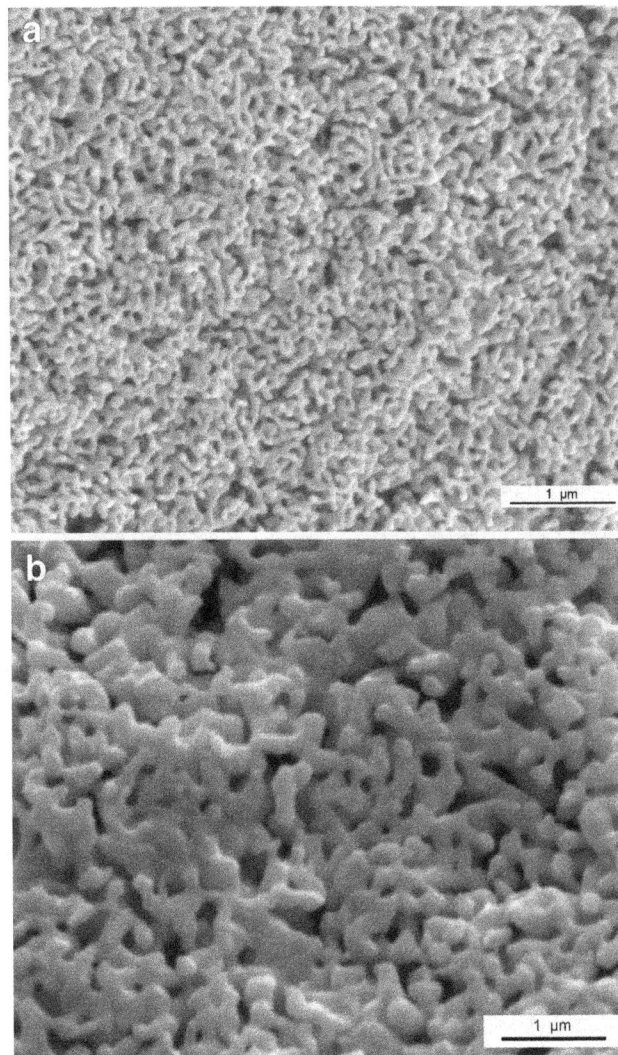

Fig. 33 The figure shows the initial nanoporous surface of the cornet (**a**) compared with the same surface after heating at 450 °C (**b**). This is the typical thermal coarsening of the nanoporous gold which preserves the starting morphology and only increases the features in size

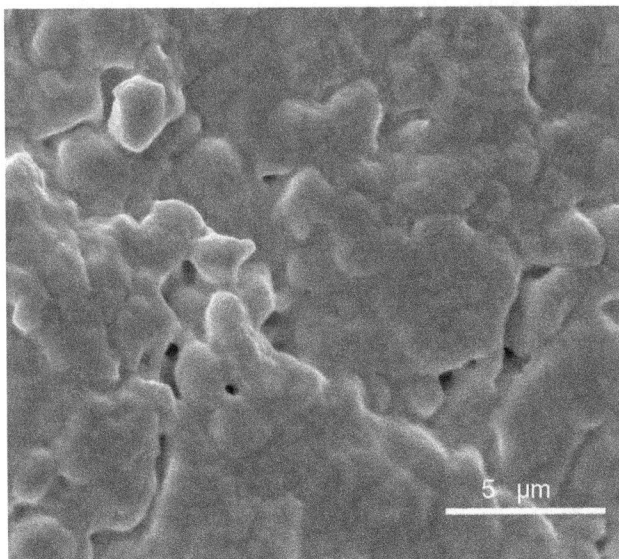

Fig. 32 SEM image. The surface morphology of the cornet after heating at 500 °C shows that the nanoporous microstructure has disappeared, at least on the surface

which leads to the coarsening of pores and ligaments [7]. This thermal coarsening preserves the starting pore structure, its only effect being increasing the size of the characteristic features (consider Fig. 33).

The mechanism of thermal coarsening is partly due to the increase in surface diffusion of gold atoms, which leads to an increase in the average pore size. The calorimetric curve (Fig. 31) shows that an exothermic reaction takes place, according to a reduction in energy of all the system. Furthermore, it must be considered that the nanoporous gold is characterized by a surface tension [32] whose release during heating may contribute to the evolution of the microstructure.

The fire assay process comes to an end now. The annealed cornet is now weighed again. The gold concentration of the initial specimen, which weight is known, is therefore computed from the two measurements.

Conclusions

This work shows that a process which has been given for granted and well established for ages reveals interesting aspects of the behaviour of precious metals such as gold and silver. Particularly, it is possible to explain how the material evolves during the fire assay and, especially, understand the reasons why the cornet changes colour and shrinks during the partition stage.

What reported here clarifies why all the stages of the procedure must be performed in order to obtain good results. Particularly, the action of lead as a scavenger of base metals and the logic of adding silver in precise amounts during the cupellation stage in order to have the nanoporous gold are understood. Finally, the microstructural analyses reveal the reason why it is necessary to make use of all the well-known working methods in the various phases of the procedure.

The cupellation process appears to be a dealloying process which can be improved thanks to the knowledge of the mechanisms of nanoporosity evolution and nanoporous gold generation, in particular, in order to avoid the cracks inside the cornet which may be responsible for loosing small pieces of material and producing errors in the gold content measurement.

Therefore, the analysts who perform the fire assay test daily can better understand what they usually do and improve the procedure further, in the light of what reported here.

Acknowledgments The authors are very grateful to Dr. Chris Corti for the useful suggestions and discussion of the results here reported. Thanks are due also to Mr. Giovanni Cascione, analyst at 8853 SpA who has prepared the specimens and taken part in the discussion of the results, giving most useful ideas for further reflection.

References

1. Biringuccio V (1540) De La Pirotechnia. Kessinger, Whitefish, print n demand
2. Bemporad E, Sebastiani M (2010) Focused ion beam and nanomechanical tests for high-resolution surface characterization: Not so far away from jewelry manufacturing. The Santa Fe Symposium on Jewelry Manufacturing Technology 2010, Eddie Bell (ed.) (Albuquerque: Met-Chem Research, 2005)
3. Smith EA (2003) The sampling and essay of the precious metals 1947, 2nd edn. Met-Chem Research, Albuquerque, Reprinted 2003
4. Raw P (1997) The assaying and refining of gold, a guide for the gold jewellery producer. World Gold Council,
5. Metalor: http://www.lbma.org.uk/assets/17_Cassagne.pdf Accessed 04 July 2013
6. Cortie MB, Maaroof AI, Stokes N, Mortari A (2007) Mesoporous gold sponge. Aust J Chem 60(70):524–527
7. Seker E, Reed ML, Begley MR (2009) Nanoporous gold: fabrication, characterization, and applications. Materials 2:2188–2215
8. Wittstock A, Biener J, Erlebacher J, Baumer M (eds.) (2012) Nanoporous gold: from an ancient technology to a high-tech material. Royal Society of Chemistry.
9. Erlebacher J, Sieradzki K (2003) Pattern formation during dealloying. Scr Mater 49:991–996
10. Erlebacher J, Aziz MJ, Karma A, Dimitrov N, Sieradzki K (2001) Evolution of nanoporosity in dealloying. Nature 410:450–453
11. Lu X, Balk TJ, Spolenak R, Arzt E (2007) Dealloying of Au-Ag thin films with a composition gradient: influence on morphology of nanoporous Au. Thin Solid Films 515:7122–7126
12. Eilks C, Elliott CM (2008) Numerical simulation of dealloying by surface dissolution via the evolving surface finite element method. J Comput Phys 227:9727–9741
13. Fujita T, Qian L, Inoke K, Erlebacher J, Chen M (2008) Three-dimensional morphology of nanoporous gold. Appl Phys Lett 92:251902
14. Zinchenko O, De Raedt HA, Detsi E, Onck PR, De Hosson JTM (2013) Nanoporous gold formation by dealloying: a Metropolis Monte Carlo study. Comput Phys Commun 184:1562–1569
15. Biener J, Hodge AM, Hamza AV (2005) Microscopic failure behavior of nanoporous gold. Appl Phys Lett 87:121908
16. Sun Y, Kucera KP, Burger SA, Balk TJ (2008) Microstructure, stability and thermomechanical behavior of crack-free thin films of nanoporous gold. Scr Mater 58:1018–1021
17. Biener J, Hodge AM, Hamza AV, Hsiung LM, Satcher JH Jr (2005) Nanoporous Au: a high yield strength material. J Appl Phys 97:024301
18. Maaroof AI, Cortie MB, Smith GB (2005) Optical properties of mesoporous gold films. J Opt A: Pure Appl Opt 7:303–309
19. Maaroof AI, Gentle A, Smith GB, Cortie MB (2007) Bulk and surface plasmons in highly nanoporous gold films. J Phys D Appl Phys 40(18):5675–5682
20. Dixon MC, Daniel TA, Hieda M, Smilgies DM, Chan MHW, Allara DL (2007) Preparation, structure, and optical properties of nanoporous gold thin films. Langmuir 23(5):2414–2422
21. Teperik TV, Popov VV, Garcia de Abajo FJ (2004) Total resonant absorption of light by plasmons on the nanoporous surface of a metal. Proceedings of the conference "Nanophotonics 2004", Nizhni Novgorod, Russia May 2–6, 2004
22. Cortie MB (2004) The weird world of nanoscale gold. Gold Bull 37(1–2):12–19
23. Zhang J, Ou J-Y, Papasimakis N, Chen Y, MacDonald KF, Zheludev NI (2011) Continuous metal plasmonic frequency selective surfaces. Opt Express 19(23):23279–23285
24. Cortie MB, Maaroof AI, Smith GB (2005) Electrochemical capacitance of mesoporous gold. Gold Bull 38(1):14–22
25. Cortie MB, Maaroof AI, Mortari A, Wuhrer R (2006) Applications of nano and mesoporous gold in electrodes and electrochemical sensors. IEEE.
26. Xu X, Gibbons TH, Cortie MB (2006) Spectrally-selective gold nanorod coatings for window glass. Gold Bull 39(4):156–165
27. Ruffato G, Romanato F, Garoli D, Cattarin S (2011) Nanoporous gold plasmonic structures for sensing applications. Opt Express 19(14):13164–13170
28. Zhang J, Ou J-Y, MacDonald KF, Zheludev NI (2012) Optical response of plasmonic relief meta-surfaces. J Optic 14:114002–114008
29. Zheng L-T, Wei Y-L, Gong H-Q, Qian L (2013) Application progress of nanoporous gold in analytical chemistry. Chinese J Anal Chem 41(1):137–144
30. Yuan L, Liu HK, Maaroof A, Konstantinov K, Liu J, Cortie M (2007) Mesoporous gold as anode material for lithium-ion cells. J New Mater Electro Syst 10:95–99
31. Jurgens B, Kubel C, Shulz C, Nowitzki T, Zielasek V, Biener J, Biener M, Hamza AV, Baumer M (2007) New gold and silver–gold catalysts in the shape of sponges and sieves. Gold Bull 40(2):142–149
32. Frenkel AI, Vasic R, Dukesz B, Li D, Chen M, Zhang L, Fujita T (2012) Thermal properties of nanoporous gold. Phys Rev B 85(195419):1–7

The role of nanogold in human tropical diseases: research, detection and therapy

Miguel Peixoto de Almeida · Sónia A. C. Carabineiro

Introduction

The Special Programme for Research and Training in Tropical Diseases, based at and executed by the World Health Organization (WHO), states that these maladies affect millions of people across the world [1–5]. They are more common in tropical and subtropical regions, as in temperate climates the cold season is able to control the insects population, which are the most common disease carriers or vectors. The flies and mosquitos may carry parasites, bacteria or viruses that can infect humans and animals through a "bite", which transmits the illness agent through a subcutaneous blood exchange. Some factors like deforestation, exploration of tropical areas, and increasing international air travel to tropical regions lead to an increase of tropical diseases. For some of them, unfortunately, vaccines are still not available.

The recent explosive growth of nanogold applications in many areas [6–9], also comprises contributions to the health field [10–14]. In particular, gold can have a direct role in treatment and detection of tropical diseases, as shown by some studies found in literature. Besides some examples of molecular gold used directly against parasites and in research dealing with animal parasitic diseases [15–17], gold nanoparticles and gold complexes can also play a significant role that goes beyond the classical approach of confirmation diagnosis,

based on detection of parasites in either blood or lymph by microscopy, as it will be explained below.

Trypanosomiases

About the diseases

There are two different types of human trypanosomiases [1, 2]: the human African trypanosomiasis and the American trypanosomiasis. The latter is more studied, but it is important to also refer the progresses made so far on the former.

The human African trypanosomiasis, also known as "sleeping sickness", is a well-known tropical disease, caused by the protozoan parasites *Trypanosoma brucei rhodesiense* or *Trypanosoma brucei gambiense*. It is spread by a bite of an infected tsetse fly, which erupts into a red sore. Within a few weeks, it causes fever, swollen lymph glands, aching muscles and joints, headaches and irritability. In advanced stages, the disease attacks the central nervous system, causing changes in personality, alteration of the circadian rhythm, confusion, slurred speech, seizures, and difficulty walking or talking. These problems can develop over many years in the *gambiense* form and some months in the *rhodesiense* form. If not treated, it can cause death. An early diagnosis is of extreme importance for a better control of this sickness.

The American trypanosomiasis, or Chagas disease, is caused by the protozoan parasite *Trypanosoma cruzi*, generically schematized in Fig. 1. It is found mainly in Latin America, where it is mostly transmitted to humans by the faeces of triatomine bugs [2]. This disease can cause death from myocarditis or meningoencephallitis during its acute phase, in less than 5–10 % of symptomatic cases. The evolution for the chronic phase (10–30 years after infection) can result in cardiac, digestive

M. P. de Almeida · S. A. C. Carabineiro (✉)
Laboratory of Catalysis and Materials (LCM),
Associate Laboratory LSRE/LCM, Faculdade de Engenharia,
Universidade do Porto, Rua Dr. Roberto Frias,
4200-465 Porto, Portugal
e-mail: scarabin@fe.up.pt

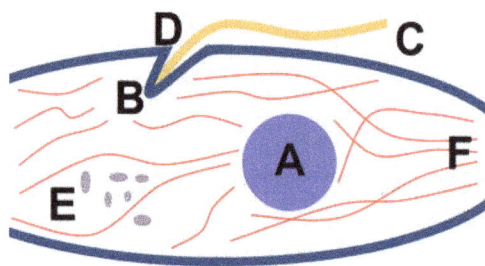

Fig. 1 *Trypanosoma cruzi* cell schematic representation (adapted from [87, 88]). **a** Nucleus, **b** flagellar pocket, **c** flagellum, **d** cytostome, **e** cytoplasmic vesicles, **f** cytoskeleton

(megaesophagus and megacolon), or cardiodigestive problems in 30–40 % of the infected patients [18].

Microscopy assays using nanogold

In 1997, Magez et al. [19], knowing that the protozoan *Trypanosoma brucei* is lysed by the cytokine tumor necrosis factor-α (TNF-α), prepared TNF-α gold nanoparticles (TNF-α-AuNP) using commercial 10 nm gold beads. These TNF-α-AuNP were endocytosed via coated pits and vesicles and are directed towards lysosome-like digestive organelles. The localization of TNF-α binding sites on intact parasites and a better understanding of intracellular uptake of TNF-α were possible precisely with the help of the conjugated colloidal gold particles. The bulk of the TNF-α gold labeling was localized in the flagellar pocket, where beads concentrated in coated pits. Sporadically, TNF-α-AuNP were found in association with the flagellum in the flagellar adhesion zone, at the entrance of the flagellar pocket, or in association with tiny filamentous material at more distant regions of the flagellum.

Soeiro et al. [20] performed ultrastructural analysis of the endocytic process (cellular uptake) using horseradish peroxidase (HRP) enzyme coupled to colloidal gold particles (HRP-AuNP). The "in vitro" findings suggest that cardiomyocytes (cells of cardiac muscle) mannose receptors, localized at the sarcolemma (muscle cell membrane), mediate *T. cruzi* recognition and can be down-modulated by parasite infection.

Okuda and co-workers [21] used ultrathin sections of *T. cruzi* epimastigote forms. These assays, which reveal gold particles at the opening of flagellar pocket, concentrated in the cytostome region, were possible through immunogold labeling (a staining technique used in electron microscopy, where colloidal gold nanoparticles are usually attached to secondary antibodies, which are attached to primary antibodies designed to bind a specific protein or other cell component). The protocol consists in cytoskeletons incubation in the first antibody, which recognizes the cytoskeletons elements; then incubated in the second antibody, which recognizes the first antibody and have attached a 15 nm gold particle. Using electron microscopy to visualize the gold nanoparticles

(AuNP), and consequently the known proximity of the elements recognized by the antibody–antibody–AuNP configuration, the relationship between the cytostome, an endocytic organelle, and the flagellum was described for the first time. Gold has high electron density which increases electron scatter enabling high contrast images.

A similar protocol allowed Monteiro et al. [22], in 2001, to describe the localization of chagasin, an endogenous tight-binding cysteine protease inhibitor in *T. cruzi*. Gold-labeled antibodies localized chagasin to the flagellar pocket and cytoplasmic vesicles of trypomastigotes and to the cell surface of amastigotes (the former are found in human blood and the later in tissues). Thin sections were incubated with affinity purified rabbit anti-chagasin followed by gold-conjugated goat anti-rabbit Immunoglobulin G (IgG).

Silva et al. [23] evaluated the presence and distribution of the Ssp4 antigen in the different amastigote *T. Cruzi* populations using gold-labeled antibodies, allowing the observation with transmission electron microscopy. Goat anti-mouse IgG was labeled with 10 nm gold particles. These were mainly located inside cytoplasmic vesicles and the flagellar pocket, suggesting that Ssp4 is released by exocytosis into the flagellar pocket.

Recently, Acosta et al. [24] used immunogold electron microscopy analysis for providing a new insight to better understand the molecular pathogenesis of Chagas heart disease. They showed that there is a common epitope between cruzipain (a lysosomal major antigen from *T. cruzi*) and either myosin or other cardiac *O*-linked *N*-acetylglucosamine containing proteins.

Eger and Soares [25] described the visualization by confocal microscopy of ingested gold (15 nm) labeled transferrin in epimastigote forms of the protozoan *T. cruzi*. This is a promising imaging tool to explore the endocytic pathway in trypanosomes (and eventually adapted for other protozoans).

Detection methods using gold

In 2003, Diniz et al. [26], prepared a gold electrode for adsorbing a polypeptide chain formed by recombinant antigens: cytoplasmic repetitive antigen (CRA) and flagellar repetitive antigen (FRA) of *T. cruzi*. The goal was to develop a biosensor for Chagas disease, based on impedance spectroscopy, with the behavior of CRA/FRA antigens adsorbed on gold and platinum electrodes being investigated. The team also used platinum electrodes, which yielded similar results as those presented by gold; however, the marked influence of oxygen dissolved in solution on the electrochemical response of platinum, requires extensive deoxygenation of solutions prior each experiment. Consequently, gold electrodes have advantages. Under proper conditions, it is possible to distinguish between serum positive and negative to Chagas disease.

Joining the last two methods (as an "immunotechnique" is applied, but gold acts as a supporting electrode for sensing purposes, instead of labeling), Ferreira et al. [27] immobilized Chagas disease antigens in a gold surface, and the anti-*T. cruzi* antibodies present in the serum sample were captured by the antigen, remaining connected to the solid phase. The human anti-IgG antibody conjugated to peroxidase (HRP) then reacted with immunocaptured anti-*T. cruzi* antibodies, if any present. The detection and quantification was ensured by peroxidase activity in presence of H_2O_2 and, in the presence of the iodide, this ion reacted with the reduced peroxidase. Under these conditions, the current intensity of I_2 reduction was proportional to the amount of anti-*T. cruzi* antibodies in the serum real sample. This assembling is schematized in Fig. 2.

In a very similar approach, Ribone and co-workers [28] prepared bioelectrodes to detect IgG antibodies occurring in sera of patients suffering from American trypanosomiasis. The main difference from the previous description [27], is the use of two different thiols, 3-mercapto-1-propionic acid (MPA), 3-mercapto-1-propanesulfonic acid (MPSA), which provide sulfonate or carboxylic residues, for the electrostatic binding of the antigen. Figure 2 also applies to the assembling described in this work, where the linkage to the gold electrode is described generically as "thiol".

Foguel et al. [29] used the same principle for the construction of an amperometric imunosensor. The novelty here was the gold based electrode being obtained from a recordable compact disc (CD-R), then modified with 4-(methylmercapto)benzaldehyde for the immobilization of Tc85 protein of the *T. cruzi*. To access the metal layer of the CD-R, concentrated HNO_3 was added on the surface and after 5–10 min the protective layers were totally removed. It is well known that organic monolayers with thiol or disulfide groups on the electrode surface are of great

interest due to the sulfur binding strongly to the gold surface. A molecule that presents this property is 4-(methylmercapto)benzaldehyde (SBZA), which has a mercapto group and maintains a free aldehydic group. In order to link the CD-R fragment and allow the amperometric process, a laminated copper wire was fixed and insulated with polytetrafluoroethylene (PTFE), so that the wire does not come into contact with the electrolyte solution, thus avoiding the oxidation and reduction of this copper.

Using a similar methodology, a microfluidic system coupled to a screen-printed carbon electrode (SPCE) was developed by Pereira and co-workers [30] for the quantitative determination of IgG specific antibodies present in serum samples of patients with Chagas disease. This time, HRP in the presence of H_2O_2 catalysed the oxidation of 4-*tert*-butylcatechol (4-TBC) and its back electrochemical reduction was detected on a modified electrode. The authors claimed that, compared with traditional IgG detection techniques, the immunosensor based on microfluidic technology showed a decrease in sample and reagent consumption, faster response times for analysis, and good reproducibility and did not require highly skilled technicians or expensive and dedicated equipment.

Deborggraeve et al. [31] developed a simple and rapid test for detection of amplified *T. brucei* DNA, the human African trypanosomiasis–polymerase chain reaction–oligochromatography (HAT–PCR–OC). In this case, gold was used essentially as a colorimetric indicator. PCR products were visualized on a dipstick through hybridization with a gold-conjugated probe (oligochromatography), occurring in 5 min. The lower detection limit of the test was 5 fg of pure *T. brucei* DNA; one parasite in 180 μl of blood is still detectable. The authors concluded that HAT–PCR–OC is a promising new tool for diagnosis of "sleeping sickness" in laboratory settings.

Nanogold chemoterapy

Nyarko et al. [32] described the effects of aqueous Au(III) and his metalloporphyrin derivative on *T. brucei brucei* growth in culture. While Au(III) porphyrin was effective against the parasites at concentrations above 4.8×10^{-6} M, aqueous Au(III) ion was toxic to the trypanosomes at concentrations as low as 2.0×10^{-7} M, due to free radicals formation. Although the parasite tested is responsible for an animal African trypanosomiasis (and this review deals mainly with human tropical diseases), the work of Nyarko et al. shows that gold might have a promising role in the battle against these noxious organisms.

A series of seven papers regarding metal-based chemotherapy against tropical diseases was published by Sánchez-Delgado and co-workers [33–39], from 1993 to 2004. Navarro, who wrote a very interesting review in 2009 about

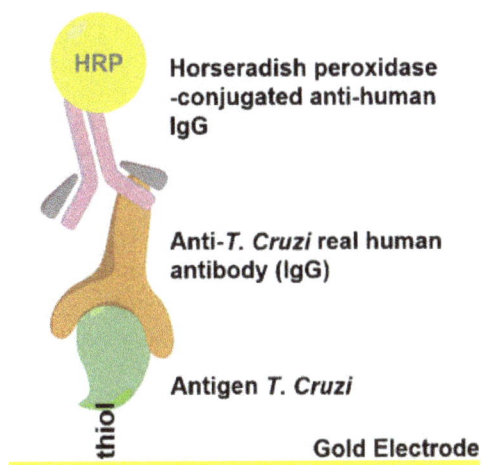

Fig. 2 Immunosensors supported on a gold electrode (adapted from [29, 30])

gold complexes as potential anti-parasitic agents [40], published since 1997, together with different co-workers, also some papers in this series [35, 36, 38, 39]. In one of those publications [36], the authors dealt with a group of metal–clotrimazole complexes evaluating their activity against *T. cruzi*, including AuCl$_3$(clotrimazole). The clotrimazole (CTZ) is an azole derivative, and this family of compounds has been developed as chemotherapeutic agents for the treatment of fungal diseases on the basis of their properties as sterol biosynthesis inhibitors, which can also affect some parasites [36]. This AuCl$_3$(CTZ) complex (Fig. 3, middle) showed an anti-*T. cruzi* epomastigotes activity only slightly higher than the free CTZ (60 % vs. 58 %), with a poor performance compared other metal–CTZ complexes [36, 40]. In another paper [38], Navarro et al. considered other azol derivative (ketoconazole [KTZ]) and other Au precursor, preparing [Au(CTZ)(PPh$_3$)]PF$_6$ (Fig. 3, top) and [Au(KTZ)(PPh$_3$)]PF$_6$·2H$_2$O (Fig. 3, bottom) complexes. Both complexes (1 µM) showed an inhibition of the proliferation of the *T. cruzi* epimastigotes around 70 %, while the free CTZ has no effect on the growth and KTZ only inhibited 39 % at the same concentration [38, 40].

It is also important to mention the role of gold in biolistic (or biological ballistics, consisting in transfecting cells by bombarding them with microprojectiles coated with DNA) immunization using a stable DNA–gold precipitate [41]. As the target population of a *T. cruzi* vaccine lives predominately in poorer rural areas in South America, the use of such DNA–gold precipitate, which does not require a cold chain, is an attractive method for vaccination.

Leishmaniasis

About the disease

Some *Leishmania* parasitic protozoa are the cause of leishmaniasis in humans. The disease is spread through the bite of phlebotomine sandflies, which breed in forest areas, caves and adobe brick houses, where most of the transmission to humans takes place. This disease can be divided in four main types [3]:

(1) Cutaneous forms, with skin ulcers usually on exposed areas, such as the face, arms and legs, that usually heal within a few months, but leaving scars.
(2) Diffuse cutaneous leishmaniasis, which produce disseminated and chronic skin lesions and it is difficult to treat.
(3) Mucocutaneous forms, where the lesions can destroy the mucous membranes of the nose, mouth, throat cavities and surrounding tissues.
(4) Visceral leishmaniasis (VL), which is characterized by high fever, considerable weight loss, anemia,

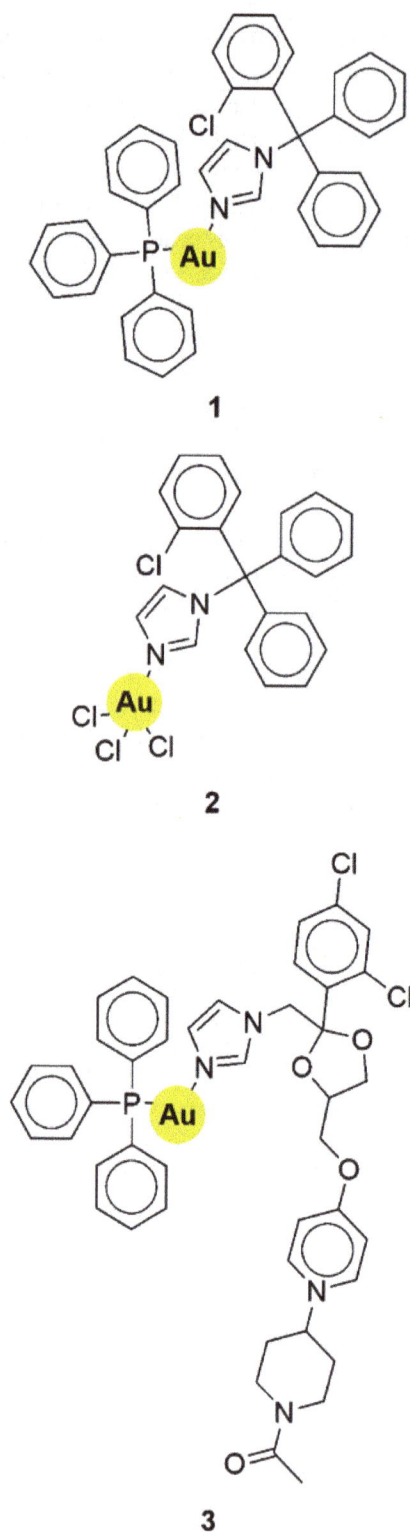

Fig. 3 Au^I-clotrimazole (**1**), Au^{III}-clotrimazole (**2**) and Au^I-ketoconazole (**3**) complexes (adapted from [40], counterions omitted for simplification)

spleen and liver swelling. If not treated, the disease can have a fatality rate, as high as 100 %, within 2 years.

Microscopy assays using nanogold

Santos et al. [42], in 1991, used transmission electron microscopy for observation of lectins associated with colloidal gold particles (15 nm). AuNP bound to concanavalin A (a lecitin protein) were found randomly and sparsely distributed on the *Leishmania donovani chagasi* surface, along the cell body and along the flagellum. The pre-treatment of promastigotes of with trypsin did not interfere with the binding of lectins to the parasite or alter the parasite ultrastructure, showing a good resistance to this protease. However, a pre-treatment with 2-mercaptoethanol increased the colloidal gold density, showing that this last compound exposed second-order concanavalin A receptors.

The intracellular fate of human transferrin (HTf) in macrophages infected by *Leishmania* was investigated by Borges et al. [43], i.e., binding the HTf to gold nanoparticles complexes for further observation in order to understand the location and processing of HTf across time. Within parasites, HTf was found in cysteine-proteinase-rich structures, suggesting that the protein can be endocytosed by intracellular amastigotes and sorted to the parasite endosomal–lysosomal compartments rather than being recycled.

Sengupta et al. [44] explored the relation between hemoglobin endocytosis and the parasite, using immunogold techniques, contributing to understand the hemoglobin destruction, a critical aspect in the disease. A specific high affinity binding site on the surface of *Leishmania donovani* promastigotes mediated rapid internalization and degradation of hemoglobin. When *Leishmania*, previously incubated with hemoglobin at 4 °C, were treated with a rabbit anti-hemoglobin antibody, followed by an anti-rabbit IgG conjugated with colloidal gold, accumulation of gold particles could be seen in the flagellar pocket. After incubation with hemoglobin–gold conjugates at 25 °C or 37 °C, the particles accumulated in discrete intracellular vesicles, suggesting the internalization of hemoglobin at both temperatures. Quantitation could be carried out by counting the number of gold particles internalized by *Leishmania* for each set of experimental conditions.

Detection methods using nanogold

Zijlstra et al. [45] evaluated a strip test employing recombinant K39 (rK39) antigen and protein-A (a surface protein originally found in the cell wall of the bacterium *Staphylococcus aureus*)/colloidal gold as read-out agents. They stated that this test has the ideal format for use in the field but also has important limitations and should be used with caution. In general, a positive test result in a patient who presents with

the classical clinical features of VL supports the diagnosis. It should be noted that a positive test result may be the result of previous (subclinical) infection and therefore not relevant to the current illness. Under field conditions, it is not always clear from the history whether a patient who reports previous treatment for VL was correctly diagnosed at the time. Such test was used recently by Carreira et al. [46] in order to study the natural infection with *Leishmania infantum* on opossums (*Didelphis aurita*), considered natural hosts of parasites, suggesting their important role in the epidemiology of VL. This work allowed the authors to present the first report of amastigotes in the tissues of *Didelphis aurita* naturally infected with *L. infantum*.

Ramos-Jesus and co-workers [47] prepared a quartz crystal gold electrode where the recombinant antigen of *Leishmania chagasi* rLci2B-NH6 was tightly immobilized on a quartz crystal gold electrode by self-assembled monolayer based on short-chain length thiol. This device was planned for the canine visceral form; however this assembling seems to be promising also for human leishmaniasis. It is important to distinguish from the similar examples given before [27, 28], as no immunotechnique is applied to sensor construction in this case. The amine groups of cysteamine provided reaction sites for covalently bind to the glutaraldehyde. The rLci2B-NH6 antigen was then immobilized through a Schiff base via glutaraldehyde by a histidine tail (Fig. 4). The Schiff base allows major exposure of epitopes and a reduced steric hindrance. The response was obtained by recognition of immobilized rLci2B-NH6 antigen.

Moreno et al. [48] designed a novel SELEX procedure (a methodology in which single stranded oligonucleotides are selected from a wide variety of sequences based on their interaction with a target molecule) using colloidal gold to select high affinity single stranded DNA aptamers that bind specifically to *L. infantum* KMP-11. Kinetoplastid membrane protein-11 (KMP-11) is a major component of the cell

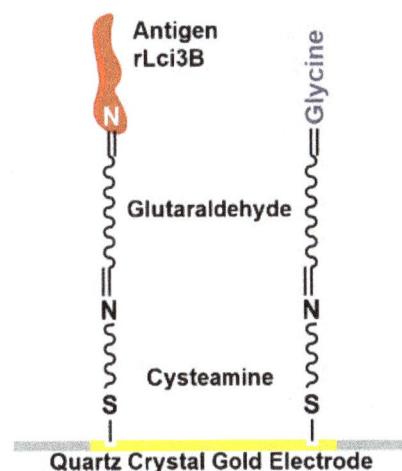

Fig. 4 Quartz crystal gold electrode supporting rLci3B-NH6 via cysteamine and glutaraldehyde (adapted from [47]). Glycine presence represents the blocking treatment in order to minimize the nonspecific binding

membrane of kinetoplastid parasites. Although its function is not known, the fact that KMP-11 is a cytoskeleton-associated protein suggests that it may be involved in mobility or in some other aspects of the flagellar structure. The method is based on the binding of the target protein to colloidal gold in order to achieve a higher amount of that protein for further purification by centrifugation. This novel methodology is very easy to use and cheaper than others that are currently being used. The same authors also developed an electrochemical biosensor based on aptamers that can recognize and specifically report the presence of *L. infantum* KMP-11. The target protein was conjugated with gold nanoparticles and the complexes electrodeposited on gold screen-printed electrodes (SPE) [49]. Using this method, Moreno et al. were able to detect 25 mg/ml of KMP-11.

Nanogold chemotherapy and thermotherapy

Recently, Ilari et al. [50] reported on the interesting antiparasitic actions of a well-known antiarthritic agent, the gold(I)-containing drug Auranofin. As trypanothione reductase, a key enzyme of *L. infantum*, contains a dithiol motif at its active site and gold(I) compounds are known to be highly thiophilic, it was found that Auranofin behaved as an effective enzyme inhibitor. It might also be a potential antileishmanial agent, as it was also found that Auranofin is able to kill the promastigote stage of *L. infantum* at micromolar concentration. These important findings will certainly contribute to the design of new drugs against this disease.

Navarro et al. [51] studied a gold complex prepared from a dypirido[3,2-*a*: 2,3-*c*]phenazine ligand, [Au(dppz)₂]Cl₃ (Fig. 5), that interacts with DNA by intercalation mode. The large leishmanicidal activity of this complex was associated to the cellular processes involving parasite DNA, constituting a new promising chemotherapeutic alternative in the search for the cure of leishmaniasis [51].

More recently, Barboza-Filho et al. [52] studied the growth of *Leishmania brasiliensis* promastigotes in culture, using natural rubber membranes with and without gold nanoparticles. It was observed that the increase of AuNP caused a decrease in the number of promastigotes

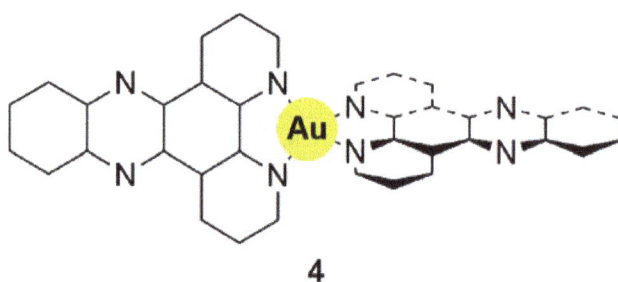

4

Fig. 5 Complex prepared by Navarro et al. from dypirido[3,2-*a*: 2,3-*c*] phenazine ligands (perpendicular to each other) (adapted from [51])

in culture medium. These results are advanced as a possible solution for developing a flexible "band-aid" for skin lesions, inhibiting the population growth of parasites in the lesions.

A recent study from Sazgarnia et al. [53] determined the efficacy of thermotherapy in the presence of AuNP and microwave radiation (2,450 MHz) on the survival of *Leishmania major* promastigotes and amastigotes. It was shown that the presence of AuNP during microwave irradiation (after cell incubation) was more lethal for promastigotes and amastigotes when compared to microwave radiation alone. This shows that AuNP are a promising new approach to treat leishmaniasis in the future.

Malaria

About the disease

Malaria is caused by the *Plasmodium* parasite, which is transmitted through the bites of infected mosquitos. After entering in the human body, the parasites multiply in the liver and infect the red blood cells. Fever, headache, and vomiting are symptoms of malaria and usually appear between 10 and 15 days after the mosquito bite. Malaria can quickly become life-threatening by disrupting the blood supply to vital organs and, in many parts of the world, the parasites have developed resistance to a number of known medicines. About 3.3 billion people (almost half of the world's population) are at risk of getting malaria. Every year, this leads to about 250 million cases and nearly 1 million deaths. Consequently, a prompt and effective treatment is crucial [4].

Microscopy assays using nanogold

In the study performed by Bhowmick et al. [54], immunogold electron microscopy was used for sub-cellular localization of *Plasmodium falciparum* enolase. This protein was detected at every stage of the parasite life cycle, in cytosol (the liquid found inside cells) and associated with nucleus, food vacuole (storage membrane bound organelle), cytoskeleton and plasma membrane. Diverse localization of enolase suggests that apart from catalysing the conversion of 2-phosphoglyceric acid into phosphoenolpyruvate in glycolysis (its typical role in cytosol), it may also be involved in other functions, namely, in red blood cell invasion, food vacuole formation and/or development and transcription. These new data seems to clarify the role of this enzyme in the parasite life.

The work of Chugh et al. [55], helped to determine the reasons why two monoclonal antibodies (AC-43 and AC-

29) significantly inhibited *Plasmodium vivax* development inside the mosquito *Anopheles culicifacies* midgut (a portion of insect's digestive system). The gold labeling showed that these two monoclonal antibodies bind to glycoproteins present in the gut epithelium (intestine wall), a known zone for its ookinetes receptors (receptors of the fertilized zygote of the parasite in the mosquito's body), which assures the crossing of the epithelial barrier and provide components for further development into oocysts (the encysted or encapsulated ookinetes in the mosquito's inside). These data point to these glycoproteins as potential candidates for a vector-directed transmission-blocking vaccine, stopping the parasite cycle in the ookinetes–oocysts transition step.

Recently, Lopes da Silva et al. [56] used bovine serum albumin (BSA) linked to 10 nm AuNP in order to follow their position in the parasite–host cell system, in the liver stage, a crucial phase of the *Plasmodium* development. The authors showed that when BSA-Au are loaded into the host cell endocytic pathway, the AuNP are later found within the parasite cytoplasm, showing the transport of materials from the host endocytic pathway towards the parasite interior, revealing a possible form of the parasite to obtain nutrients.

Detection methods using nanogold

Stevens et al. [57] described a different approach using AuNP. The Immunoglobulin M (IgM) antibody bounded to the assay membrane acts as the capture molecule. The gold–antibody conjugate acts as the label, generating a visible increase in optical density proportional to the concentration of analyte present, that is, the malarial antigen *P. falciparum* histidine-rich protein II (PfHRP2). If this antigen (which acts as a bridge between the membrane–IgM and Au–IgG systems) is not present, Au–IgG will disappear after washing and no changes in optical density will be observed. A positive test is schematized in Fig. 6.

Amperometric immunosensors can also be used for monitoring this disease. Sharma et al. [58] presented recently a sensor based on AuNP/alumina sol–gel modified SPE for antibodies to PfHRP2. The AuNP electrode (AuNP/Al$_2$O$_3$ sol–gel/SPE) had much larger amperometric current (390 nA) than the bare SPE (120 nA) and Al$_2$O$_3$ sol–gel/SPE (154 nA). AuNPs/Al$_2$O$_3$ sol–gel/SPE was three times more sensitive compared to unmodified/bare SPE for the same concentration of analyte. Recently, Fu et al. [59] developed a device consisting on a two-dimensional paper network signal-amplified immunoassay for malaria protein PfHRP2 detection. The paper card contains reagents stored in dry form, including an antibody conjugated to a gold particle label, and the user only needs to add water and the sample. Factors like the signal/noise ratio and test sensitivity are always a concern with AuNP based (and other) tests. Therefore, improvements in this area are also important, besides the test principle itself, as demonstrated for the PfHRP2 malaria biomarker [60].

Potipitak et al. [61] created a device specifically for *P. falciparum* that can be adapted for other parasites. A biotinylated probe was linked to the gold electrode of quartz crystal microbalance (QCM) surface based on the specific interaction between avidin protein and biotin (vitamin B7). Prior to this immobilization, the QCM surface was pre-treated (e.g., with MPA) to add the ester group on the surface, in order to bind with the amine group of avidin. This technique was based on DNA piezoelectric biosensor using the Au electrode of QCM. The deposited mass, due to DNA hybridization on the QCM surface, resulted in a shift of the quartz resonance frequency. Moreover, the new sensor is cost-effective since both sides of the quartz Au surface can be used separately, reducing the sensor price to 50 %.

Very recently, Guirgis et al. [62] described a homogeneous assay based on the fluorescence quenching of cyanine 3B (Cy3B)-labeled recombinant *P. falciparum* heat shock protein 70 (PfHsp70) upon binding to AuNP functionalized with an anti-Hsp70 monoclonal antibody. Upon competition with the free antigen, the Cy3B-labeled recombinant PfHsp70 is released to solution resulting in an increase of fluorescence intensity. Concerning test kits involving colloidal gold, Piper et al. [63] recently showed that a panel of monoclonal antibodies against *Plasmodium* lactate dehydrogenase can be used in various combinations to uniquely identify all species of malaria parasites that infect humans. Their results should help the development of new test with greater specificity, sensitivity and ability to differentiate among malaria parasite species, adding to some solutions already in the market [64].

Nanogold chemoterapy

Concerning antimalarial gold compounds, and in context of the already mentioned series of seven papers related to metal-based chemotherapy against tropical diseases [33–39], Navarro et al. described the reaction of AuPPh$_3$Cl with chloroquine (CQ) and KPF$_6$ leading to the new

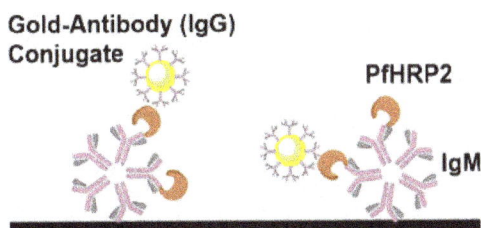

Fig. 6 PfHRP2 positive assay (adapted from [57])

complex [Au(PPh$_3$)(CQ)]PF$_6$. This compound was found to be considerably more active than CQ diphosphate and other previously reported metal–CQ complexes (Ru or Rh [34]) against two chloroquine-resistant strains of *P. falciparum* in vitro [35]. Navarro et al. also prepared a series of new Au(I) and Au(III) complexes containing CQ in combination with other ligands, which display activity against CQ-sensitive but also against CQ-resistant strains of *P. falciparum* [39]. The highest activity for this series was obtained for [Au(CQ)(PEt$_3$)]PF$_6$. More recently, Navarro et al. [65] studied the mechanism of antimalarial action of [Au(CQ)(PPh$_3$)]PF$_6$. This compound seems to avoid heme aggregation (which allows accumulation of toxic levels of compound, resulting in parasite death). The high activities observed against parasites resistant to chloroquine are probably due to the structural modification of CQ introduced by the presence of the gold-triphenylphosphine fragment. The scheme of these complexes can be found in Fig. 7.

Blackie et al. [66] described the synthesis of [Au(**R**)(PPh$_3$)]NO$_3$ and [Au(C$_6$F$_5$)(**R**)] complexes, where **R** denotes a CQ group or a ferrocenyl-4-amino-7-chloroquinoline, and their use as anti-malarial agents. Rh complexes were also tested. The CQ complexes showed better efficacy against CQ-resistant strains of the *P. falciparum* when comparing to free CQ. For the CQ-resistant strain, there is a considerable drop in efficacy. Concerning to the compounds with the ferrocenyl-4-amino-7-chloroquinolines, those containing ferroquine were considered the most efficient, but all showed improved efficacy with respect to CQ in both sensitive and resistant strains.

Fricker et al. [67] used a set of six Au(III) complexes, showing that all of them were able to inhibit cathepsin B, with a half maximal inhibitory concentration (IC$_{50}$) values in the range of 0.2–1.4 μM. Cysteine proteases cathepsin B play multiple roles in the parasite life cycle like nutrition, host invasion, protein processing, and evasion of the host immune response, so their inhibition can be extremely important.

Also well-known gold-based drugs like Auranofin (the antiarthritic medicine that also demonstrated potential antileishmanial activity, as referred above), showed recently very pronounced antiplasmodial effects in vitro [68]. Auranofin proved to be a potent inhibitor of mammalian thioredoxin reductases causing severe intracellular oxidative stress. Given the high sensitivity of *P. falciparum* to oxidative stress, the authors thought that Auranofin might act as an effective antimalarial agent and showed that this compound (and a few other related gold complexes) strongly inhibits *P. falciparum* growth in vitro.

Recently, Bjelosevic et al. [69] prepared a set of gold(I) complexes based on 1,10-bis(diphenylphosphino) metallocene derivatives (Fig. 8) and evaluated their activity against malaria (W2 chloroquine-resistant strain of *P. falciparum*) and other diseases. Although the IC$_{50}$ values (half-maximal inhibitory concentrations, which indicate how much are needed to induce inhibitions of 50 %) are low, they are still much higher than those for chloroquine. The biological activities of ruthenocenyl-based gold compounds are superior to their ferrocenyl analogues and thus provide directions as to which of these ligands can be used for medicinal applications.

Very recently, Hemmert et al. [70] described the use of gold(I) and gold(III) complexes with *N*-heterocyclic carbene (NHC) ligands as antimalarial agents, specifically against the chloroquine-resistant *P. falciparum* strain FcM29-Cameroon. A group of four dinuclear gold(I) complexes plus three dinuclear gold(III) complexes (Fig. 9, top) was studied, followed by another group of three mononuclear gold(I) complexes (Fig. 9, bottom). The gold(I) dinuclear complexes were found to have moderate antimalarial effect (IC$_{50}$ never lower than 9 μM), but less efficient than the silver(I) counterparts. However, all gold complexes were found to not cause hemolysis (red blood cells membrane bursting with subsequent release of hemoglobin), in contrast to some of the silver analogues. The second group of gold complexes (mononuclear gold(I)) revealed an improved

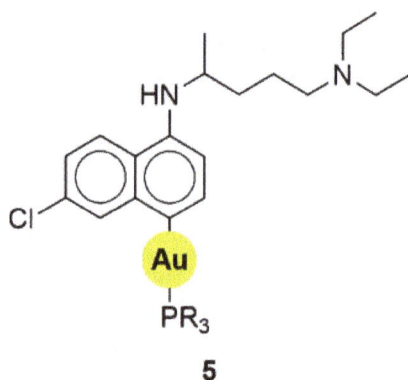

Fig. 7 [Au(chloroquine)(PR$_3$)]PF$_6$ complexes (R = Me, Et or Ph) prepared by Navarro et al. (adapted from [39, 65])

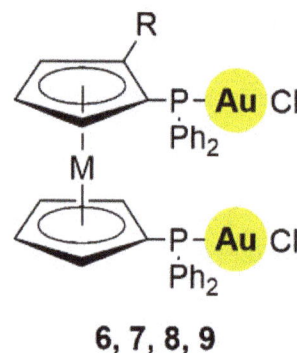

Fig. 8 Au^I complexes based on metallocene derivatives prepared by Bjelosevic et al. and tested as antimalarial agents (**6**: M = Ru, R = CHCH$_3$N(CH$_3$)$_2$; **7**: M = Fe, R = CHCH$_3$OCOCH$_3$; **8**: M = Ru, R = CHCH$_3$OCOCH$_3$; **9**: M = Ru, R = CHCH$_3$NHCO(CH$_2$)$_2$COOH) (adapted from [69])

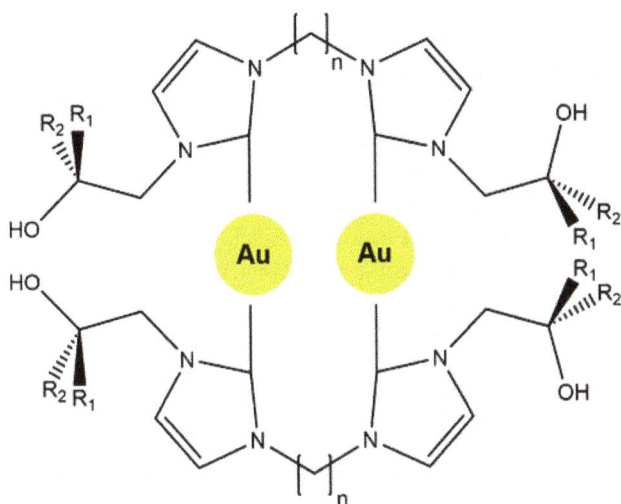

10, 11, 12, 13, 14, 15, 16

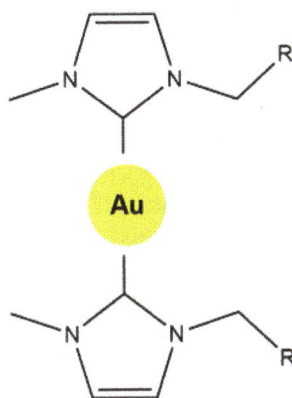

17, 18, 19

Fig. 9 Dinuclear Au^I (**10**: $n=1$, $R_1=R_2=CH_3$; **11**: $n=2$, $R_1=R_2=CH_3$; **12**: $n=2$, $R_1=Ph$, $R_2=H$; **13**: $n=3$, $R_1=Ph$, $R_2=H$), dinuclear Au^{III} (**14**: $n=1$, $R_1=R_2=CH_3$; **15**: $n=2$, $R_1=Ph$, $R_2=H$; **16**: $n=3$, $R_1=Ph$, $R_2=H$) and mononuclear Au^I (**17**: $R = CONHPh$; **18**: R = quinoline; **19**: $R=$ 2,2′-bipyridine) complexes (adapted from [70]). Note: for simplicity, the following are omitted in the figure: two PF$_6^-$ counterions for each one of the **10–16** complexes, two Br atoms coordinated to each Au atom for the complexes **14–16** and one Cl$^-$, PF$_6^-$ or Br$^-$ counterion for complexes **17–19**, respectively

performance compared to the first group, achieving IC$_{50}$ values as low as 0.33 μM for one of the compounds.

Soni and Prakash [71] described the larvicidal effect of AuNP (synthesized using biomass derived from the *Chrysosporium tropicum* fungus and HAuCl$_4$) against *Anopheles stephensi* larvae, a primary mosquito vector of malaria. The exposure to AuNP solutions can lead up to 100 % mortality, depending on the larval instar (developmental stage). Although this study is not a therapy protocol for the disease itself, the larvicidal effect obtained is significant and can lead a better, environmentally safer and greener approach to control this malaria vector in early stages.

Dengue

About the disease

Dengue is a febrile illness that affects infants, young children and adults, in tropical and sub-tropical areas of the world. It is transmitted through the bite of an *Aedes* mosquito, infected with any of the four types of dengue viruses. The symptoms include fever, with severe headache, pain behind the eyes, muscle and joint pain, and rash. They usually appear 3 to 14 days after the insect bite. Severe dengue (plasma leakage, strong hemorrhages, organ failure) is a potentially lethal complication, affecting both children and adults. Naturally, early clinical diagnosis can increase survival of patients [5].

Microscopy assays using nanogold

The reports on the use of microscopy in connection with nanogold, dealing with dengue virus itself, are scarce, probably due to the viral nature of the pathogen, in contrast to the cellular protozoans agents of the other diseases described above. Two works, performed about two decades ago, used immunogold applied to intracellular localization of a dengue antigen:

Thet and Thein [72] showed the potential use of protein-A–Au complex as a marker in the detection of dengue antigen by immunogold labeling for light microscopy. Dengue 4 prototype virus H241, propagated in C6/36 mosquito cell cultures, was used for in vitro experiments. The dengue antigen was detected by using polystyrene beads coated with anti-dengue IgG. The presence of dengue antigen was shown by the formation of pink color (seen with the naked eye) on the beads when incubated with antigen, antibody and protein-A–Au complex. The intensity of the pink color was found to increase with the degree of infection on the infected coverslip cultured cells. The results obtained demonstrated that the use of the protein-A–Au complex had increased sensitivity and was able to detect antigen in the tissue culture at the post infection as early as the first day.

Chen et al. [73] also used immunogold labeling, emphasizing its high sensitivity. A protein-A–Au–Ag staining was used to detect the virus antigens in cultured dengue inoculated C6/36 clone of *Aedes albopictus* cells and human endothelial cells. Data from direct immunofluorescence antibody (DFA) test were compared to immunogold labeling. The study revealed that all DFA-positive specimens were also found positive for immunogold labeling, but not vice versa. This showed that the method using nanogold was more sensitive than DFA.

In a recent study, Vancini et al. [74] analysed the early events in the infection process of dengue and other *Flavivirus*, using electron microscopy and immunogold labeling of viral particles during cell entry. The obtained data supports the hypothesis of cells infection by a mechanism that involves direct penetration of the host cell plasma membrane.

Detection methods using nanogold

Like in other biosensors described for the previously mentioned diseases, gold can be in the bulk (surface) form, being a support for the sensor assembling active part, as described by Kumbhat et al. [75] for serological diagnosis of dengue virus infection, using surface plasmon resonance (SPR) as analytical technique.

Gold can also be in the form of nanoparticles, as in the following examples: Hsu and co-workers [76] described a sensitive AuNP based inductively coupled plasma mass spectrometry (ICP-MS) amplification and magnetic separation method for the detection of oligonucleotide virus-specific RNA sequences. There is a sandwich-type binding of two designed probe sequences (as illustrated in Fig. 10) that specifically recognize the target regions. They have attached (1) magnetic beads for easy separation and (2) AuNP based beads for ICP-MS amplification detection. Compared with the standard methodology (plaque assay) for the quantification of dengue, the method described allows early detection of the virus in complicated and small-volume samples, with high specificity, good analytical sensitivity, and superior time-effectiveness. The process can be described in five main steps: (1) virus lysing, (2) incubation with magnetic probe and washing, (3) incubation with AuNP probe and washing, (4) releasing and dissolving AuNP for (5) ICP-MS analysis.

Nascimento et al. [77] described the preparation and characterization of a novel gold nanoparticle–polyaniline hybrid composite (AuNpPANI, schematized in Fig. 11), containing SH terminal groups with the ability of immobilizing dengue serotype-specific primers (ST1, ST2 and ST3). Electrochemical impedance spectroscopy (EIS) and cyclic voltammetry (CV) were performed. The authors showed that AuNpPANI-ST(1–3) systems (Fig. 12, left) are capable of detecting dengue serotypes with high specificity and reproducibility at picomolar concentrations. The AuNpPANI-ST system

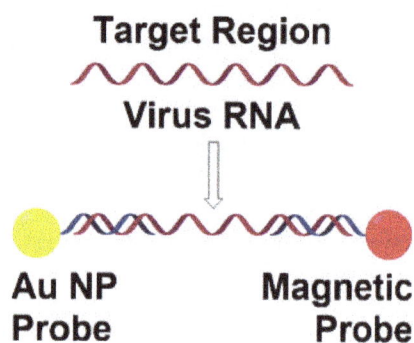

Fig. 10 Sandwich-type binding of two probes to the target region of the RNA virus (adapted from [76])

exhibited a highly selectivity response to the complementary target of human patient's dengue genome (Fig. 12, right). According to the authors, this can be a step in development of dengue serotype biosensors that are functional even in the presence of small volumes and low concentrations of the analyte. Even in those conditions, the CV and EIS results showed unequivocal evidence of an existing interaction between dengue serotype-specific primers and their complementary genomic DNA targets.

The dengue IgM capture ELISA is the immunoenzymatic system recommended by the Pan American Health Organization and the World Health Organization for the serological diagnosis of dengue virus infection, due to its high sensitivity, ease of performance, and use of a single acute-phase serum sample [78]. However, tests with the ELISA system are time-consuming and require equipment for washing, incubation, and reading of the results. In 2003, Vazquez et al. [78] described the use of AuBioDOT, a multistep visual diagnostic immunoassay that uses technology based on the IgM capture ELISA principle. This system uses white polyethylene opaque plates as the solid phase, colloidal gold as the marker, and silver ion amplification. It does not require special equipment, it is totally manually operated, and it can be performed in less than 1 h. The application of AuBioDOT for the detection of anti-dengue virus IgM antibodies is recommended as an alternative method for the diagnosis of dengue virus infection, both for clinical diagnosis and for seroepidemiological surveillance. The system is useful under field conditions and in laboratories and requires little equipment. The AuBioDOT IgM capture test for the detection of anti-dengue virus antibodies is a multistep immunoassay that uses manual operation, visual reading, and colloidal gold-labeled conjugated monoclonal antibodies. Colloidal AuNP were used to make a microfluidics-based bioassay that is able to recognize specific DNA sequences via conformational change-induced fluorescence quenching. In this method, a self-assembled monolayer of AuNP was fabricated on the channel wall of a microfluidic chip, and DNA probes were bonded to the monolayer via thiol groups. This test was applied for the detection of the PCR product of dengue virus and results indicate that the assay is specific for the target gene [79].

Oliveira et al. [80, 81] immobilized concanavalin A lecitin on gold electrode using AuNP and polyvinyl butyral and put the biosensor in contact to sera from patients infected by dengue. Changes passible to be detected by cyclic voltammetry and electrochemical impedance spectroscopy occurred.

As Foguel et al. [29] suggested for *T. Cruzi* detection, Cavalcanti et al. [82] also used a gold film electrode obtained from a recordable compact disk, but for the detection of non-structural protein 1 (NS1) of the dengue virus. This protein is abundantly present in blood during the acute phase of the dengue infection, in a correlation with viremia

Fig. 11 Schematic representation of AuNpPANI synthesis (adapted from [77])

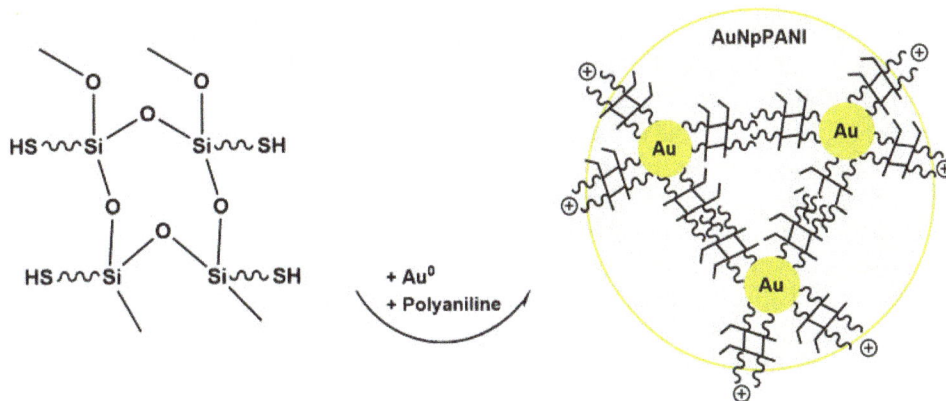

levels and, consequently, can be used to early diagnostic of the dengue hemorrhagic fever. These disposable gold electrodes with anti-NS1, successfully immobilized onto gold film surface via protein-A, present a high sensitivity (linear response from 1 to 100 ng_{NS1}/ml) with a relatively low detection limit (0.33 ng_{NS1}/ml) and selectivity in a electrochemical label-free detection technique. This same protein (NS1) was also the object of the study of Muller et al. [83], using microprojection arrays, a surface where are distributed micrometric cones or other shapes with sharp points, capable of easy skin penetration (with more than 20,000 projections/cm^2). They are usually applied to skin for the purpose of delivering high molecular weight compounds which cannot be delivered efficiently by other means, but used here for directly extracting circulating protein biomarkers from the skin. The use of gold on microprojection arrays deals with their coating in order to allow a better grafting (eventually with help of other small linking molecules) of biomolecules like anti-NS1 antibody for NS1 capture in the subject skin epithelia extracellular fluid. The detection of captured NS1 is made using an anti-NS1 antibody conjugated to HRP, using the principle described in Fig. 2, however with a support–antibody–antigen–antibody HRP assembling instead of the support–antigen–antibody–antibody HRP. This test surpasses others commonly used (like ELISA), which only accept serum/plasma samples, thus requiring significant laboratory-based processing and more invasive procedures for blood collecting.

There are other processes involving nanogold for dengue research and detection, but only slightly different than the several techniques already mentioned: Andrade et al. [84] showed an assembly consisting of a AuNpPANI on a gold

electrode, as already shown in Fig. 12; however, this system recognizes dengue glycoproteins instead of DNA specific regions. Chen et al. [85] used a QCM similar to a malaria test mentioned above, but supporting DNA dengue specific probes and also using a signal amplification technique with AuNP probes as those shown in Fig. 10. Oliveira et al. [86] also used the principles described in the last paragraph, but employed a different lecitin (from *Cratylia mollis* seeds) and Fe_3O_4 nanoparticles.

Unfortunately, until the time this review was written, to the best of our knowledge, there were no reports on the use of gold for dengue therapy, being this an important area of future research.

Conclusions

The results reported above strongly evidence the important role of gold in tropical diseases research. In terms of *imaging*, gold has high electron density which increases electron scatter enabling high contrast images, allowing a new understanding into the molecular immune pathogenesis of these maladies. Immunogold techniques contributed so far to understand the hemoglobin destruction, a critical aspect in the leishmaniasis disease, and also to the localization of an important malaria parasite protein. In terms of dengue, the combination of immunogold labeling and electron microscopy allowed the detection of virus antigens, and to obtain data supporting a mechanism of cells infection by direct penetration of the host cell plasma membrane.

To what concerns the *detection* of tropical diseases, important findings have also been achieved with gold. The research has been growing and several sensors can now assure easier diagnosis, namely in places where the comfort and resources of the laboratory are absent. As described above, gold can be in the bulk (surface) form, being a support for the sensor assembling active part, or in nanoparticles. Au electrodes showed advantages (as they are not influenced by oxygen dissolved in solution) for

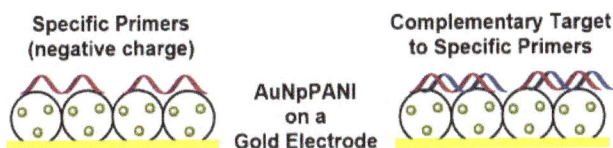

Fig. 12 Schematic representation the AuNpPANI on a gold electrode biosensor assembling (adapted from [77])

Chagas disease detection, compared to Pt analogues. For sleeping sickness, one parasite in 180 μL of blood was still detectable in laboratory settings using gold as a colorimetric indicator, which is a remarkable result. Promising test kits using colloidal gold were reported for leishmaniasis and malaria. For the former, a novel cheaper and simplified SELEX procedure using colloidal gold has been reported. Interesting amperometric immunosensors containing gold were used for monitoring malaria. For dengue detection, a fast, fully manual, visual diagnostic immunoassay comprising the IgM capture ELISA principle, with colloidal gold as marker, was reported as an alternative to conventional ELISA systems, which are time-consuming and require a large quantity of equipment. The use of gold on microprojection arrays also surpasses ELISA. Several sensitive AuNP-based tests have also been reported for the efficient early detection of the dengue virus in complicated and small-volume samples.

In terms of disease *therapy*, gold complexes, like the antiarthritic drug Auranofin, showed potential antileishmanial and antimalarial activity. Other gold complexes are promising chemotherapeutic alternatives in the search for the cure of several tropical diseases, namely, $AuCl_3(CTZ)$ against Chagas disease, $[Au(dppz)_2]Cl_3$ for leishmaniasis, $[Au(PPh_3)(CQ)]PF_6$ and other Au(I) and Au(III) complexes containing CQ and NHC ligands for malaria. The latter were found to not cause hemolysis, in contrast to some silver analogues. AuNP also have been reported to have interesting roles in therapy, namely against *Leishmania* promastigotes in culture medium, becoming promising for flexible "band-aids" to be used skin lesions. AuNP also showed an interesting larvicidal effect for a mosquito vector of malaria, which can be a encouraging approach to control the malaria vector in early stages.

In terms of *vaccines*, the role of gold in biolistic immunization using a stable DNA–gold precipitate, not needing a cold chain, is also an attractive method for vaccination against Chagas disease in rural areas. Gold labeling showed that monoclonal antibodies bind to glycoproteins, pointing to these as potential candidates for a vector-directed transmission-blocking malaria vaccine, stopping the parasite cycle in the ookinetes–oocysts transition step.

After all that has been investigated on this topic, the overall conclusion is that the potential of gold for stimulating research in human tropical diseases is considerable. New results will certainly come out soon and lead to even more practical and commercial applications, the full extent of which has still to be envisaged.

Acknowledgments The authors are grateful to the financial support for this work provided by projects PTDC/QUI-QUI/100682/2008 and PEst-C/EQB/LA0020/2011 financed by FEDER through COMPETE, and by Fundação para a Ciência e a Tecnologia (FCT), and for CIENCIA 2007 program (for SACC).

References

1. WHO (2012) TDR — Diseases and topics — African trypanosomiasis. http://www.who.int/tdr/diseases-topics/african-trypanosomiasis/en/index.html. Accessed 20th August 2012
2. WHO (2012) TDR — Diseases and Topics — Chagas. http://www.who.int/tdr/diseases-topics/chagas/en/index.html. Accessed 20th August 2012
3. WHO (2012) TDR — Diseases and topics — Leishmaniasis. http://www.who.int/tdr/diseases-topics/leishmaniasis/en/index.html. Accessed 20th August 2012
4. WHO (2012) TDR — Diseases and topics — Malaria. http://www.who.int/tdr/diseases-topics/malaria/en/index.html. Accessed 20th August 2012
5. WHO (2012) TDR — Diseases and topics — Dengue. http://www.who.int/tdr/diseases-topics/dengue/en/index.html. Accessed 20th August 2012
6. Bond GC, Louis C, Thompson DT (2006) Catalysis by gold, vol 6. Catalytic science series. Imperial College Press, London
7. Carabineiro SAC, Thompson DT (2007) Catalytic applications for gold nanotechnology. In: Heiz EU, Landman U (eds) Nanocatalysis. Springer-Verlag, Berlin, pp 377–489
8. Carabineiro SAC, Thompson DT (2010) Gold catalysis. In: Corti C, Holliday R (eds) Gold: science and applications. CRC Press, Taylor and Francis Group, Boca Raton, pp 89–122
9. de Almeida MP, Carabineiro SAC (2012) The best of two worlds from the gold catalysis universe: making homogeneous heterogeneous. ChemCatChem 4:18–29.
10. Fricker SP (1996) Medical uses of gold compounds: past, present and future. Gold Bull 29:53–60
11. Bawarski WE, Chidlowsky E, Bharali DJ, Mousa SA (2008) Emerging nanopharmaceuticals. Nanomed Nanotechnol Biol Med 4:273–282.
12. Boisselier E, Astruc D (2009) Gold nanoparticles in nanomedicine: preparations, imaging, diagnostics, therapies and toxicity. Chem Soc Rev 38:1759–1782.
13. Arvizo R, Bhattacharya R, Mukherjee P (2010) Gold nanoparticles: opportunities and challenges in nanomedicine. Expert Opin Drug Deliv 7:753–763.
14. Berners-Price SJ, Filipovska A (2011) Gold compounds as therapeutic agents for human diseases. Metallomics: Integr Biometal Sci 3:863–873.
15. Hemphill A, Ross CA (1995) Flagellum-mediated adhesion of *Trypanosoma congolense* to bovine aorta endothelial cells. Parasitol Res 81:412–420.
16. Abdo J, Kristersson T, Seitzer U, Renneker S, Merza M, Ahmed J (2010) Development and laboratory evaluation of a lateral flow device (LFD) for the serodiagnosis of *Theileria annulata* infection. Parasitol Res 107:1241–1248.
17. Alvarez I, Gutierrez G, Barrandeguy M, Trono K (2010) Immunochromatographic lateral flow test for detection of antibodies to Equine infectious anemia virus. J Virol Methods 167:152–157.
18. Rassi A Jr, Rassi A, Marin-Neto JA (2010) Chagas disease. Lancet 375:1388–1402.

19. Magez S, Geuskens M, Beschin A, del Favero H, Verschueren H, Lucas R, Pays E, de Baetselier P (1997) Specific uptake of tumor necrosis factor-α is involved in growth control of *Trypanosoma brucei*. J Cell Biol 137:715–727.

20. Soeiro MN, Paiva MM, Barbosa HS, Meirelles MN, Araújo-Jorge TC (1999) A cardiomyocyte mannose receptor system is involved in *Trypanosoma cruzi* invasion and is down-modulated after infection. Cell Struct Funct 24:139–149

21. Okuda K, Esteva M, Segura EL, Bijovsky AT (1999) The cytostome of *Trypanosoma cruzi* epimastigotes is associated with the flagellar complex. Exp Parasitol 92:223–231.

22. Monteiro ACS, Abrahamson M, Lima A, Vannier-Santos MA, Scharfstein J (2001) Identification, characterization and localization of chagasin, a tight-binding cysteine protease inhibitor in *Trypanosoma cruzi*. J Cell Sci 114:3933–3942

23. Silva EO, Saraiva EMB, de Souza W, Souto-Padron T (1998) Cell surface characterization of amastigotes of *Trypanosoma cruzi* obtained from different sources. Parasitol Res 84:257–263.

24. Acosta DM, Soprano LL, Ferrero M, Landoni M, Esteva MI, Couto AS, Duschak VG (2011) A striking common *O*-linked *N*-acetylglucosaminyl moiety between cruzipain and myosin. Parasite Immunol 33:363–370.

25. Eger I, Soares MJ (2012) Endocytosis in *Trypanosoma cruzi* (Euglenozoa: Kinetoplastea) epimastigotes: visualization of ingested transferrin–gold nanoparticle complexes by confocal laser microscopy. J Microbiol Methods 91:101–105.

26. Diniz FB, Ueta RR, Pedrosa AMD, Areias MD, Pereira VRA, Silva ED, da Silva JG, Ferreira AGP, Ferreira AGP, Gomes YM, Gomes YM (2003) Impedimetric evaluation for diagnosis of Chagas' disease: antigen–antibody interactions on metallic eletrodes. Biosens Bioelectron 19:79–84.

27. Ferreira AAP, Colli W, da Costa PI, Yamanaka H (2005) Immunosensor for the diagnosis of Chagas' disease. Biosens Bioelectron 21:175–181.

28. Ribone ME, Belluzo MS, Pagani D, Macipar MS, Lagier CM (2006) Amperometric bioelectrode for specific human immunoglobulin G determination: optimization of the method to diagnose American trypanosomiasis. Anal Biochem 350:61–70.

29. Foguel MV, dos Santos GP, Pupim Ferreira AA, Yamanaka H, Benedetti AV (2011) Amperometric immunosensor for Chagas' disease using gold CD-R transducer. Electroanalysis 23:2555–2561.

30. Pereira SV, Bertolino FA, Fernandez-Baldo MA, Messina GA, Salinas E, Sanz MI, Raba J (2011) A microfluidic device based on a screen-printed carbon electrode with electrodeposited gold nanoparticles for the detection of IgG anti-*Trypanosoma cruzi* antibodies. Analyst 136:4745–4751.

31. Deborggraeve S, Claes F, Laurent T, Mertens P, Leclipteux T, Dujardin JC, Herdewijn P, Buescher P (2006) Molecular dipstick test for diagnosis of sleeping sickness. J Clin Microbiol 44:2884–2889.

32. Nyarko E, Hara T, Grab DJ, Habib A, Kim Y, Nikolskaia O, Fukuma T, Tabata M (2004) In vitro toxicity of palladium(II) and gold(III) porphyrins and their aqueous metal ion counterparts on *Trypanosoma brucei brucei* growth. Chemico Biol Interact 148:19–25.

33. Sánchez-Delgado RA, Lazardi K, Rincón L, Urbina JA, Hubert AJ, Noels AN (1993) Toward a novel metal-based chemotherapy against tropical diseases: 1. Enhancement of the efficacy of clotrimazole against *Trypanosoma cruzi* by complexation to ruthenium in RuCl$_2$(clotrimazole)$_2$. J Med Chem 36:2041–2043

34. Sánchez-Delgado RA, Navarro M, Pérez H, Urbina JA (1996) Toward a novel metal-based chemotherapy against tropical diseases: 2. Synthesis and antimalarial activity *in vitro* and *in vivo* of new ruthenium– and rhodium–chloroquine complexes. J Med Chem 39:1095–1099

35. Navarro M, Pérez H, Sánchez-Delgado RA (1997) Toward a novel metal-based chemotherapy against tropical diseases: 3. Synthesis and antimalarial activity in vitro and in vivo of the new gold–chloroquine complex [Au(PPh$_3$)(CQ)]PF$_6$. J Med Chem 40:1937–1939.

36. Sánchez-Delgado RA, Navarro M, Lazardi K, Atencio R, Capparelli M, Vargas F, Urbina JA, Bouillez A, Noels AF, Masi D (1998) Toward a novel metal-based chemotherapy against tropical diseases: 4. Synthesis and characterization of new metal-clotrimazole complexes and evaluation of their activity against *Trypanosoma cruzi*. Inorg Chim Acta 275–276:528–540

37. Navarro M, Lehmann T, Cisneros-Fajardo EJ, Fuentes A, Sánchez-Delgado RA, Silva P, Urbina JA (2000) Toward a novel metal-based chemotherapy against tropical diseases: Part 5. Synthesis and characterization of new Ru(II) and Ru(III) clotrimazole and ketoconazole complexes and evaluation of their activity against Trypanosoma cruzi. Polyhedron 19:2319–2325.

38. Navarro M, Cisneros-Fajardo EJ, Lehmann T, Sánchez-Delgado RA, Atencio R, Silva P, Lira R, Urbina JA (2001) Toward a novel metal-based chemotherapy against tropical diseases: 6. Synthesis and characterization of bew copper(II) and gold(I) clotrimazole and ketoconazole complexes and evaluation of their activity against *Trypanosoma cruzi*. Inorg Chem 40:6879–6884.

39. Navarro M, Vasquez F, Sanchez-Delgado RA, Perez H, Sinou V, Schrevel J (2004) Toward a novel metal-based chemotherapy against tropical diseases: 7. Synthesis and in vitro antimalarial activity of new gold–chloroquine complexes. J Med Chem 47:5204–5209.

40. Navarro M (2009) Gold complexes as potential anti-parasitic agents. Coord Chem Rev 253:1619–1626.

41. Bryan M, Guyach S, Norris K (2013) Biolistic DNA vaccination against trypanosoma infection. In: Sudowe S, Reske-Kunz AB (eds) Biolistic DNA delivery, vol. 940. Methods in molecular biology. Humana Press, pp 305–315.

42. Santos MAM, de Andrade PP, de Andrade CR, Padovan PA, de Souza W (1991) Effect of trypsin and 2-mercaptoethanol on the exposure of sugar residues on the surface of *Leishmania donovani chagasi*. Parasitol Res 77:553–557

43. Borges VM, Vannier-Santos MA, de Souza W (1998) Subverted transferrin trafficking in *Leishmania*-infected macrophages. Parasitol Res 84:811–822.

44. Sengupta S, Tripathi J, Tandon R, Raje M, Roy RP, Basu SK, Mukhopadhyay A (1999) Hemoglobin endocytosis in *Leishmania* is mediated through a 46-kDa protein located in the flagellar pocket. J Biol Chem 274:2758–2765.

45. Zijlstra EE, Nur Y, Desjeux P, Khalil EAG, El-Hassan AM, Groen J (2001) Diagnosing visceral leishmaniasis with the recombinant K39 strip test: experience from the Sudan. Trop Med Int Health 6:108–113.

46. Carreira JCA, da Silva AVM, de Pita PD, Brazil RP (2012) Natural infection of *Didelphis aurita* (Mammalia: Marsupialia) with *Leishmania infantum* in Brazil. Parasites Vectors 5:111–115.

47. Ramos-Jesus J, Carvalho KA, Fonseca RAS, Oliveira GGS, Barrouin Melo SM, Alcantara-Neves NM, Dutra RF (2011) A piezoelectric immunosensor for *Leishmania chagasi* antibodies in canine serum. Anal Bioanal Chem 401:917–925.

48. Moreno M, Rincon E, Pineiro D, Fernandez G, Domingo A, Jimenez-Ruiz A, Salinas M, Gonzalez VM (2003) Selection of aptamers against KMP-11 using colloidal gold during the SELEX process. Biochem Biophys Res Commun 308:214–218.

49. Moreno M, Gonzalez VM, Rincon E, Domingo A, Dominguez E (2011) Aptasensor based on the selective electrodeposition of protein-linked gold nanoparticles on screen-printed electrodes. Analyst 136:1810–1815.

50. Ilari A, Baiocco P, Messori L, Fiorillo A, Boffi A, Gramiccia M, Di Muccio T, Colotti G (2012) A gold-containing drug against parasitic polyamine metabolism: the X-ray structure of trypanothione reductase from *Leishmania infantum* in complex with auranofin reveals a dual mechanism of enzyme inhibition. Amino Acids 42:803–811.

51. Navarro M, Hernandez C, Colmenares I, Hernandez P, Fernandez M, Sierraalta A, Marchan E (2007) Synthesis and characterization of [Au(dppz)$_2$]Cl$_3$. DNA interaction studies and biological activity against *Leishmania (L) mexicana*. J Inorg Biochem 101:111–116.

52. Barboza-Filho CG, Cabrera FC, Dos Santos RJ, De Saja Saez JA, Job AE (2012) The influence of natural rubber/Au nanoparticle membranes on the physiology of *Leishmania brasiliensis*. Exp Parasitol 130:152–158.

53. Sazgarnia A, Taheri AR, Soudmand S, Parizi AJ, Rajabi O, Darbandi MS (2013) Antiparasitic effects of gold nanoparticles with microwave radiation on promastigots and amastigotes of *Leishmania major*. Int J Hyperth 29:79–86.

54. Bhowmick IP, Kumar N, Sharma S, Coppens I, Jarori GK (2009) *Plasmodium falciparum* enolase: stage-specific expression and sub-cellular localization. Malaria Journal 8. doi:10.1186/1475-2875-8-179

55. Chugh M, Gulati BR, Gakhar SK (2010) Monoclonal antibodies AC-43 and AC-29 disrupt *Plasmodium vivax* development in the Indian malaria vector *Anopheles culicifacies* (Diptera: culicidae). J Biosci 35:87–94.

56. Lopes da Silva M, Thieleke-Matos C, Cabrita-Santos L, Ramalho JS, Wavre-Shapton ST, Futter CE, Barral DC, Seabra MC (2012) The host endocytic pathway is essential for *Plasmodium berghei* late liver stage development. Traffic 13:1351–1363.

57. Stevens DY, Petri CR, Osborn JL, Spicar-Mihalic P, McKenzie KG, Yager P (2008) Enabling a microfluidic immunoassay for the developing world by integration of on-card dry reagent storage. Lab Chip 8:2038–2045.

58. Sharma MK, Agarwal GS, Rao VK, Upadhyay S, Merwyn S, Gopalan N, Rai GP, Vijayaraghavan R, Prakash S (2010) Amperometric immunosensor based on gold nanoparticles/alumina sol–gel modified screen-printed electrodes for antibodies to *Plasmodium falciparum* histidine rich protein-2. Analyst 135:608–614.

59. Fu E, Liang T, Spicar-Mihalic P, Houghtaling J, Ramachandran S, Yager P (2012) Two-dimensional paper network format that enables simple multistep assays for use in low-resource settings in the context of malaria antigen detection. Anal Chem 84:4574–4579.

60. Nash MA, Waitumbi JN, Hoffman AS, Yager P, Stayton PS (2012) Multiplexed enrichment and detection of malarial biomarkers using a stimuli-responsive iron oxide and gold nanoparticle reagent system. ACS Nano 6:6776–6785.

61. Potipitak T, Ngrenngarmlert W, Promptmas C, Chomean S, Ittarat W (2011) Diagnosis and genotyping of *Plasmodium falciparum* by a DNA biosensor based on quartz crystal microbalance (QCM). Clin Chem Lab Med 49:1367–1373.

62. Guirgis BSS, Sa e Cunha C, Gomes I, Cavadas M, Silva I, Doria G, Blatch GL, Baptista PV, Pereira E, Azzazy HME, Mota MM, Prudencio M, Franco R (2012) Gold nanoparticle-based fluorescence immunoassay for malaria antigen detection. Anal Bioanal Chem 402:1019–1027.

63. Piper RC, Buchanan I, Choi YH, Makler MT (2011) Opportunities for improving pLDH-based malaria diagnostic tests. Malaria Journal 10.

64. Peng Y, Wu J, Wang J, Li W, Yu S (2012) Study and evaluation of Wondfo rapid diagnostic kit based on nanogold immunochromatography assay for diagnosis of *Plasmodium falciparum*. Parasitol Res 110:1421–1425.

65. Navarro M, Castro W, Martinez A, Sanchez-Delgado RA (2011) The mechanism of antimalarial action of [Au(CQ)(PPh$_3$)]PF$_6$: structural effects and increased drug lipophilicity enhance heme aggregation inhibition at lipid/water interfaces. J Inorg Biochem 105:276–282.

66. Blackie MAL, Beagley P, Chibale K, Clarkson C, Moss JR, Smith PJ (2003) Synthesis and antimalarial activity in vitro of new heterobimetallic complexes: Rh and Au derivatives of chloroquine and a series of ferrocenyl-4-amino-7-chloroquinolines. J Organomet Chem 688:144–152.

67. Fricker SP, Mosi RM, Cameron BR, Baird I, Zhu Y, Anastassov V, Cox J, Doyle PS, Hansell E, Lau G, Langille J, Olsen M, Qin L, Skerlj R, Wong RSY, Santucci Z, McKerrow JH (2008) Metal compounds for the treatment of parasitic diseases. J Inorg Biochem 102:1839–1845.

68. Sannella AR, Casini A, Gabbiani C, Messori L, Bilia AR, Vincieri FF, Majori G, Severini C (2008) New uses for old drugs. Auranofin, a clinically established antiarthritic metallodrug, exhibits potent antimalarial effects *in vitro*: mechanistic and pharmacological implications. FEBS Lett 582:844–847.

69. Bjelosevic H, Guzei IA, Spencer LC, Persson T, Kriel FH, Hewer R, Nell MJ, Gut J, Van Rensburg CEJ, Rosenthal PJ, Coates J, Darkwa J, Elmroth SKC (2012) Platinum(II) and gold(I) complexes based on 1,1'-bis(diphenylphosphino)metallocene derivatives: synthesis, characterization and biological activity of the gold complexes. J Organomet Chem 720:52–59.

70. Hemmert C, Fabié A, Fabre A, Benoit-Vical F, Gornitzka H (2013) Synthesis, structures, and antimalarial activities of some silver(I), gold(I) and gold(III) complexes involving *N*-heterocyclic carbene ligands. Eur J Med Chem 60:64–75.

71. Soni N, Prakash S (2012) Entomopathogenic fungus generated nanoparticles for enhancement of efficacy in *Culex quinquefasciatus* and *Anopheles stephensi*. Asian Pac J Trop Dis 2:S356–S361.

72. Thet W, Thein T (1987) Rapid and sensitive detection of dengue viral antigen using immunogold in light microscopy and solid phase gold immunoassay (SPGIA). Microbiol Immunol 31:183–188

73. Chen WJ, Chen SL, Fang AH, Wang MT (1993) Detection of dengue virus antigens in cultured cells by using protein A–gold–silver staining (pAgs) method. Microbiol Immunol 37:359–363

74. Vancini R, Kramer LD, Ribeiro M, Hernandez R, Brown D (2013) Flavivirus infection from mosquitoes in vitro reveals cell entry at the plasma membrane. Virology 435:406–414.

75. Kumbhat S, Sharma K, Gehlot R, Solanki A, Joshi V (2010) Surface plasmon resonance based immunosensor for serological diagnosis of dengue virus infection. J Pharm Biomed Anal 52:255–259.

76. Hsu IH, Chen W-H, Wu T-K, Sun Y-C (2011) Gold nanoparticle-based inductively coupled plasma mass spectrometry amplification

and magnetic separation for the sensitive detection of a virus-specific RNA sequence. J Chromatogr A 1218:1795–1801.

77. Nascimento HPO, Oliveira MDL, de Melo CP, Silva GJL, Cordeiro MT, Andrade CAS (2011) An impedimetric biosensor for detection of dengue serotype at picomolar concentration based on gold nanoparticles–polyaniline hybrid composites. Colloids Surf B: Biointerfaces 86:414–419.

78. Vazquez S, Lemos G, Pupo M, Ganzon O, Palenzuela D, Indart A, Guzman MG (2003) Diagnosis of dengue virus infection by the visual and simple AuBioDOT immunoglobulin M Capture system. Clin Diagn Lab Immunol 10:1074–1077.

79. Li YT, Liu HS, Lin HP, Chen SH (2005) Gold nanoparticles for microfluidics-based biosensing of PCR products by hybridization-induced fluorescence quenching. Electrophoresis 26:4743–4750.

80. Oliveira MDL, Correia MTS, Diniz FB (2009) Concanavalin A and polyvinyl butyral use as a potential dengue electrochemical biosensor. Biosens Bioelectron 25:728–732.

81. Oliveira MDL, Correia MTS, Diniz FB (2009) A novel approach to classify serum glycoproteins from patients infected by dengue using electrochemical impedance spectroscopy analysis. Synth Met 159:2162–2164.

82. Cavalcanti IT, Guedes MIF, Sotomayor MDPT, Yamanaka H, Dutra RF (2012) A label-free immunosensor based on recordable compact disk chip for early diagnostic of the dengue virus infection. Biochem Eng J 67:225–230.

83. Muller DA, Corrie SR, Coffey J, Young PR, Kendall MA (2012) Surface modified microprojection arrays for the selective extraction of the dengue virus NS1 protein as a marker for disease. Anal Chem 84:3262–3268.

84. Andrade CAS, Oliveira MDL, de Melo CP, Coelho LCBB, Correia MTS, Nogueira ML, Singh PR, Zeng X (2011) Diagnosis of dengue infection using a modified gold electrode with hybrid organic–inorganic nanocomposite and *Bauhinia monandra* lectin. J Colloid Interface Sci 362:517–523.

85. Chen S-H, Chuang Y-C, Lu Y-C, Lin H-C, Yang Y-L, Lin C-S (2009) A method of layer-by-layer gold nanoparticle hybridization in a quartz crystal microbalance DNA sensing system used to detect dengue virus. Nanotechnology 20:215501.

86. Oliveira MDL, Nogueira ML, Correia MTS, Coelho LCBB, Andrade CAS (2011) Detection of dengue virus serotypes on the surface of gold electrode based on *Cratylia mollis* lectin affinity. Sensors Actuators B Chem 155:789–795.

87. Vaughan S, Gull K (2003) The trypanosome Flagellum. J Cell Sci 116:757–759.

88. Chaves CR, Fontes A, Farias PMA, Santos BS, de Menezes FD, Ferreira RC, Cesar CL, Galembeck A, Figueiredo RCBQ (2008) Application of core–shell PEGylated CdS/Cd(OH)$_2$ quantum dots as biolabels of *Trypanosoma cruzi* parasites. Appl Surf Sci 255:728–730.

5

Sensitive electrogenerated chemiluminescence peptide-based biosensor for the determination of troponin I with gold nanoparticles amplification

Meng Shan · Min Li · Xiaoying Qiu · Honglan Qi · Qiang Gao · Chengxiao Zhang

Abstract A sensitive electrogenerated chemiluminescence (ECL) peptide-based biosensor was fabricated for the determination of troponin I (TnI) by employing gold nanoparticles as amplification platform. Two specific peptides including peptide1 with a sequence of CFYSHSFHENWPS and peptide2 with a sequence of FYSHSFHENWPSK were employed as capture peptide and report peptide, respectively. The peptide2 was labeled with ruthenium bis(2,2′-bipyridine) (2,2′-bipyridine-4,4′-dicarboxylic acid)-N-hydroxysuccinimide ester (Ru(bpy)$_2$(dcbpy)NHS) at NH$_2$-containing lysine via acylation reaction and utilized as the ECL probe. Gold nanoparticles were electrodeposited onto gold electrode and used as an amplification platform. The peptide-based biosensor was fabricated by self-assembling peptide1 onto the surface of gold nanoparticles-modified gold electrode through a thiol-containing cysteine at the end of the peptide1. When the biosensor reacted with target TnI, and then incubated with the ECL probe, a strong ECL response was electrochemically generated. The ECL intensity is directly proportional to the logarithm of the concentration of TnI in the range from 1 to 300 pg/mL. The biosensor employing gold nanoparticles as amplification platform shows high sensitivity for the detection of TnI with a detection limit of 0.4 pg/mL (S/N=3). Moreover, the biosensor is successfully applied to analysis of TnI in human serum sample. This work demonstrates that the combination of a highly binding peptide with nanoparticle amplification is a great promising approach for the design of ECL biosensor.

Keywords Electrogenerated chemiluminescence · Gold nanoparticles · Peptide · Troponin I

Introduction

Acute myocardial infarction is a major cause of human death and responsible for one third of deaths in the world. Measurement of cardiac markers is critical in assisting the diagnosis of acute myocardial infarction [1, 2]. Cardiac troponin I (TnI), with a molecular weight of 24 kDa, a part of the troponin complex that is present in cardiac muscle tissues, has been known as a reliable biomarker of cardiac muscle tissue injury and was widely used in the early diagnosis of acute myocardial infarction [3, 4]. The concentration of TnI in blood rises rapidly within 4–6 h after the onset of an acute myocardial infarction and reaches to the maximum at approximately 12 h. After 6–8 days, the TnI level returns to normal, and thus, the concentration of TnI in blood can provide a long diagnostic window for detecting cardiac injury [3]. A variety of methods such as colorimetric [5], electrochemical [2, 3], fluorescent [6, 7], and chemiluminescence [8] methods have been developed to determine TnI. Despite the extensive development of these methods, the major limitation in currently used methods for the determination of TnI assays is low sensitivity at the time of a patient's presentation, owing to a delayed increase in circulating levels of cardiac troponin [2]. Therefore, it is still a critical demand on sensitive and specific methods for the determination of TnI, especially in the point-of-care applications.

Electrogenerated chemiluminescence (also called electrochemiluminescence and abbreviated ECL) method

M. Shan · M. Li · X. Qiu · H. Qi (✉) · Q. Gao · C. Zhang (✉)
Key Laboratory of Analytical Chemistry for Life Science of Shaanxi Province, School of Chemistry and Chemical Engineering, Shaanxi Normal University, Xi'an 710062, People's Republic of China
e-mail: honglanqi@snnu.edu.cn
e-mail: cxzhang@snnu.edu.cn

has attracted considerable interest due to its high sensitivity, rapidity, easy controllability, and wide dynamic range [9, 10]. Several ECL methods have been developed for the determination of TnI [11–15]. Cui designed ECL immunosensor for the detection of human cardiac troponin I by using luminol [11] and *N*-(aminobutyl)-*N*-(ethylisoluminol) [12] -functionalized gold nanoparticles as labels. Smith developed an ECL immunoassay for detection of rat TnI in serum [13]. The commonly used ECL immunoassays are normally conducted by employing antibodies as molecular recognition elements. However, the antibody drawbacks associated with the production and stability. Short linear binding peptides, which are obtained using phage display, have received considerable interest in protein analysis due to the advantages, such as stable, resistant to harsh environments, and more amenable to engineering at the molecular level than antibodies [16]. Recently, Park et al. reported a new peptide (FYSHSFHENWPS) that selectively bound to TnI with a disassociation constant of the complex in nanomolar level [17]. We developed two homogeneous ECL peptide-based methods for the determination of TnI using this peptide as a molecular recognition element [14, 15]. In previously work [14], one peptide (FYSHSFHENWPSK), as a molecular recognition element, was labeled with ruthenium complex through NH_2-containing lysine on the peptide via acylation reaction and utilized as an ECL probe. In the presence of TnI, a decrease in ECL signal was observed upon the binding event between the ECL probe and target TnI. The binding of small peptide with large target protein results in a sensitive ECL detection of protein compared with homogeneous immunoassay employing antibody as recognition element. However, the detection limit of previously homogeneous ECL method (1.2×10^{-10} g/mL) is limited due to the high background and only one ECL molecule is attached directly to each peptide.

The elaboration of ECL biosensors is probably one of the most promising ways to solve some of the problems concerning sensitivity, speediness, and stability. And much effort has been devoted to improve the sensitivity of ECL biosensors, such as employment of the nanoparticles-based signal amplification strategy. Nanoparticles including carbon nanotube, metal nanoparticles were employed as the amplification platform for the immobilization of molecular recognition elements [18], such as antibody [19], aptamer [20], carbohydrate [21], or peptide[22]. Gold nanoparticles (GNPs), with unique properties such as their fascinating electrocatalytic activity, large surface area, excellent conductivity, and stability, have been widely used in designing ECL biosenesors [23, 24]. Generally, the ECL biosensors utilize GNPs for the modification of the substrate electrodes, which provide large electrode area and also facilitate the electron transfer between the ECL signals and the electrodes, thus affording the possibility of the improvement of ECL

performance. The unique properties of GNPs-modified electrode interfaces lead to novel ECL biosensors with high sensitivity and good stability in immunoassay [25], DNA bioassay [26], and glycan biosensor [27]. In an alternative way, GNPs can work as carriers of conventional ECL signals such as luminol [11] and ruthenium complex [28] and, thus, afford substantial ECL signal amplification. We developed an ECL immunoassay for the determination of human immunoglobulin G at gold nanoparticles-modified paraffin-impregnated graphite electrode [29], an ECL DNA biosensor at gold nanoparticles-modified gold electrode [30] and a signal off aptasensor for the determination of thrombin at gold nanoparticles-modified gold electrode [31] with high sensitivity.

The aim of this work is to develop a highly sensitive ECL peptide-based method for the determination of protein, on basis of the idea of encompassing gold nanoparticles as amplification platform and peptide as molecular recognition element. The principle scheme is demonstrated in Fig. 1. Two specific peptides including peptide1 with a sequence of CFYSHSFHENWPS, in which a thiol-containing cysteine residue was incorporated at the end of the specific peptide to facilitate self-assembly onto the surface of gold, peptide2 with a sequence of FYSHSFHENWPSK, in which a NH_2-containing lysine residue was incorporated at the end of the peptide to covalently couple with ECL signal, were designed according to ref. [17] and employed as capture peptide and report peptide, respectively. The peptide2 was labeled with ruthenium bis(2,2'-bipyridine) (2,2'-bipyridine-4,4'-dicarboxylic acid)-*N*-hydroxysuccinimide ester (Ru(bpy)$_2$(dcbpy)NHS) via acylation reaction through NH_2-containing lysine at the end of the peptide to form the ECL probe Ru-peptide2. Gold nanoparticles were electrodeposited onto gold electrode and used as an amplification platform. The ECL peptide-based biosensor was fabricated by self-assembling the peptide1 onto a gold nanoparticles-modified gold electrode surface through a thiol-containing cysteine at the end of the peptide1. When the biosensor reacted with target TnI, and then incubated with the ECL probe, a strong ECL response was electrochemically generated. In this paper, the characteristics of the ECL probe and the ECL peptide-based biosensor and the analytical performance for TnI are presented.

Experimental

Reagents and apparatus

Two peptides chemically synthesized, including peptide1 with a sequence of CFYSHSFHENWPS (13 mer, MW=1,640.77), peptide2 with a sequence of FYSHSFHENWPSK (13 mer, MW=1,665.80), were designed according to ref. [17] and purchased from Sinoasis Pharmaceuticals, Inc

Fig. 1 Schematic representation
of the ECL peptide-based
biosensor for the determination
of TnI

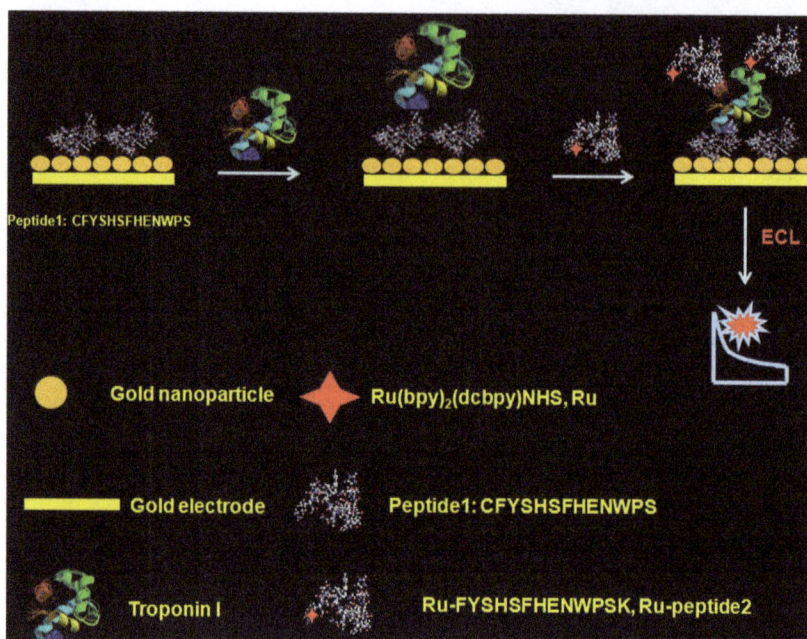

(China). Cardiac troponin I (TnI, human heart) was obtained from Abcam, Inc. (Cambridge, UK). Mercaptohexanol (MCH), Bis(2,2′-bipyridine)-4,4′-dicarboxybipyridine-ruthenium di(N-succinimidyl ester) bis(hexafluorophosphate) $(Ru(bpy)_2(dcbpy-NHS)(PF_6)_2$, abbreviated as Ru) and $HAuCl_4$ were obtained from Sigma Aldrich (USA). Human alpha-fetoprotein (AFP) and prostate specific antigen (PSA) were obtained from Fitzgerald Industries International, Inc. (USA). Human immunoglobulin G (IgG) and bovine serum albumin (BSA) were obtained from Beijing Biosynthesis Biotechnology Co., Ltd. (China). Albumin chicken egg protein was obtained from Sino-American Biotechnology Co., Ltd. (China).

The serum samples were provided by Xianyang Central Hospital (China). Phosphate buffered saline (PBS) (0.1 M) consisted of 0.1 M NaH_2PO_4, 0.1 M Na_2HPO_4, and 0.1 M KCl (pH 7.4). The other reagents used in this work were of analytical grade and directly used without additional purification. Millipore Milli-Q water (18.2 MΩ cm) was used to prepare all solutions.

ECL measurements were performed with a MPI-A ECL detector (Xi'an Remax Electronics, China). A commercial cylindroid glass cell was used as an ECL cell, which contained a conventional three-electrode system that consisted of a gold electrode (2.0 mm diameter) as the working electrode, a platinum wire as the counter electrode, and an Ag/AgCl (saturated KCl) as the reference electrode. ECL emissions were detected with a photomultiplier tube (PMT) that was biased at −900 V unless otherwise stated. Electrochemical experiments were performed with a CHI 660 electrochemical workstation (Chenhua Instruments Co., China).

Preparation of ECL probes

The ECL probe, $Ru(bpy)_2(dcbpy-NHS)(PF_6)_2$ labeled peptide2 (Ru-peptide2), were synthesized according to literatures with some modifications [14, 15]. Briefly, 1 mg of the peptide2 (0.0006 mol) was dissolved in 0.5 mL of 0.1 M phosphate buffer (PB) consisted of 0.1 M NaH_2PO_4 and 0.1 M Na_2HPO_4 (pH 7.4). Then, a 25 μL 0.02 M $Ru(bpy)_2(dcbpy-NHS)(PF_6)_2$ was added into 0.5 mL 10-diluted peptide2 solution under stirring, followed by an overnight incubation. The Ru-peptide2 was purified by dialysis for 12 h at 4 °C using MD34-2 Da molecular weight cutoff membrane with 0.1 M PBS (pH 7.4). The concentration of Ru-peptide2 solution was estimated to be $1.5×10^{-5}$ M, on the basis of UV absorbance of $Ru(bpy)_2(dcbpy-NHS)(PF_6)_2$ at 457 nm [32, 33].

Fabrication of the ECL peptide-based biosensor

The biosensor was fabricated by two steps including an electrochemical deposition step and an immobilization step. The procedure for the deposition of gold nanoparticles onto gold electrode was adapted from the ref. [34]. Prior to the experiment, the gold electrode was polished with 0.3 μm alumina slurry and then ultra-sonicated in water for 5 min. The polished gold electrode was immersed in 0.1 M $HClO_4$ containing 0.1 % $HAuCl_4$, which was degassed with N_2 stream for at least 20 min before the electrochemical deposition. Gold nanoparticles were electrodeposited on the gold electrode by holding a constant potential of +1.1 V (vs. Ag/AgCl, sat. KCl) for 60 s and then a constant potential of

−0.1 V for 5 min in 0.1 M $HClO_4$ containing 0.1 % $HAuCl_4$ to form gold nanoparticles-modified gold electrode (GNPs/Au electrode).

The peptide1 (CFYSHSFHENWPS) was immobilized onto the surface of GNPs/Au electrode by dipping the electrode into 11.3 μM peptide1 solution for 2 h at 4 °C and rinsing with 10 mM PB (pH 7.4) to remove the unbinding peptide1. The peptide-modified GNPs/Au electrode was then immersed in 1 mM mercaptohexanol solution for 30 min to block the uncovered surface of the electrode. The resulting electrode (peptide1/GNPs/Au electrode) was washed with water and used as the ECL peptide-based biosensor.

ECL measurement

The ECL peptide-based biosensor fabricated was immersed in 100 μL 10 mM PBS (pH 7.4) containing different concentrations of TnI for 1 h at room temperature. Then, the resulting ECL peptide-based biosensor was dipped into 100 μL of 1.5 μM ECL probe for 1 h at room temperature. After each incubation, the biosensor was rinsed thoroughly with 10 mM PBS (pH 7.4) to remove adsorption components. The ECL measurement was performed at a constant potential of +0.90 V in 2.0 mL of 0.10 M PBS (pH 7.4) containing 50 mM tripropylamine (TPA). The concentration of TnI was quantified by an increased ECL intensity ($\Delta I = I_S - I_0$), where I_S was the ECL intensity of ECL peptide-based biosensor reacted with TnI and I_0 was the blank ECL intensity of ECL peptide-based biosensor. All experiments were carried out at room temperature.

Results and discussion

Characterization of the ECL probe

The ECL probe Ru-peptide2 synthesized was characterized by UV–vis spectroscopy and ECL. Figure 2a shows UV–vis spectra of the peptide2, Ru(bpy)$_2$(dcbpy)NHS and Ru-peptide2. The characteristic peaks of peptide2 appear at 263 nm (line a) and the characteristic peaks of Ru(bpy)$_2$(dcbpy)NHS appear at 201, 276, and 457 nm (line b). The characteristic absorption peaks of Ru-peptide2 appear at 207, 276, and 457 nm (line c), corresponding to the characteristic peaks of Ru(bpy)$_2$(dcbpy)NHS at 450, 276, 207 nm and peptide at 263 nm, respectively. This indicates that Ru(bpy)$_2$(dcbpy)NHS is labeled to the peptide2.

Figure 2b shows ECL intensity–potential profiles of Ru(bpy)$_2$(dcbpy-NHS)(PF$_6$)$_2$ (line a) and Ru-peptide2 (line b) in 0.10 M PBS containing 50 mM TPA. From Fig. 2b, it can be seen that both ECL peaks appear at near + 900 mV, which is similar with that (+900 mV) in ref. [35],

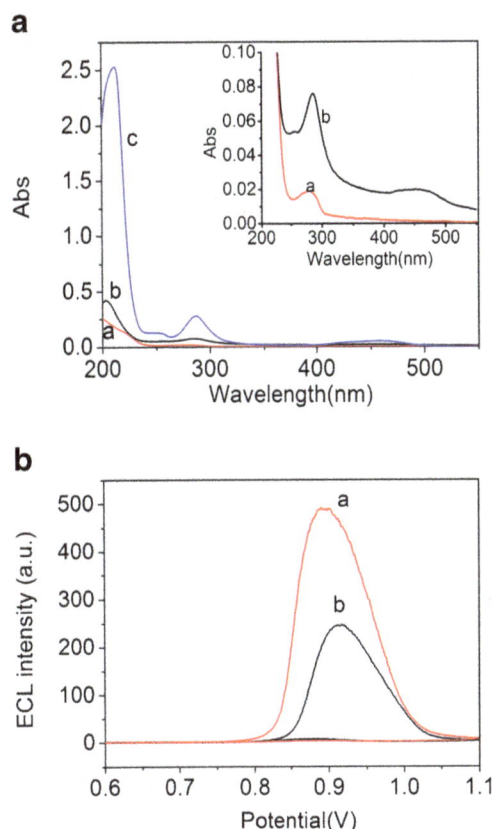

Fig. 2 **a** UV–vis spectra of peptide (*a*), Ru(bpy)$_2$(dcbpy-NHS)(PF$_6$)$_2$ (*b*) and Ru-peptide2 (*c*). **b** ECL intensity-potential profiles of 1.5× 10^{-7} M Ru(bpy)$_2$(dcbpy-NHS)(PF$_6$)$_2$ (*a*) and 1.5×10^{-7} M Ru-peptide2 (*b*) in 0.10 M PBS (pH 7.4) containing 50 mM TPA. Scan rate, 50 mV/s

indicating that the ECL behavior of Ru-peptide2 is similar to that of Ru(bpy)$_2$(dcbpy)NHS and Ru(bpy)$_2$(dcbpy)NHS is attached to peptide2.

Feasibility of ECL peptide-based biosensor for the determination of TnI

The fabrication process of the peptide-based biosensor was characterized by cyclic voltammetry in the presence of the ferri/ferrocyanide redox couple as redox probe (see Fig. S1 A in supporting information). As expected, K$_3$[Fe(CN)$_6$]/K$_4$[Fe(CN)$_6$] showed the reversible behavior on a bare gold electrode and on a gold nanoparticles-modified electrode with a peak-to-peak separation ΔE_p of 76 mV (Fig. S1 A, line a–b). After gold nanoparticles were electrodeposited onto gold electrode, the peak current increased from 23.89 to 29.21 μA (Fig. S1 A, line b), ascribing to the increase of electrode surface area, which is confirmed by the CV results in 0.1 M H$_2$SO$_4$ (Fig. S1 B). The immobilization of peptide1 on the surface of gold nanoparticles-modified gold electrode led to a significant increase in the peak-to-peak separation (ΔE_p= 143 mV) (Fig. S1 A, line c) and decrease of peak current (21.22 μA), indicating that peptide1 is self-assembled on the electrode. This is mainly attributed to the fact that the peptide1 modified on the surface of the electrode prohibits the mass transfer of $[Fe(CN)_6]^{3-/4-}$ from the solution to the surface of electrode. After reacted with TnI and Ru-peptide2, the peak-to-peak separation further increased (ΔE_p=196 mV) and the peak current decreased to 17.22 μA (Fig. S1A, line d). This indicates that the peptide is self-assembled onto the electrode surface and the sandwich conjugates is formed onto the surface electrode.

Figure 3a shows the ECL intensity vs potential profiles of the ECL peptide-based biosensor reacted with 1.0×10^{-11} and 1.0×10^{-10} g/mL TnI, respectively. Compared line a and line b, it can be seen that the ECL peptide-based biosensor shows a low ECL signal (line a, 1,048) while the ECL peptide-based biosensor reacted with 1.0×10^{-11} g/mL TnI displays a higher ECL signal (line b, 4,955) in 0.10 M PBS (pH 7.4) containing 50 mM TPA. Compared line b with line c, it can be clearly observed that the ECL peak intensity increases from 4,955 to 8,648 as the concentration of TnI is elevated from 1.0×10^{-11}

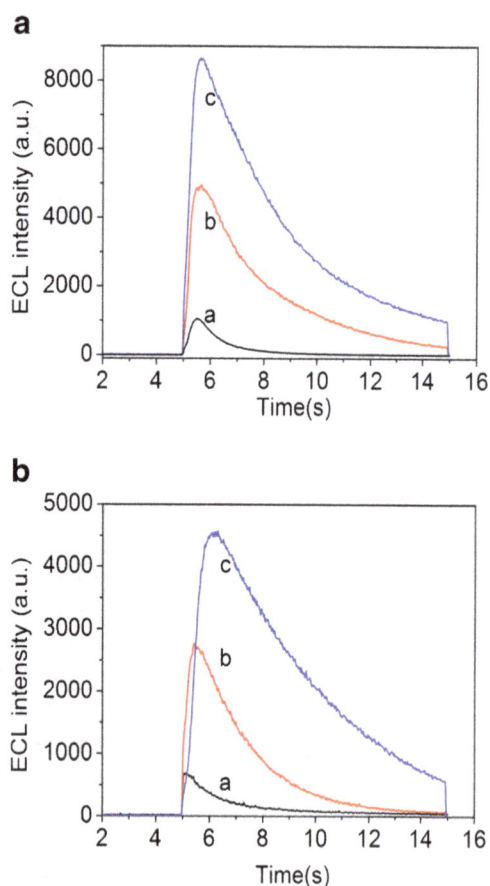

Fig. 3 ECL intensity-potential profiles of the ECL peptide-based biosensor fabricated on GNPs-modified gold electrode (**a**) and the ECL peptide-based biosensor fabricated on bare gold electrode (**b**) in 0.10 M PBS (pH 7.4) containing 50 mM TPA at a constant potential of 0.90 V, before (*a*) and after reaction with 1.0×10^{-11} g/mL TnI (*b*) and 1.0×10^{-10} g/mL TnI (*c*), respectively

to 1.0×10^{-10} g/mL. The results indicate that the ECL method is feasible for the determination of TnI.

In order to illustrate the function of gold nanoparticles, another kind of the peptide-based biosensor was fabricated by self-assembling peptide1 onto bare gold electrode surface through a thiol-containing cysteine at the end of the peptide1 and evaluated according to the protocol described in experimental section. Figure 3b shows the ECL intensity-potential profiles of the ECL peptide-based biosensor employing gold electrode as platform for the determination of TnI. The ECL peptide-based biosensor shows a low ECL signal (line a, 697). The ECL intensity were 2,779 (line b) and 4,576 (line c) after reacting with 1.0×10^{-11} and 1.0×10^{-10} g/mL TnI, respectively. Comparing Fig. 3a, b, the ECL intensity at the peptide1/GNPs/Au electrode was 1.5–1.9 times higher than that obtained at a peptide1/Au electrodes at same condition. The packing density of the peptide1 on the different electrodes surface is estimated on basis of the electrode surface area and the amount of peptide1. The amount of peptide1 was related with the charge associated with the electrode desorption reaction arising from the one-electron reduction of peptide1 layer on gold surface according Faraday law [36]. The peptide densities at the bare electrode and gold nanoparticles-modified electrode were 3.58×10^{-10} and 11.2×10^{-10} molecules.cm^{-2}, respectively, corresponding the effective electrode area of 0.037 mm^2 for bare gold electrode and 0.05 mm^2 for gold nanoparticles-modified gold electrode (as seen in supporting information Figure S2). The packing density of the peptide1 on gold nanoparticles-modified electrode was 3.1-folds larger than that of the bare gold electrode. In summer, gold nanoparticles not only facilitate the electron transfer at the electrode interface and catalyze the ECL process of ruthenium complex/TPA system [25, 37], but also increase the interface area of the electrode to capture more molecular recognition element for recognition of targets, and then immobilize numerous signal-generating molecules. The signal enhancement of the gold nanoparticles for the ECL peptide-based biosensor designed is evident.

Optimization of conditions

The applied potential is an important parameter because it decides the sensitivity of the ECL peptide-based biosensors. The dependence of the ECL intensity of the ECL peptide-based biosensor on applied potential was checked for 1.0×10^{-10} g/mL TnI. Figure 4a shows that the ECL intensity increases when the applied potential is raised from +0.8 to +0.9 V and reaches a maximum at +0.9 V. Therefore, the constant potential of +0.9 V was chosen in following experiments.

Figure 4b shows the effect of binding time between the peptide1 and TnI on the ECL intensity. The ECL intensity sharply increases with increasing binding time from 20 to 60 min and reaches a maximum at about 60 min. When further

a

b

c

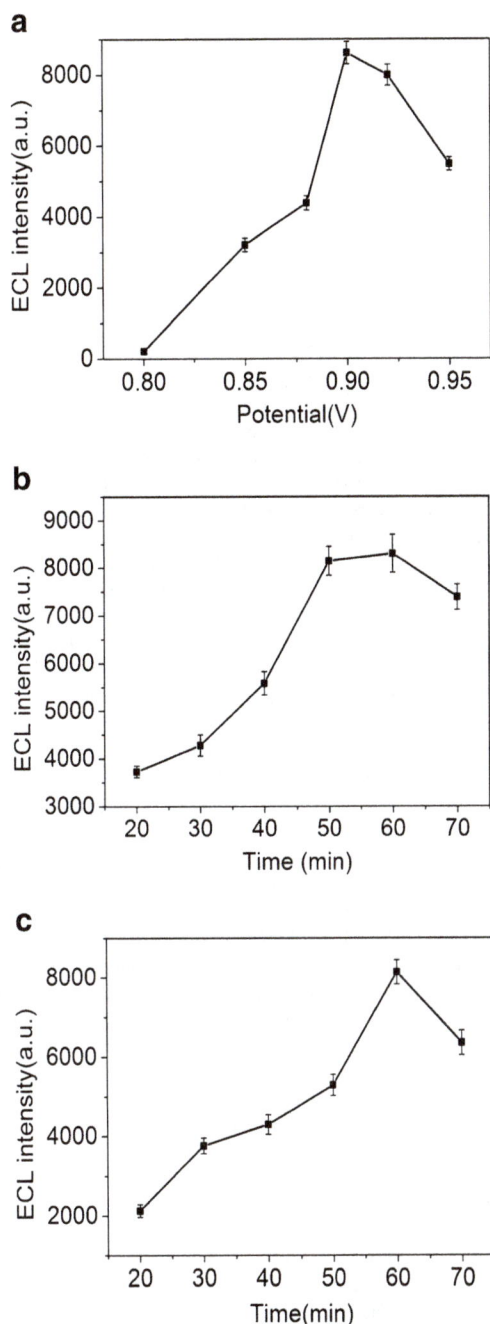

Fig. 4 a Effect of applied potential on the ECL intensity; **b** effect of binding time between peptide1 and 1.0×10^{-10} g/mL TnI on the ECL intensity. **c** Effect of binding time between 1.0×10^{-10} g/mL TnI and Ru-peptide2 on the ECL intensity. In 0.10 M PBS (pH 7.4) containing 50 mM TPA

increasing the binding time, the ECL intensity decreases slightly. This is attributed to steric and electrostatic hindrance arising from the more tightly packed TnI monolayer. Figure 4c demonstrates the effect of the binding time between TnI and the Ru-peptide2 on the ECL intensity. The results showed that the ECL intensity sharply increased with increasing binding time from 20 to 60 min, and then, reached a maximum at 60 min. Considering both of the binding time between

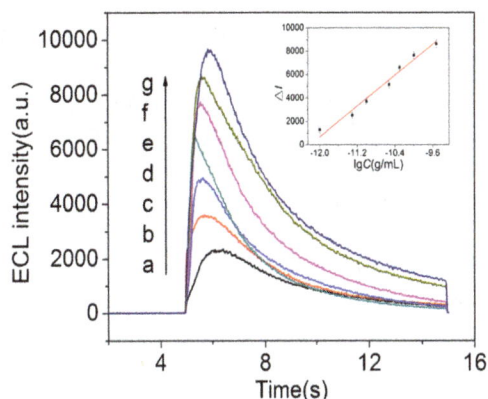

Fig. 5 ECL intensity vs time profiles for different concentrations of TnI in 0.10 M PBS (pH=7.4) containing 50 mM TPA: **a** 1.0×10^{-12} g/mL, **b** 5.0×10^{-12} g/mL, **c** 1.0×10^{-11} g/mL, **d** 3.0×10^{-11} g/mL, **e** 5.0×10^{-11} g/mL, **f** 1.0×10^{-10} g/mL, **g** 3.0×10^{-10} g/mL. *Insert*, calibration curve of TnI. Experimental conditions, applied potential, 0.90 V; binding time, 60 min; 0.10 M PBS (pH 7.4) containing 50 mM TPA

peptide1 and TnI with the binding time of Ru-peptide2 and TnI, we chosen 60 min as binding time for both of incubation process in following experiments.

Analytical performance of ECL peptide-based biosensor

The quantitative behavior of the ECL peptide-based biosensor fabricated was assessed under the optimized conditions. Figure 5 shows the ECL intensity vs time profiles of the ECL peptide-based biosensors for the determination of TnI. The ECL intensity increases with an increase of the concentration of TnI. The increased ECL intensity has a linear relationship with the logarithm of the concentration of TnI in the range from 1.0×10^{-12} to 3.0×10^{-10} g/mL. The linear regression equation is $\Delta I = 3,303 \lg C + 40,305$ (unit of C is in gram per milliliter) and the correlation coefficient was 0.9662. The relative standard derivation for 1.0×10^{-11} g/mL TnI was 2.8 %. The detection limit (DL) is 0.4 pg/mL, which is 300-fold lower than that obtained by homogenous ECL method using Ru-labeled peptide [14] and 3-fold lower than that obtained by homogenous ECL method using Ru-

Table 1 Analytical results of TnI in clinical serum samples

Sample number	CL method (ng/mL)[a]	This method ($n=3$, $P=0.9$, ng/mL)[b]	Relative error (%)
1	5.09	4.85±0.70	−4.7
2	2.94	2.95±0.06	0.4
3	1.02	1.01±0.01	−1.0

[a] The CL results of TnI in serum samples from clinical reports provided by Xianyang Central Hospital.

[b] Confidence interval

encapsulated liposomes labeled peptide as the ECL probe in our previous report [15]. The employment of gold nanoparticles as amplification platform and peptide as molecular recognition element to improve the sensitivity is evident.

The evaluation of the selectivity of the ECL peptide-based biosensor was performed by examining 1.0×10^{-10} g/mL $(8.3 \times 10^{-11}$ M) TnI or 1.5×10^{-7} M other proteins including AFP, PSA, albumin chicken egg protein and IgG, respectively. A significant increase in ECL intensity (72.5 %) by the interaction with TnI was observed (as shown in Fig. S3). On the other hand, very slight increases in ECL intensity were found for the other tested proteins including AFP (7.2 %), PSA (6.3 %), albumin chicken egg protein (8.4 %), and IgG (7.9 %), respectively. This indicates that the developed strategy has good selectivity for TnI.

Sample analysis

To evaluate a potential application of the ECL peptide-based biosensor, serum samples obtained from Xianyang Central Hospital were assayed using the proposed method in this work. The results are presented in Table 1. The obtained results show an acceptable agreement with the data provided by Xianyang Central Hospital (China) using a standard chemiluminescence (CL) method with an Abbott Immunoanalyzer (Abbott Axsym, i1000, USA), no statistical significance (P value=0.9) is obtained between the result using the ECL method in this work and that CL method, therefore, signifying the feasibility of the ECL method in clinical sample.

Conclusion

In this work, we fabricated a high sensitive electrogenerated chemiluminescence peptide-based biosensor for the detection of TnI. Great signal amplification was achieved with an extremely low detection limit of 0.4 pg/mL attributed to the combination of gold nanoparticles as amplification platform and peptide as molecular recognition element. The strategy presented in this study could be easily extended to the detection of other biomarkers.

Acknowledgements We gratefully acknowledge the financial support from The National Science Foundation of China (nos. 21375084, 21275095, 21027007) and the Natural Science Basic Research Plan in Shaanxi Province of China (no. 2013KJXX-73) and the Fundamental Research Funds for the Central Universities (no.GK261001185).

References

1. Qureshi A, Gurbuz Y, Niazi JH (2012) Biosensors for cardiac biomarkers detection: a review. Sens Actuators B 171–172: 62–76
2. McDonnell B, Hearty S, Leonard P, O'Kennedy R (2009) Cardiac biomarkers and the case for point-of-care testing. Clin Biochem 42: 549–561
3. Wu J, Cropek DM, West AC, Banta S (2010) Development of a troponin I biosensor using a peptide obtained through phage display. Anal Chem 82:8235–8243
4. Akanda MR, Aziz MA, Jo K, Tamilavan V, Hyun MH, Kim S, Yang H (2011) Optimization of phosphatase- and redox cycling-based immunosensors and its application to ultrasensitive detection of troponin I. Anal Chem 83:3926–3933
5. Guo H, Yang D, Gu C, Bian Z, He N, Zhang J (2005) Development of a Low density colorimetric protein array for cardiac troponin I detection. J Nanosci Nanotechnol 5:2161–2166
6. Nandhikonda P, Heagy MD (2011) An abiotic fluorescent probe for cardiac troponin I. J Am Chem Soc 133:14972–14974
7. Rusling JR, Kumar CV, Gutkind JS, Patel V (2010) Analyst 135: 2496–2511
8. Cho IH, Paek EH, Kim YK, Kim JH, Paek SH (2009) Chemiluminometric enzyme-linked immunosorbent assays (ELISA)-on-a-chip biosensor based on cross-flow chromatography. Anal Chim Acta 632:247–255
9. Hu L, Xu G (2010) Applications and trends in electro-chemiluminescence. Chem Soc Rev 39:3275–3304
10. Miao W (2008) Electrogenerated chemiluminescence and its biorelated applications. Chem Rev 108:2506–2553
11. Li F, Yu Y, Cui H, Yang D, Bian Z (2013) Label-free electrochemiluminescence immunosensor for cardiac troponin I using luminol functionalized gold nanoparticles as a sensing platform. Analyst 138(6):1844–1850
12. Shen W, Tian D, Cui H, Yang D, Bian Z (2011) Nanoparticle-based electrochemiluminescence immunosensor with enhanced sensitivity for cardiac troponin I using N-(aminobutyl)-N-(ethylisoluminol)-functionalized gold nanoparticles as labels. Biosens Bioelectron 27: 18–24
13. Sun D, Hamlin D, Butterfield A, Watson DE, Smith HW (2010) Electrochemiluminescent immunoassay for rat skeletal troponin I (Tnni2) in serum. J Pharmacol Toxicol Methods 61:52–58
14. Wang C, Qi H, Qiu X, Gao Q, Zhang C (2012) Homogeneous electrogenerated chemiluminescence peptide-based method for determination of troponin I. Anal Methods 4:2469–2474
15. Qi H, Qiu X, Xie D, Ling C, Gao Q, Zhang C (2013) Ultrasensitive electrogenerated chemiluminescence peptide-based method for the determination of cardiac troponin I incorporating amplification of signal reagent-rncapsulated liposomes. Anal Chem 85:3886–3894
16. Petrenko VA, Vodyanoy VJ (2003) Phage display for detection of biological threat agents. J Microbiol Methods 53:253–262
17. Park JP, Cropek DM, Banta S (2010) High affinity peptides for the recognition of the heart disease biomarker troponin I identified using phage display. Biotechnol Bioeng 105:678–686
18. Li Y, Schluesener HJ, Xu S (2010) Gold nanoparticle-based biosensors. Gold Bull 43:29–41
19. Jiang X, Chai Y, Yuan R, Cao Y, Chen Y, Wang H, Gan X (2013) An ultrasensitive luminol cathodic electrochemiluminescence immunosensor based on glucose oxidase and nanocomposites: graphene-carbon nanotubes and gold-platinum alloy. Anal Chim Acta 783:49–55
20. Deng S, Cheng L, Lei J, Cheng Y, Huang Y, Ju H (2013) Label-free electrochemiluminescent detection of DNA by hybridization with a molecular beacon to form hemin/G-quadruplex architecture for signal inhibition. Nanoscale 5:5435–41

21. Han E, Ding L, Jin S, Ju H (2011) Electrochemiluminescent biosensing of carbohydrate-functionalized CdS nanocomposites for in situ label-free analysis of cell surface carbohydrate. Biosens Bioelectron 26:2500–2555

22. Liu L, Zhao F, Ma F, Zhang L, Yang S, Xia N (2013) Electrochemical detection of β-amyloid peptides on electrode covered with N-terminus-specific antibody based on electrocatalytic O_2 reduction by Aβ(1–16)-heme-modified gold nanoparticles. Biosens Bioelectron 49:231–235

23. Qi H, Peng Y, Gao Q, Zhang C (2009) Applications of nanomaterials in electrogenerated chemiluminescence biosensors. Sensors 9: 674–695

24. Cao X, Ye Y, Liu S (2011) Gold nanoparticle-based signal amplification for biosensing. Anal Biochem 417:1–16

25. Yin XB, Qi B, Sun X, Yang X, Wang E (2005) 4-(Dimethylamino)butyric acid labeling for electrochemiluminescence detection of biological substances by increasing sensitivity with gold nanoparticle amplification. Anal Chem 77:3525–3530

26. Yao W, Wang L, Wang H, Zhang X, Li L, Zhang N, Pan L, Xing N (2013) An electrochemiluminescent DNA sensor based on nano-gold enhancement and ferrocene quenching. Biosens Bioelectron 40:356–361

27. Chen Z, Liu Y, Wang Y, Zhao X, Li J (2013) Dynamic evaluation of cell surface N-glycan expression via an electrogenerated chemiluminescence biosensor based on concanavalin A-integrating gold-nanoparticle-modified $Ru(bpy)_3^{2+}$-doped silica nanoprobe. Anal Chem 85:4431–4438

28. Duan R, Zhou X, Xing D (2010) Electrochemiluminescence biobarcode method based on cysteamine-gold nanoparticle conjugates. Anal Chem 82:3099–3103

29. Qi H, Zhang Y, Peng Y, Zhang C (2008) Homogenous electrogenerated chemiluminescence immunoassay for human immunoglobulin G using N-(aminobutyl)-N-ethylisoluminol as luminescence label at gold nanoparticles modified paraffin-impregnated graphite electrode. Talanta 75:684–690

30. Li Y, Qi H, Yang J, Zhang C (2009) Detection of DNA immobilized on bare gold electrodes and gold nanoparticle-modified electrodes via electrogenerated chemiluminescence using a ruthenium complex as a tag. Microchim Acta 164:69–76

31. Li Y, Qi H, Gao Q, Yang J, Zhang C (2010) Nanomaterial-amplified "signal off/on" electrogenerated chemiluminescence aptasensors for the detection of thrombin. Biosens Bioelectron 26:754–759

32. Shimidzu T, Iyoda T, Izaki K (1985) Photoelectrochemical properties of bis(2,2′-bipyridlne) (4,4′-dlcarboxy-2,2′-bipyrldlne)ruthenlum(II) chloride. J Phys Chem 89:642–645

33. Li Y, Qi H, Peng Y, Yang J, Zhang C (2007) Electrogenerated chemiluminescence aptamer-based biosensor for the determination of cocaine. Electrochem Commun 9:2571–2575

34. Martín H, Carro P, Creus AH, Gonzá lez S, Andreasen G, Salvarezza RC, Arvia AJ (2000) The influence of adsorbates on the growth mode of gold islands electrodeposited on the basal plane of graphite. Langmuir 16:2915–2923

35. Zhang J, Qi H, Li Y, Yang J, Gao Q, Zhang C (2008) Electrogenerated chemiluminescence DNA biosensor based on hairpin DNA probe labeled with ruthenium complex. Anal Chem 80:2888–2894

36. El-Deabl MS, Ohsaka T (2004) Molecular-level design of binary self-assembled monolayers on polycrystalline gold electrodes. Electrochim Acta 49:2189–2194

37. Chen Z, Zu Y (2007) Gold nanoparticle-modified ITO electrode for electrogenerated chemiluminescence: well-preserved transparency and highly enhanced activity. Langmuir 23: 11387–11390

6

The effect of nonhomogeneous silver coating on the plasmonic absorption of Au–Ag core–shell nanorod

Jian Zhu · Fan Zhang · Jian-Jun Li · Jun-Wu Zhao

Abstract Plasmonic light absorption properties of bimetallic Au–Ag core–shell nanorod with nonhomogeneous Ag coating are investigated both theoretically and experimentally. When the Ag coating on the Au nanorod is not homogeneous, the overall aspect ratio and Ag composition greatly depends on the coating uniform, which strongly affects intensity changing and wavelength shift of the plasmonic absorption. When the transverse Ag coating is faster than longitudinal coating, the Au–Ag core–shell nanorod could present two intense longitudinal plasmonic absorption peaks with equal intensity and small wavelength gap as the Ag shell thickness has a critical value. Furthermore, this critical Ag shell thickness could be decreased when the Au nanorod core has a small aspect ratio. When the longitudinal Ag coating is faster than transverse coating, the increasing intensity of longitudinal peak corresponding to outer Ag surface is always weaker than the longitudinal peak corresponding to Au–Ag interface. Thus, we cannot obtain two equal intense plasmonic absorption peaks when the longitudinal Ag coating is thicker than the transverse coating. However, the transverse peak corresponding to Au–Ag interface could be enhanced by decreasing the aspect ratio of the Au nanorod core. Thus, we can always find three distinct absorption peaks as the aspect ratio of Au nanorod is decreased and the longitudinal Ag coating is greater than transverse coating.

Keywords Au–Ag bimetallic nanoparticles · Nonhomogeneous silver coating · Core–shell structure · Nanorod · Plasmonic absorption

J. Zhu (✉) · F. Zhang · J.-J. Li · J.-W. Zhao (✉)
The Key Laboratory of Biomedical Information Engineering of Ministry of Education, School of Life Science and Technology, Xi'an Jiaotong University, Xi'an, 710049, People's Republic of China
e-mail: jianzhusummer@163.com
e-mail: nanoptzhao@163.com

Introduction

In recent years, noble metallic nanoparticles have received considerable attention owing to their novel optical properties based on localized surface plasmon resonance (LSPR). Although noble metal gold and silver have the same face-centered cubic crystal structure and similar lattice constants, silver nanoparticles display stronger LSPR absorption with higher plasmon energy (about 400 nm for Ag nanosphere and 520 nm for Au nanosphere) and more intense local electric field enhancement [1–3]. Therefore, Ag nanoparticles have been often used as substrates for surface-enhanced Raman scattering (SERS) [4] and sensing and imaging based on resonance light scattering [5, 6].

Because symmetry breaking induced redistribution of surface charges, the plasmonic optical properties are strongly dependent on particle shape. Thus, producing metallic nanoparticles with low symmetry (nonspherical symmetry) could improve plasmon tunability such as resonance wavelength and local field intensity [7, 8]. However, the synthesis of complex morphological silver nanoparticles such as Ag nanorod is much harder than gold [9]. One of the effective methods to achieve nanoparticles with well-defined morphology and silver surface is coating an Ag nanolayer on the premade gold nanoparticles with nonspherical symmetry [9–12].

By using the microwave-polyol method, Au–Ag core–shell nanocrystals have been successfully synthesized [10]. The corresponding crystal structures and their growth mechanisms had also been studied. Furthermore, the transmission electron microscope (TEM) images observed by Tsuji et al. demonstrated that the shapes of initiated Au seeds greatly affect the shapes of formed Ag shells [13]. The Au–Ag core–shell nanocubes with varying shaped cores have been reported by Gong et al. [14]. The prepared Au–Ag core–shell nanoparticles display very abundant and distinct LSPR characteristics, which are dependent on core shape and core size. By using iodide ions, the growth direction of the Ag shell on the gold nanodisk core could also be tuned, and then the shape of the coated Ag shell could also be controlled [15]. The LSPR and SERS response in Au dumb bells with silver coating have also been reported [9]; this type of core–shell Au–Ag nanostructure is expected to serve as excellent SERS substrates because of the higher enhancement factors for silver as compared to gold. Ma et al. reported a facile method for generating Au–Ag core–shell nanocubes with controllable edge lengths [16]. By varying the ratio of Ag ions to Au seeds, the thickness of the Ag shells could be finely tuned from 1.2 to 20 nm, and the plasmon excitation of the Au cores would be completely screened when the Ag shell has a critical value of 3 nm. The refractive index sensitivity of gold–silver core–shell nanoparticles has also been studied [17]. It has been found that coating a layer of silver brings about a higher refractive index sensitivity in comparison to the pure Au nanobars [18].

It is known that the Ag and Au nanorods have two LSPR bands corresponding to transverse and longitudinal resonance, respectively. The longitudinal plasmon bands are sensitive to the aspect ratio and could be tuned from visible to infrared region. An interesting topic arises from the effect of Ag coating on the LSPR of Au–Ag core–shell nanorods. In our previous theoretical study, two transverse LSPR bands from outer Ag surface and Au–Ag interface have been observed in Au–Ag core–shell nanowires [19]. In the report of Yu et al. [20], the optical properties of bimetallic Au–Ag core–shell nanorods were characterized by using a steady-state extinction spectra and ultrafast transient absorption spectroscopy. They have experimentally observed two longitudinal LSPR bands in Au–Ag core–shell nanorods corresponding to the Ag shell and Au core. In the study of Liu et al., the silver coating induced blue shift, and enhancement of longitudinal plasmon mode was observed [21]. And the blue shift has been attributed to the changing of effective dielectric function and decreasing of the overall aspect ratio induced by silver coating.

In a recent report, it is interesting to find that the silver coating is not always homogeneous [3]. In the preparation method of Jiang et al., the thickness of the Ag shell at the side increases faster than that at the ends as the amounts of $AgNO_3$ is increased [3]. How about the effect of nonhomogeneous silver coating on the plasmonic absorption properties of gold nanorod? In this report, we studied the plasmonic absorption properties of Au–Ag core–shell nanostructure with nonhomogeneous Ag coating. It has been found that the anisotropy of Ag coating greatly affects intensity changing, wavelength shift, and peak number of the plasmonic absorption. Furthermore, the physical origin of the nonhomogeneous Ag coating-dependent LSPR tunability has also been discussed.

The model

The geometry of the core–shell structure Au–Ag ellipsoidal nanorods with rotation symmetry is shown in Fig. 1. The Au ellipsoid core is modeled as a prolate (cigar shaped) spheroid, which is generated by rotating an ellipse about its major axis. Because of the rotation symmetry, the Au ellipsoid core has a semimajor axis c_1 and equal semiminor axis $a_1=b_1$. The outer surface of the Ag shell has semimajor axis c_2 and semiminor axis $a_2=b_2$. And the coated Ag shell has a transverse thickness of $t_T=a_2-a_1$ and longitudinal thickness of $t_L=c_2-c_1$. When the silver is homogeneously coated on the gold nanorod, t_T and t_L have the same value. In this model, the dielectric constant of the Au nanorod core, Ag coating shell, and environmental medium is given by ε_1, ε_2, and ε_3, respectively. It is important to note that the dielectric function ε_1 and ε_2 have real and imaginary wavelength (λ)-dependent components [22]. In our analysis, the size of the Au–Ag nanostructure is much smaller than the incident wavelength. Thus, the nanoparticles are subjected to an almost uniform field and oscillate like a simple dipole. Therefore, the quasistatic approximation can be employed in the calculation. The basic equations of the polarizabilities along the principal axes and the plasmonic absorption cross-sections could be derived from the Laplace's equation and have been reported in [21, 23]. By plotting the absorption cross-sections with different wavelength, we could obtain the absorption spectrum.

Results and discussion

Absorption spectra of Au–Ag nanorods with homogeneous silver coating

Figure 2a shows the calculated absorption spectra of bare Au nanorods and Au–Ag core–shell nanorods with homogeneous Ag coating. In this calculation, the semiminor axis of the inner Au nanorod is set as $a_1=5$ nm, the aspect ratios of the Au nanorod is set as $p=c_1/a_1=4$, the environmental dielectric constant is set as $\varepsilon_3=1.0$. For pure Au nanorods, there are two LSPR absorption peaks corresponding to transverse and longitudinal resonance appearing at about 515 and 686 nm, respectively. As the coating Ag shell thickness is increased from 1 to 11 nm, both the transverse (denoted as peak 2) and longitudinal (denoted as peak 3) peaks blue shift and get intense. However, the shifting and intensity increase of the longitudinal peak are more intense, and then peak 2 has been merged gradually by peak 3. What is more, the Ag coating results in a new plasmonic absorption peak taking place at shorter wavelength, which is denoted as peak 1. As the coating Ag shell thickness is increased, the peak 1 red shifts slightly and gets intense greatly. All these Ag coating-dependent absorption properties are in good agreement with the

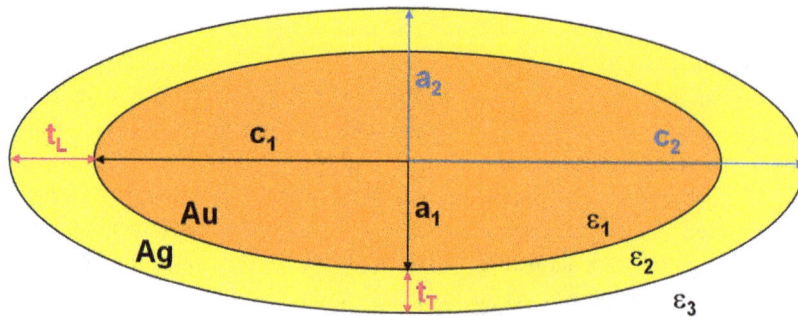

Fig. 1 Geometry of core–shell structure Au–Ag ellipsoidal nanorods: ε_1, ε_2, ε_3 are the dielectric functions for the Au core, Ag nanoshell, and embedding regions respectively, c denotes the semimajor axis and $a=b$ the semiminor axis. t_T denotes the Ag shell thickness in transverse direction and t_L denotes the Ag shell thickness in longitudinal direction

Fig. 2 **a** Calculated absorption spectra of Au–Ag core–shell nanorods with homogeneous Ag coating, the Ag shell thickness is changed from 0 to 11 nm. **b** Calculated absorption spectrum of Au–Ag core–shell nanorods with nonhomogeneous Ag coating, $t_T=7.5$ nm and $t_L=3.25$ nm

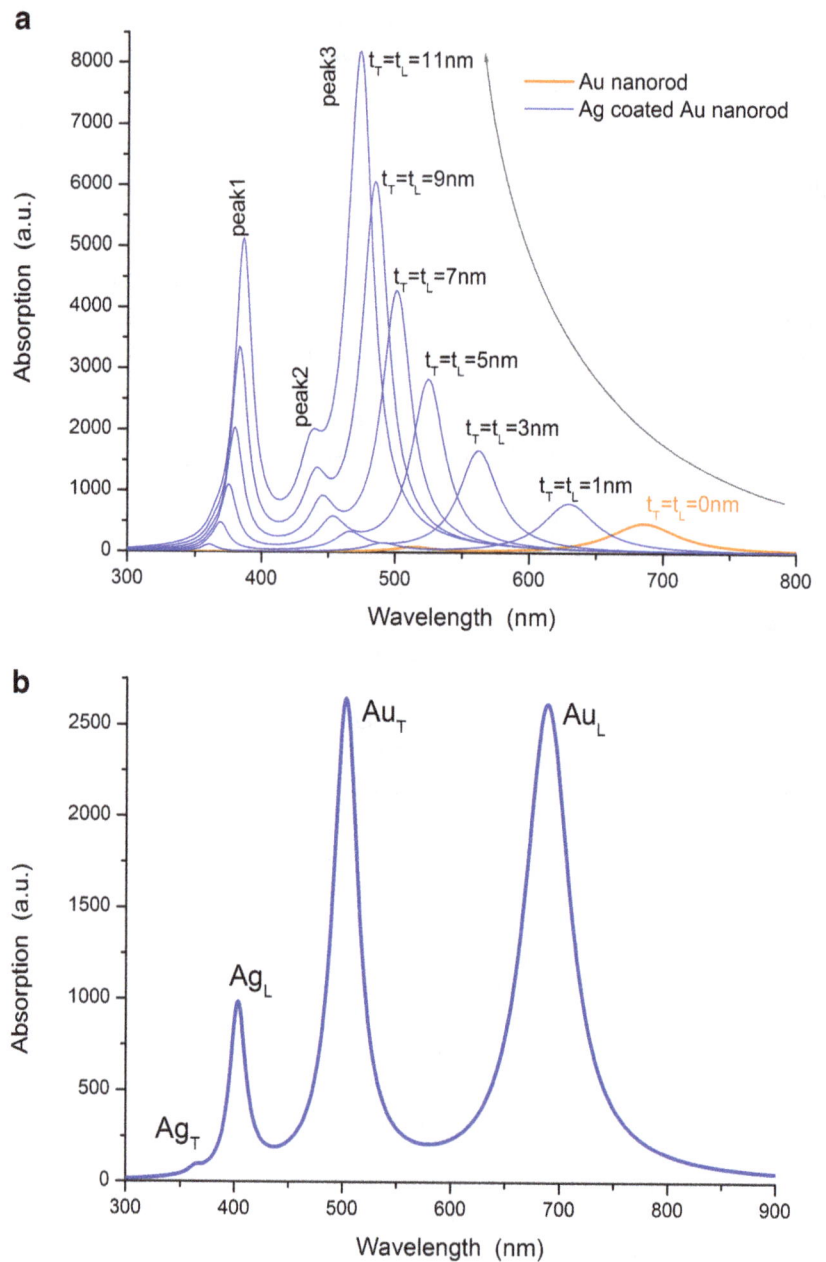

experimental results [21]. Because the Ag nanoparticles have more intense plasmonic absorption, the physical mechanism of the intensity increasing of peaks 2 and 3 could be resulted from the increase composition of Ag in the Au–Ag bimetallic nanoparticles [21]. Because the Ag nanoparticles have shorter plasmon resonance wavelength, the intense blue shift of peaks 2 and 3 should also be attributed to the increase composition of Ag in the bimetallic nanoparticles. On the other hand, a homogeneous Ag layer coating lowers the overall aspect ratio of the core–shell nanostructure, which provides the minor reason of the blue shift of peak 3.

Although the surface plasmons are coherent excitation and are affected by the coupling between metallic surfaces, different surface electron oscillations may take major effect on different plasmon modes. In our previous study of the transverse LSPR in Au–Ag and Ag–Au core–shell structure nanowires, the two plasmonic peaks have been assigned to outer surface of wall metal and the interface between core and wall metals, respectively [19]. Therefore, there should be four LSPR peaks in the Au–Ag core–shell nanorod. However, one could only find three distinct plasmon peaks in Fig. 2a. The peaks 2 and 3 correspond to the transverse and longitudinal plasmon from Au–Ag interface (denoted as Au_T and Au_L). The peak 1 corresponds to the longitudinal plasmon from outer Ag surface (denoted as Ag_L). Because the transverse plasmonic absorption is always weaker than the longitudinal mode, the fourth peak corresponding to the transverse plasmon from outer Ag surface (denoted as Ag_T) is usually too weak to be observed and could only be found under certain conditions. For example, when the Ag coating is nonhomogeneous (t_T=7.5 nm and t_L=3.25 nm) and the environmental dielectric constant is increased to 3.5, one can find four LSPR peaks in the absorption spectrum as shown in Fig. 2b. In the absorption spectral testing, the gold nanorods are usually suspended in the aqueous solution with a surrounding dielectric constant of $\varepsilon_{surrounding}$=1.7. Thus, it has been observed that the longitudinal absorption intensity is usually larger than the transverse intensity. In Fig. 2a, the surrounding dielectric constant is set as ε_3=1.0, and the longitudinal absorption intensity is also larger than the transverse intensity. However, when the surrounding dielectric constant is increased, the polarization of the dielectric media will affect the surface plasmon resonance. Because of the nonspherical symmetry, the effect on the transverse and longitudinal SPR is different. Previous experimental reports indicate that the transverse absorption increases and the longitudinal decreases when the surrounding dielectric constant is increased [24–26]. For example, by comparing the absorption spectra of bare Au nanorods and silica-coated Au nanorods, one can find that the transverse absorption increases and the longitudinal decreases as the silica coat thickness increased [24]. In Fig. 2b, in order to observe the fourth peak corresponding to the transverse plasmon from outer Ag surface, the

surrounding dielectric constant is increased to ε_3=3.5. Therefore, the transverse plasmon absorption from Au–Ag interface increases greatly, whereas the longitudinal plasmon absorption from Au–Ag interface fades down.

Absorption spectra of Au–Ag nanorods with nonhomogeneous silver coating

In Fig. 3, we studied the effect of nonhomogeneous Ag coating on the absorption spectral properties of Au–Ag core–shell nanorods. For the case of t_T=2t_L, i.e., the Ag shell coating in the transverse direction is faster than that of longitudinal direction, the overall aspect ratio decreases rapidly and the composition of Ag in the bimetallic particle increases rapidly. Therefore, the peak 3 blue shifts intensely and the peak 1 gets intense rapidly as the Ag shell thickness is increased. As shown in Fig. 3a, the intensity increase of peak 1 is faster than that of peak 3, which is similar to the experimental observation [3]. Thus, the peak 1 at shorter wavelength becomes more intense than the peak 3 at longer wavelength when the Ag coating is thick, which is different from the calculation results of homogeneous coating case in Fig. 2a and the experimental result of [21]. Furthermore, because of this nonhomogeneous silver coating, the bimetallic nanostructures could present two intense plasmonic absorption peaks with equal intensity and small wavelength gap ca. 30 nm as the Ag shell thickness has a critical value, as shown in the inset of Fig. 3a. This plasmonic property provides potential for double-channel optical sensing based on plasmonic absorption [27]. In Fig. 3a, the disappearance of peak 2 in absorption spectrum is due to the following reasons. Firstly, the increase of outer Ag shell thickness leads to the great enhancement of peaks 1 and 3, which masks the weak transverse plasmon resonance of inner Au core. This trend could be clearly reflected in Fig. 2a. Secondly, the red shift of peak 1 and blue shift of peak 3 make the peak 2 merged together with peaks 1 and 3. Meanwhile, peak 2 only becomes distinct as the surrounding dielectric constant is large as shown in Fig. 2b. However, the surrounding dielectric constant in Fig. 3a is set as ε_3=1, which is much smaller than that of Fig. 2b. Therefore, the peak 2 in Fig. 3a is too weak to be observed.

In order to make the Ag coating-dependent intensity changing clearer, we also plotted the peak intensity of the absorption as the function of Ag coating thickness in Fig. 3c. When t_T= 2t_L, the intensity of peak 3 is always a little greater than that of peak 1 as the Ag shell thickness is less than 8 nm. The increasing speed of peak 3 fades down as the Ag coating is further increased. And then peaks 3 and 1 have the same intensity when Ag shell thickness reaches 9 nm. However, the intensity difference between peaks 1 and 3 increases again when the Ag shell thickness is greater than 9 nm. For the case of t_L=2t_T, i.e., the Ag shell coating in the longitudinal direction is faster than that of transverse direction (but the ratio t_L/

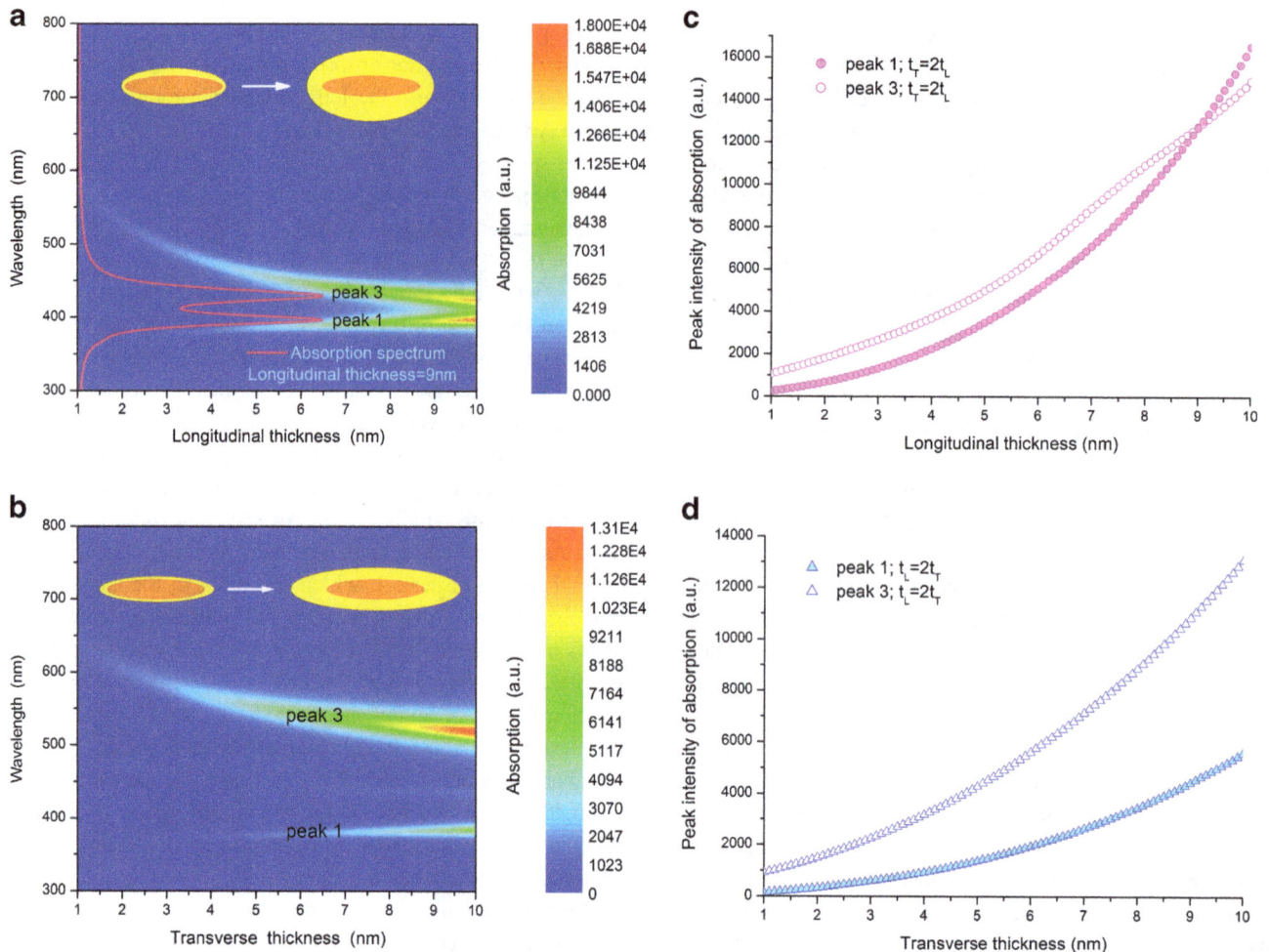

Fig. 3 Calculated absorption spectra of Au–Ag core–shell nanorods with nonhomogeneous silver coating, the Au nanorod core has a large aspect ratio of $p=4$. **a** Transverse Ag shell thickness is larger, $t_T=2t_L$. **b** Longitudinal Ag shell thickness is larger, $t_L=2t_T$. **c** Peak intensity of the absorption as a function of longitudinal Ag coating thickness. **d** Peak intensity of the absorption as a function of transverse Ag coating thickness, $\varepsilon_3=1.0$

$t_T<p$), the overall aspect ratio decreases slowly and the Ag composition of the bimetallic particle also increases slowly. Therefore, the blue shift of peak 3 is gentle and the wavelength gap between peaks 1 and 3 is wide as shown in Fig. 3b. Furthermore, the intensity increase of peak 1 is also gentle and is always weaker than peak 3. As shown in Fig. 3c, d, the intensity difference between peaks 1 and 3 has been monotonously enlarged as the Ag coating is increased. And we cannot obtain two intense plasmonic absorption peaks with equal intensity when the longitudinal Ag coating is thicker as shown in Fig. 3d.

In order to find the effect of aspect ratio of the Au nanorod on the nonhomogeneous coating-dependent plasmonic absorption, we also plotted the absorption spectra of Au–Ag core–shell nanorods when the aspect ratio of the Au nanorod core is decreased to $p=2$. As shown in Fig. 4, the way of the absorption intensity changing and wavelength shifting is similar to the cases of $p=4$. However, because of the decrease of the aspect ratio and the volume of the Au nanorod, the Ag

coating-dependent Ag composition increase becomes intense. Therefore, the Ag coating-induced absorption intensity increase becomes rapid. Thus, we can obtain two intense plasmonic absorption peaks with equal intensity and small wavelength gap when the Ag coating thickness is thin. As shown in the inset of Fig. 4a, the Au–Ag core–shell nanorods present two intense plasmonic absorption peaks with equal intensity and small wavelength gap ca. 45 nm as the longitudinal Ag shell thickness is 4 nm. The peak intensity of the absorption as the function of longitudinal Ag coating thickness has been plotted in Fig. 4c. When $t_T=2t_L$, the intensity of peak 3 is always greater than that of peak 1 as the Ag shell thickness is less than 4 nm. However, the increasing speed of peak 3 gets intense first and then fades down as the Ag coating is increased. At last, peaks 3 and 1 have the same intensity when Ag shell thickness reaches 4 nm. In Fig. 4b, because the intensity discrepancy between transverse and longitudinal plasmonic absorption is decreased, the relative intensity of peak 2 is much greater than that of the absorption spectra in

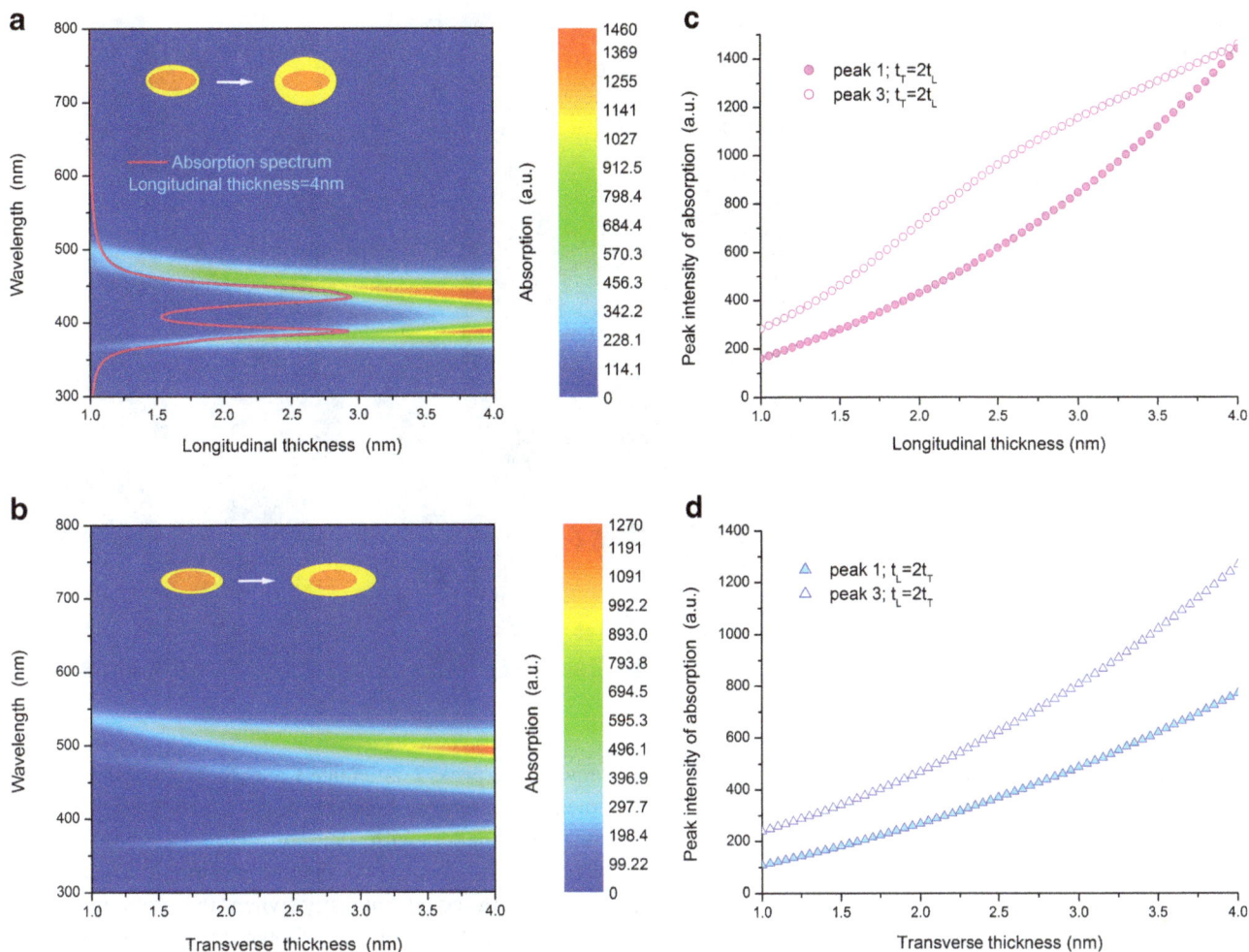

Fig. 4 Calculated absorption spectra of Au–Ag core–shell nanorods with nonhomogeneous silver coating, the Au nanorod core has a small aspect ratio of $p=2$. **a** Transverse Ag shell thickness is larger, $t_T=2t_L$. **b** Longitudinal Ag shell thickness is larger, $t_L=2t_T$. **c** Peak intensity of the absorption as a function of longitudinal Ag coating thickness. **d** Peak intensity of the absorption as a function of transverse Ag coating thickness, $\varepsilon_3=1.0$

Fig. 3b. We can always find three distinct absorption peaks as the Au nanorod is short and the longitudinal Ag coating is thicker.

Experimental study of the absorption spectra of Au–Ag nanorods with silver coating (transverse coating is larger than longitudinal coating)

The preparation of Au nanorods

Au nanorods were prepared according to a seed-mediated growth protocol developed by Sau and Murphy with some slight modification [28]. Initially, Au seed solution was prepared by quickly injecting NaBH$_4$ (0.01 M, 0.6 mL) into the mixture of HAuCl$_4$ (0.01 M, 0.25 mL) and CTAB (0.1 M, 7.5 mL) solution in a test tube of 15 mL, followed by 1 min of violent stirring. The color of the solution changed from yellow to brownish yellow, which indicates the start formation of the Au seed. Then, the resultant solution was kept at

27 °C for 2 h for the next use. Secondly, to grow Au nanorods, 0.01 M HAuCl$_4$ (4 mL), 0.01 M AgNO$_3$, 0.1 M AA (0.64 mL), and 1 M HCl were injected into 0.1 M CTAB (90 mL) in order. With gentle stirring, the color of the solution changed from brownish yellow to transparent. Different aspect ratios of gold nanorods could be obtained by tuning the volume of AgNO$_3$ and HCl. Finally, 0.2 mL Au seed solution was injected to initiate growth reaction. Then the ultimate solution was kept at a constant temperature of 27 °C undisturbed overnight. During this time, the transparent solution progressively changed to purple or deep blue, depending on the aspect ratio of the Au nanorods. When the volume of AgNO$_3$ is relatively small, we obtained the shorter Au nanorods with an aspect ratio of 3.3. Figure 5a depicted the synthesized shorter Au nanorods with an average length of about 40 nm and width of about 12 nm. In order to get longer Au nanorods, the AgNO$_3$ volume has been increased and HCl was added at the same time. As shown in Fig. 5b, longer Au nanorods with an aspect ratio

Fig. 5 TEM images of **a** short
Au nanorods with an aspect ratio
of 3.3. **b** Long Au nanorods with
an aspect ratio of 4.8. **c** Au–Ag
core–shell nanorod with thin Ag
shell thickness. **d** Au–Ag core–
shell nanorod with thick Ag shell
thickness

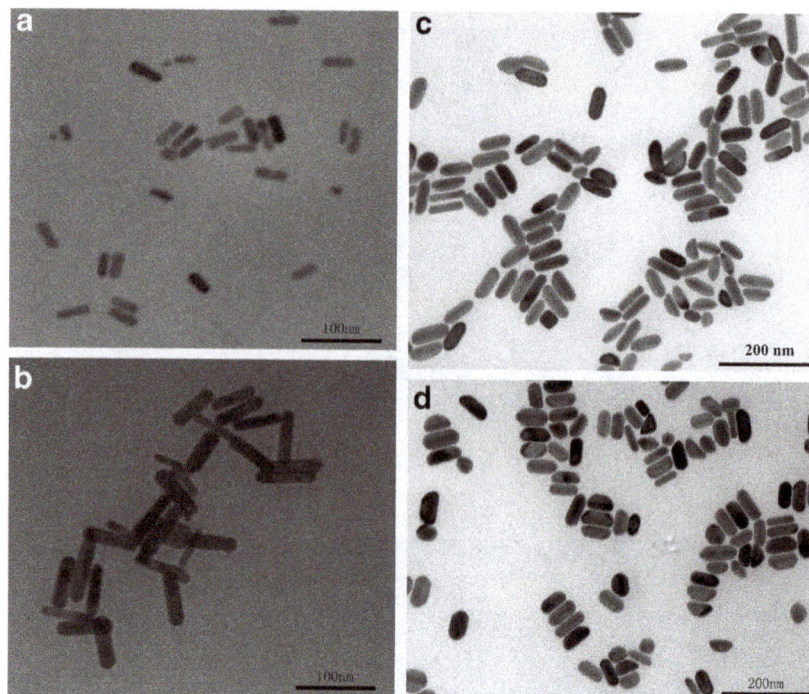

of 4.8 (with an average length of about 62 nm and width of
about 13 nm) were obtained.

The synthesis of Au–Ag core–shell nanorods

Au nanorods with Ag coating were prepared by the method in
previous literature with some minor modification [3, 16, 29,
30]. In summary, 45 mL prepared Au nanorods were centrifuged for 20 min at 10,000 rpm, then the supernatant was
carefully removed and the precipitate was redispersed into the
same volume of 0.08 M CTAB. One minute of the ultrasonic
solution is needed. Afterwards, the solution is divided into
nine portions. For each portion, AgNO$_3$ (0.01 M), AA (0.1 M,
0.4 mL), and NaOH (0.1 M, 1 mL) was added in that order. By
adding the volume of AgNO$_3$ from 0.1 to 0.9 mL, we obtained
Au–Ag core–shell nanorods with different Ag shell thickness.
The solution was then kept at 65 °C for 4 h. The typical TEM
images of the synthesized Au–Ag core–shell nanorods with
thin Ag coating and thick Ag coating are compared. As can be
observed in Fig. 5c, when the volume of AgNO$_3$ is 0.5 mL,
the Ag shell at the side facet is thin, whose average thickness
is about 5 nm. And thicker Ag shell at the side facet of about
10 nm could be observed when the volume of AgNO$_3$ reaches
0.8 mL as shown in Fig. 5d.

Figure 6a depicts the absorption spectra of Au–Ag
core–shell nanorods with longer Au nanorods core. As
can be seen in Fig. 6a, for longer gold nanorods, as the
volume of AgNO$_3$ is increased, peak 1 slightly red
shifts with intensity increasing quickly, while peak 3
blue shifts greatly with intensity increasing slowly. At

the same time, peak 2 attenuated and then disappeared.
The intensity difference between peaks 1 and 3 becomes
smaller and smaller with the continuous increase of
AgNO$_3$. We can obtain two intense plasmonic absorption peaks with equal intensity when the volume of
AgNO$_3$ reaches 0.85 mL, as shown in Fig. 6a. For
shorter gold nanorods, two intense plasmonic absorption
peaks with equal intensity appeared when the volume of
AgNO$_3$ reaches 0.4 mL as shown in Fig. 6b. As the
volume of AgNO$_3$ is further increased, peak 1 exceeds
peak 3 and leaves only one peak at about 450 nm at
last. As a matter of fact, longer and shorter gold
nanorods represent aspect ratios of large and small, the
volume of AgNO$_3$ corresponds to the thickness of silver
coating. The silver coating gets thicker as the amount of
AgNO$_3$ is increased. A thin silver layer was developed
at first, as shown in line A of Fig. 6b. Owing to its
nonhomogeneous coating, the increase of the side of the
nanorod's thickness is faster than the end as shown in
line B of Fig. 6b. It is obvious that its side facet of
silver coating is thicker than line A of Fig. 6b. At last,
the nanorods turned to nearly spherical structure as
shown in line C of Fig. 6b. Figure 6c, d depicts the
dependences of the plasmon peak wavelengths and absorption intensities on the volume of the AgNO$_3$ solution. Because the Ag shell thickness is controlled by the
amount of AgNO$_3$, the amount of AgNO$_3$ corresponds
to the Ag shell thickness, and the Ag shell thickness is
increased as the volume of the AgNO$_3$ solution is
increased. Increasing the silver coating results in the

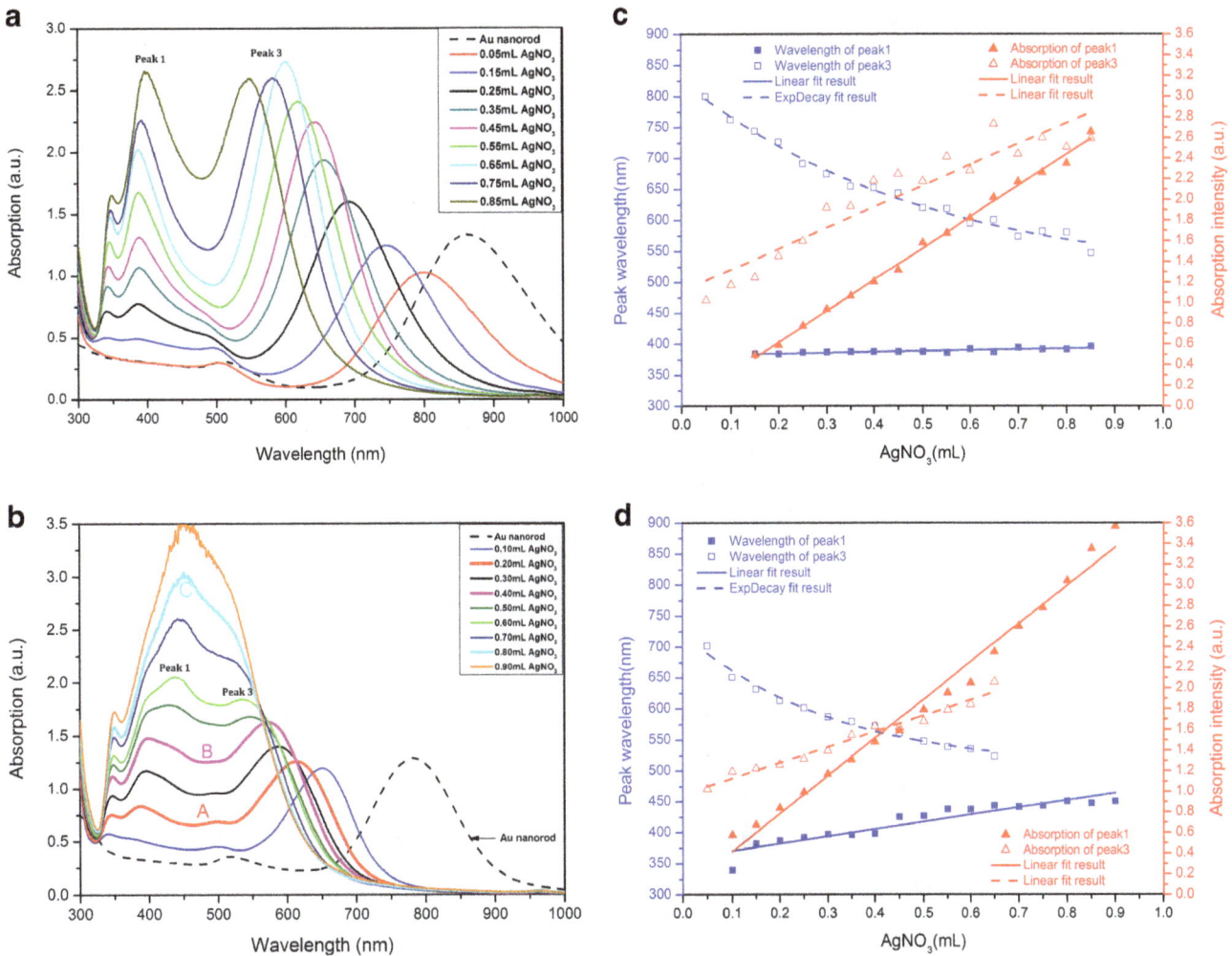

Fig. 6 Absorption spectra of Au nanorod samples and Au–Ag core–shell nanorod samples with different volume of the 0.01 M AgNO₃ solution. **a** The Au nanorods have large aspect ratio and the AgNO₃ volume is increased from 0.05 to 0.85 mL with an increment of 0.1 mL. **b** The Au nanorods have small aspect ratio and the AgNO₃ volume is increased from 0.1 to 0.9 mL with an increment of 0.1 mL. Dependences of the plasmon peak wavelengths and absorption intensities on the volume of the AgNO₃ solution, the Au nanorods have a **c** large aspect ratio and **d** small aspect ratio

absorption intensity of both peaks 1 and 3, getting intense linearly. The linear relation could also be found for the silver coating-dependent plasmon shifting of peak 1. However, the peak 3 blue shift exponentially as the silver coating is increased. Furthermore, one can find the silver coating-dependent intensity increasing of peak 1 is more intense as the gold nanorods are short. Therefore, two absorption peaks with equal intensity could be obtained with thinner silver coating. These experimental observations are in good agreement with our theoretical calculation above.

Conclusions

In conclusion, nonhomogeneous Ag coating of the Au nanorod and aspect ratio of Au nanorod core strongly affect

the intensity change and wavelength shift of the plasmonic absorption. Theoretical calculations show that, when the transverse Ag coating is faster, the Au–Ag core–shell nanorod could present two intense longitudinal plasmonic absorption peaks with equal intensity and small wavelength gap as the Ag shell thickness has a critical value. When the longitudinal Ag coating is faster, the intensity increasing of longitudinal peak corresponding to outer Ag surface is always weaker than the longitudinal peak corresponding to Au–Ag interface. The critical value of Ag shell thickness could be decreased when Au nanorod is short, at the same time the transverse peak corresponding to Au–Ag interface could be enhanced. It is worth noting that the two equal intensity of absorption peaks exhibit Au–Ag core–shell nanorod obtained from the nonhomogeneous silver coating may have great potential for many biological and chemical applications such as design and fabrication of dual-channel SPR sensor.

Acknowledgments This work was supported by the Program for New Century Excellent Talents in University under grant no. NCET-10-0688, the Fundamental Research Funds for the Central Universities under grant no. 2011jdgz17, and the National Natural Science Foundation of China under grant nos. 11174232, 61178075, and 81101122.

References

1. Moskovits M, Srnová-Šloufová I, Vlčkova B (2002) Bimetallic Ag–Au nanoparticles: extracting meaningful optical constants from the surface-plasmon extinction spectrum. J Chem Phys 116:10435–10446

2. Xu HX (2005) Multilayered metal core–shell nanostructures for inducing a large and tunable local optical field. Phys Rev B 72:0734051–0734055

3. Jiang R, Chen H, Shao L, Li Q, Wang J (2012) Unraveling the evolution and nature of the plasmons in (Au core)–(Ag shell) nanorods. Adv Mater 24:OP200–OP207

4. Zhang Q, Moran CH, Xia XH, Rycenga M, Li NX, Xia YN (2012) Synthesis of Ag nanobars in the presence of single-crystal seeds and a bromide compound, and their surface-enhanced Raman scattering (SERS) properties. Langmuir 28:9047–9054

5. Li Y, Jing C, Zhang L, Long YT (2012) Resonance scattering particles as biological nanosensors in vitro and in vivo. Chem Soc Rev 41:632–642

6. Wu LP, Li YF, Huang CZ, Zhang Q (2006) Visual detection of Sudan dyes based on the plasmon resonance light scattering signals of silver nanoparticles. Anal Chem 78:5570–5577

7. Tao AR, Habas S, Yang PD (2008) Shape control of colloidal metal nanocrystals. Small 4:310–325

8. Chen HJ, Shao L, Woo KC, Ming T, Lin HQ, Wang JF (2009) Shape-dependent refractive index sensitivities of gold nanocrystals with the same plasmon resonance wavelength. J Phys Chem C 113:17691–17697

9. Cardinal MF, Rodríguez-González B, Alvarez-Puebla RA, Pérez-Juste J, Liz-Marzán LM (2010) Modulation of localized surface plasmons and SERS response in gold dumbbells through silver coating. J Phys Chem C 114:10417–10423

10. Tsuji M, Miyamae N, Lim S, Kimura K, Zhang X, Hikino S, Nishio M (2006) Crystal structures and growth mechanisms of Au@Ag core–shell nanoparticles prepared by the microwave-polyol method. Cryst Growth Des 6:1801–1807

11. Xue C, Millstone JE, Li S, Mirkin CA (2007) Plasmon-driven synthesis of triangular core–shell nanoprisms from gold seeds. Angew Chem Int Ed 46:8436–8439

12. Duan J, Park K, MacCuspie RI, Vaia RA, Pachter R (2009) Optical properties of rodlike metallic nanostructures: insight from theory and experiment. J Phys Chem C 113:15524–15532

13. Tsuji M, Matsuo R, Jiang P, Miyamae N, Ueyama D, Nishio M, Hikino S, Kumagae H, Kamarudin KSN, Tang XL (2008) Shape-dependent evolution of Au@Ag core–shell nanocrystals by PVP-assisted *N,N*-dimethylformamide reduction. Cryst Growth Des 8:2528–2536

14. Gong JX, Zhou F, Li ZY, Tang ZY (2012) Synthesis of Au@Ag core–shell nanocubes containing varying shaped cores and their localized surface plasmon resonances. Langmuir 28:8959–8964

15. Hong SC, Choi YJ, Park SH (2011) Shape control of Ag shell growth on Au nanodisks. Chem Mater 23:5375–5378

16. Ma YY, Li WY, Cho EC, Li ZY, Yu TY, Zeng J, Xie ZX, Xia YN (2010) Au@Ag core–shell nanocubes with finely tuned and well-controlled sizes, shell thicknesses, and optical properties. ACS Nano 4:6725–6734

17. Deng JJ, Du J, Wang Ye TYF, Di JW (2011) Synthesis of ultrathin silver shell on gold core for reducing substrate effect of LSPR sensor. Electrochem Commun 13:1517–1520

18. Lee YH, Chen H, Xu QH, Wang J (2011) Refractive index sensitivities of noble metal nanocrystals: the effects of multipolar plasmon resonances and the metal type. J Phys Chem C 115:7997–8004

19. Zhu J (2009) Surface plasmon resonance from bimetallic interface in Au–Ag core–shell structure nanowires. Nanoscale Res Lett 4:977–981

20. Yu K, You GJ, Polavarapu L, Xu QX (2011) Bimetallic Au/Ag core–shell nanorods studied by ultrafast transient absorption spectroscopy under selective excitation. J Phys Chem C 115:14000–14005

21. Liu MZ, Guyot-Sionnest P (2004) Synthesis and optical characterization of Au/Ag core/shell nanorods. Phys Chem B 108:5882–5888

22. Perenboom JAAJ, Wyder P, Meier F (1981) Electronic properties of small metallic particles. Phys Rep 78:173–292

23. Liu MZ, Guyot-Sionnest P (2006) Preparation and optical properties of silver chalcogenide coated gold nanorods. J Mater Chem 16:3942–3945

24. Yi DK (2011) A study of optothermal and cytotoxic properties of silica coated Au nanorods. Mater Lett 65:2319–2321

25. Huang H, Liu X, Zeng Y, Yu X, Liao B, Yi P, Chu PK (2009) Optical and biological sensing capabilities of Au2S/AuAgS coated gold nanorods. Biomaterials 30:5622–5630

26. Marinakos SM, Chen S, Chilkoti A (2007) Plasmonic detection of a model analyte in serum by a gold nanorod sensor. Anal Chem 79:5278–5283

27. Chakravadhanula VSK, Elbahri M, Schürmann U, Takele H, Greve H, Zaporojtchenko V, Faupel F (2008) Equal intensity double plasmon resonance of bimetallic quasi-nanocomposites based on sandwich geometry. Nanotechnology 19:225302–225306

28. Sau TK, Murphy CJ (2004) Seeded high yield synthesis of short Au nanorods in aqueous solution. Langmuir 20:6414–6120

29. Fu Q, Zhang DG, Yi MF, Wang XX, Chen YK, Wang P, Ming H (2012) Effect of shell thickness on a Au–Ag core–shell nanorods-based plasmonic nano-sensor. J Opt 14:085001–085005

30. Zhu J, Zhang F, Li JJ, Zhao JW (2013) Optimization of the refractive index plasmonic sensing of gold nanorods by non-uniform silver coating. Sensor Actuat B-Chem 183:556–564

7

Electrocatalytic activity of ligand-protected gold particles: formaldehyde oxidation

Kun Luo · Haiming Wang · Xiaogang Li

Abstract Tris(hydroxymethyl)phosphine oxide (THPO) and triphenyl phosphine oxide (PPh$_3$O) were introduced onto the surface of colloidal gold nanoparticles (Au NPs), and the effect of capping ligands on the catalytic electrooxidation of formaldehyde was studied voltammetrically by using colloidal Au-NP-modified glassy carbon electrodes (GCEs). This was compared with polycrystalline Au and another Au-NP-modified GCE without a capping molecule. We found that PPh$_3$O causes a larger decrease in the catalytic activity of the Au NPs in liquid than THPO does, indicating that the catalytic activity of the Au NPs is closely associated with the capping ligands. The effect of capping ligands is discussed based on the available surface ratio (ASR), which is defined as the ratio of the total surface area measured electrochemically to the calculated value based on the number and geometry of the Au NPs. These were determined to be 70.6 % for THPO and 0.23 % for PPh$_3$O, respectively. The significant blocking of formaldehyde is probably due to the structure and hydrophobicity of the benzene rings in the PPh$_3$O molecule, which is responsible for the decrease in catalytic activity of the Au NPs.

Keywords Gold nanoparticle · Ligand · Catalytic activity · Formaldehyde · Voltammetry

Introduction

Since the catalytic property of gold nanoparticles (Au NPs) was revealed by Haruta et al. [1] and Hutchings et al. [2], the effect of the particulate size of gold has been extensively investigated [3–7]. Nanoparticles between 2 and 50 nm have been determined to be heterogeneous catalysts [8–10]. Gold-based catalysts are normally composites of Au NPs and insulating oxide supports, and these catalysts exhibit the best low-temperature activity for CO oxidation among all catalysts [11]. Their catalytic activity exceeds the catalytic activity of platinum group metals by a factor of about five [12, 13]. Gold catalysts have also been reported to be uniquely selective for the partial oxidation of propene to propene oxide [4]. However, disagreement on the mechanisms of the catalytic processes remains [8].

The large surface energy of Au NPs makes them prone to aggregation, and capping ligands or macromolecules need to be used to stabilize these colloidal metal systems [14], which normally results in a decrease in catalytic activity of the Au NPs. Au also undergoes deactivation upon poisoning as a result of chemisorption, as do other noble metal catalysts. For example, chloride is regarded as a potent poison of gold catalysts, which readily adsorbs onto the surface of gold when HAuCl$_4$ is used as a starting material [8, 15]. Calcination is often employed to remove capping molecules from Au NPs, which leads to the growth of Au particulates. Electrochemical deposition or stripping of PbO$_2$ films offers another route for the removal of organic species on a gold surface [16, 17]. However, it is not feasible for use with Au-based catalysts because the Au NPs are normally deposited on insulating oxides. It is thus necessary to understand the effect of surface chemistry on the electrocatalytic activity of Au-NP catalysts.

The catalytic activity of noble metals can be studied electrochemically in solutions of strong reducing agents such as hydrazine, borohydride, dimethylamine borane, and formaldehyde. Au was found to be more active in formaldehyde alkaline solutions [18] than in other solutions. Formaldehyde dissolves in water as a weakly acidic gem-diol, which deprotonates in alkaline solutions to form the electroactive enolate anion [19]. The enolate anion is oxidized on gold and

K. Luo (✉) · H. Wang · X. Li
Key Laboratory of New Processing Technology for Nonferrous Metals and Materials, Ministry of Education, College of Materials Science and Engineering, Guilin University of Technology, Guilin 541004, China
e-mail: luokun@glut.edu.cn

follows the reported scheme [20, 21] to form HCOO⁻ and CO_2, and this is accompanied by hydrogen evolution [22]. On platinum and palladium, formaldehyde oxidation occurs without hydrogen generation.

The catalytic activity of Au NPs can be studied using Au-NP-modified glassy carbon electrodes (GCEs). The signal-to-noise ratio during electroanalysis is often closely associated with the loading and the surface chemistry of the Au NPs on the modified GCEs. A few methods are available for the preparation of Au-NP-modified GCEs such as electrodeposition [23], the spontaneous reduction of $AuCl_4^-$ on carbon [24], or the direct adsorption of Au-NP colloids [24]. To increase the Au-NP loading on modified electrodes without serious agglomeration, the self-assembly of Au-NP colloids at immiscible liquid/liquid (L/L) interfaces can also be used [25–30]. A spontaneous interfacial reaction was recently reported by Rao et al. [31, 32] between chloro(triphenylphosphine)gold(I) ($Au(PPh_3)Cl$) in toluene and tetrakis(hydroxymethyl)phosphonium chloride (THPC) in an aqueous alkaline solution. This reaction directly generates robust Au-NP ultrathin films at the L/L interface. Our previous work [33] indicated that the spontaneous L/L interfacial reaction is actually composed of two simultaneous steps: Au particle formation and self-assembly that leads to compositionally similar ultrathin Au-NP films in the interfacial reaction and the self-assembly methods. Organophosphorus molecules that form during the interfacial reaction play an important role in determining the morphology of the interfacial deposit. In this paper, a further voltammetric investigation was performed using Au-NP-modified GCEs in alkaline formaldehyde solutions to study the effect of ligand molecules on the catalytic activity of Au NPs in liquids.

Experimental

Chemicals

Formaldehyde (CH_2O, 36 % in water, Aldrich), sodium tetrachloroaurate (III) dihydrate ($NaAuCl_4$, 99.99 %, Alfa Aesar), tetrakis(hydroxymethyl)phosphonium chloride (THPC, 80 %, Aldrich), triphenyl phosphine (PPh_3, 99 %, Aldrich), hydrochloric acid (HCl, 37 %, Aldrich), nitric acid (HNO_3, 69 %, BDH), and phosphate buffer solution (pH=7.32, Fisher Scientific) were used as received.

Preparation of the Au colloid

A total of 165 µl of 50 mM THPC solution was injected into 8 ml of 6.25 mM NaOH solution. After stirring for 5 min, 3 ml of a 3.3-mM $NaAuCl_4$ solution was added dropwise, and this is a variant of the approach described by Duff et al. [34]. The appearance of a dark red solution indicated the formation of an Au colloid.

Modification of glassy carbon electrodes

Glassy carbon electrodes (GCEs, 3 mm diameter, CH Instruments Inc.) were employed as catalytically inert substrates. The GCEs were carefully polished with diamond spray (0.1 µm, Kemet International Ltd.) and sonicated in deionized water before use.

Modification 1   A total of 5 ml of the as-prepared Au colloid was mixed and shaken vigorously with 5 ml of pure toluene as described by Binks et al. [25] where the resultant interfacial deposit was transferred onto a clean GCE leading to a modified electrode designated "GCE-AuTHPO" after drying in air.

Modification 2   A total of 5 ml of the as-prepared Au colloid was mixed with 5 ml of a 5-mM PPh_3 toluene solution. After vigorous shaking, a deposit was formed at the toluene/water interface, which was transferred onto a clean GCE, and a modified electrode designated "GCE-AuPPhO" was prepared after drying in air.

Modification 3   A clean GCE was immersed in a 3.3-mM $NaAuCl_4$ solution for 5 min, and a modified electrode designated "GCE-AuCl" was prepared by the spontaneous interaction between $AuCl_4^-$ and carbon as suggested by Vaskelis et al. [22].

All the modified GCEs were rinsed with deionized water before use. Transmission electron microscopy (TEM) was carried out to characterize the Au NPs on the modified GCEs with a Tecnai F30 FEG-TEM system operating at 300 kV.

Electrochemistry

A three-electrode system under the control of a PGSTAT30 Potentiostat (Autolab, Eco Chemie B.V.) was used. A platinum mesh and a silver/silver chloride wire (Ag/AgCl/3 M KCl) were used as counter and reference electrodes, respectively. The working electrodes were gold polycrystalline discs, unmodified GCE, and the modified GCEs described above. All the alkaline formaldehyde solutions were freshly prepared and were degassed in pure argon before commencing voltammetry. The amount of Au NPs on the modified GCEs was determined by anodic stripping in a dilute aqua regia solution [35] containing 0.1 M HCl and 0.32 M HNO_3 (in a molar ratio of roughly 1:3). Linear sweep voltammetry was carried out from 0 to 1.4 V (vs. Ag/AgCl) repeatedly until the stripping peaks were invisible. The sum of the serial oxidation peaks was taken to correspond to the amount of Au deposited as per Faraday's laws. The total available surface area of the Au NPs on the modified GCEs was measured by the ratio of the reduction peak area to a value of 482 μC/cm^2 in a phosphate buffer solution (pH 7.32), according to the method of Hoogvliet and Oesch [35, 36]. The potentials in the figures are quoted versus the Ag/AgCl reference electrode.

Results and discussion

Electrocatalytic oxidation of formaldehyde

Figure 1 compares cyclic voltammetry (CV) curves of the polycrystalline Au electrode, the unmodified GCE, and the Au-NP-modified GCEs. In Fig. 1a, the catalytic electrooxidation of CH_2O on the bulk Au electrode generates two oxidation peaks at ca. -0.27 and $+0.68$ V (vs. Ag/AgCl) in a solution containing 0.1 M CH_2O and 0.1 M KOH (solid). These peaks are not visible in the absence of CH_2O in the alkaline solution (dotted). This result is in agreement with previous literature [20]. In contrast, there is only one peak present in the voltammograms of the Au-NP-modified GCEs where the curve of the bare GCEs

appears to be catalytically inert toward the reaction (dotted lines in Fig. 1b–d). This is indicative of the strong catalytic activity of the Au NPs toward the anodic oxidation of formaldehyde. The decrease in the peak at -0.27 V in Fig. 1b–d implies that the anodic oxidation of formaldehyde on Au NPs follows a different pathway compared with bulk Au. According to Yang et al. [37], the electrooxidation of formaldehyde on Au-NP-modified GCEs produces $HCOO^-$ because of GCE substrate restrictions while the reaction on bulk Au normally generates CO_2 instead of $HCOO^-$, which also highlights the product selectivity characteristic of Au-based nanocatalysts. The mechanistic difference was thus examined again and verified by the results in Fig. 1.

The peak potentials in Fig. 1b–d were also found to vary with surface molecules and were around $+0.35$ V for GCE-AuCl, $+0.40$ V for GCE-AuTHPO, and $+0.50$ V for GCE-AuPPhO. The order is thus: GCE-AuCl < GCE-AuTHPO < GCE-AuPPhO. However, the peak current density based on the geometry of the supporting GCEs has a different sequence as the values are 0.0047, 0.0016, and 0.0022 A cm^{-2} for GCE-AuCl, GCE-AuTHPO, and GCE-AuPPhO, respectively. This order is thus: GCE-AuCl < GCE-AuPPhO < GCE-AuTHPO. To understand this phenomenon, the effects of morphology and capping ligand molecules of the Au NPs on the catalytic property have to be further investigated.

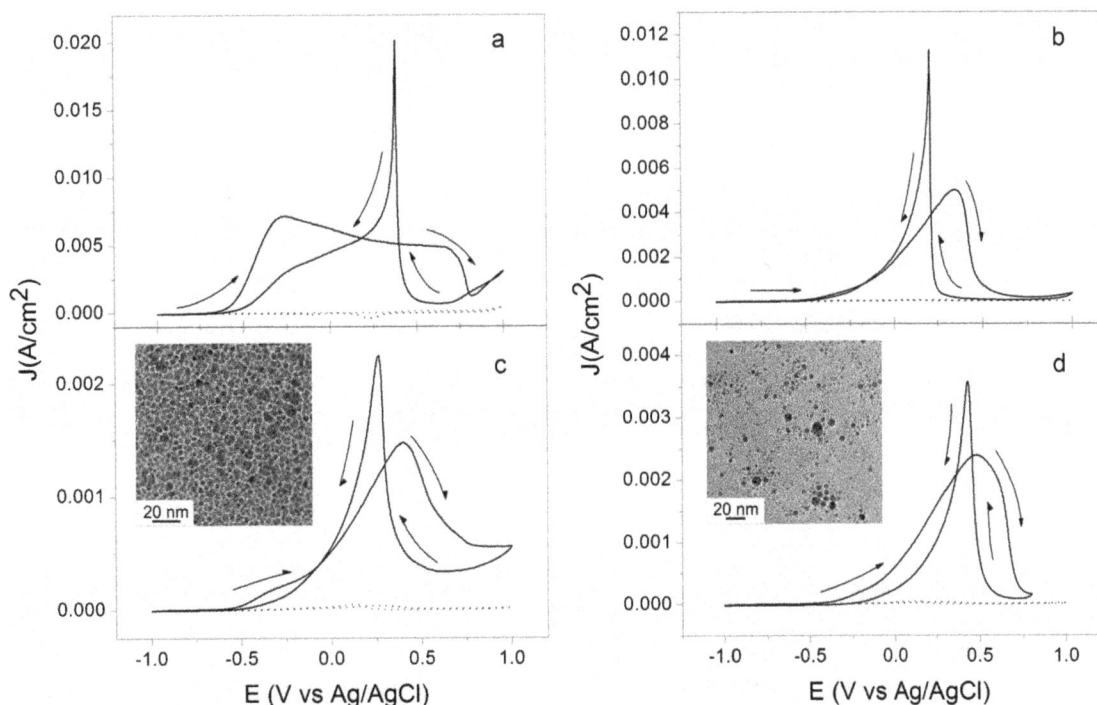

Fig. 1 Cyclic voltammograms of unmodified GCE (*dotted line*) in a 0.1-M CH_2O and 0.1-M KOH solution at a potential scan rate of 50 mV s^{-1} for *a*, gold electrode; *b*, GCE-AuCl; *c*, GCE-AuTHPO; and *d*, GCE-AuPPhO. The third scans for each electrode were obtained for comparison. *Insets* are the TEM images of the Au NPs on GCE-AuTHPO and GCE-AuPPhO, respectively

Morphology of the Au NPs

TEM micrographs of the Au NPs on the modified GCEs are shown in the insets of Fig. 1c, d. The average diameters were measured to be 2.1±0.7 nm (N=544) for GCE-AuTHPO and 2.6±1.5 nm (N=331) for GCE-AuPPhO, which is indicative of the similar sizes of the two Au deposits. We have previously determined the nature of the functional groups on the two Au-NP deposits by XPS analysis [33]. These included tris(hydroxymethyl)phosphine oxide (THPO) for GCE-AuTHPO and tri(phenyl)phosphine oxide (PPh$_3$O) plus THPO for GCE-AuPPhO. THPO is the cleavage product of THPC in alkaline solution, and it adsorbs onto the surface of the colloidal Au NPs as a protecting ligand during the reduction of NaAuCl$_4$. The THPO-functionalized Au-NP colloid undergoes self-assembly when it is in contact with pure toluene leading to GCE-AuTHPO without agglomerated Au NPs. When PPh$_3$ is present in toluene, a competitive ligand exchange process probably occurs between THPO and PPh$_3$. Most THPO is replaced by PPh$_3$O (PPh$_3$ can be converted into PPh$_3$O in a basic environment) although a small amount of THPO still remains. For GCE-AuCl deposition, previous reports on the catalytic oxidation of formaldehyde and stripping experiments [24, 38] have indicated the presence of metallic Au NPs on the GCE surface because of the spontaneous interaction between AuCl$_4^-$ and carbon. A similar reduction of silver ions by carbon has also been described previously [38]; however, to date, no morphological evidence has been available to support these conclusions. In our experiments, we attempted to find Au particles in the TEM images, but only a few Au NPs were available for measurement on the GCE-AuCl possibly because of the limited loading of Au. However, deductions can still be made through voltammetric investigations.

The amount of Au that deposited on the GCE-AuCl, GCE-AuTHPO, and GCE-AuPPhO electrodes was determined by anodic stripping, where linear sweep voltammetry was repeated from 0 to 1.4 V in a mixed acid solution (0.1 M HCl and 0.32 M HNO$_3$) at a scan rate of 200 mV s^{-1} until the Au stripping peak was invisible. Additionally, all the stripping peaks were at ca. 1.1 V (not shown). The sum of the serial oxidation peaks corresponds to the amount of Au deposited according to Faraday's laws:

$$m = QM_{Au}/zF, \tag{1}$$

where Q is the sum of the measured charge of the stripping peaks, M_{Au} is the atomic mass of gold, z is the charge transferred per Au atom (taken as 3 for the stripping process), and F is the Faraday constant. The mass of Au NPs on the modified GCE was then calculated separately using Eq. 1, giving 0.15, 11.2, and 1,150 ng for the GCE-AuCl, GCE-AuTHPO, and GCE-AuPPhO, respectively. The loading of

the GCE-AuCl is much smaller than that reported previously [24], possibly because of the weaker interaction between AuCl$_4^-$ and carbon in basic solutions compared with low-pH solutions. The current density values based on the loading of Au NPs for the electrooxidation of formaldehyde in Fig. 1 were calculated to be 3.3×10^7, 1.00×10^5, and 2.0×10^3 A g^{-1} for GCE-AuCl, GCE-AuTHPO, and GCE-AuPPhO, respectively, which is in agreement with the sequence of peak potentials shown in Fig. 1. Therefore, both the thermodynamic and kinetic data suggest that the presence of organophosphorus ligand molecules tends to decrease the catalytic activity of Au NPs toward the electrooxidation of formaldehyde where PPh$_3$O exhibits more influence than THPO. A further discussion about the available surface area of the Au NPs may help to determine the reason for PPh$_3$O being better at blocking formaldehyde than THPO.

Effect of ligand molecules

The effect of organophosphorus molecules on the catalytic activity of the Au NPs on the modified GCEs can be determined by the available surface ratio (ASR), which is defined as the ratio of the measured total surface area to the calculated area based on the number and geometry of Au NPs. Assuming that the Au NPs are spherical, the available surface ratio can be expressed as follows:

$$ASR = \frac{S_m}{S} = \frac{S_m}{N\pi d^2}, \tag{2}$$

where ASR represents the available surface ratio, and S_m, N, and d are the measured total surface area, particle number, and diameter of the Au NPs, respectively. The total mass (m) and surface area (s) based on the geometry of the Au NPs can be expressed as follows:

$$m = N\rho_{Au}V = \frac{1}{6}N\pi d^3 \rho_{Au}, \tag{3}$$

$$S = N\pi d^2, \tag{4}$$

where V and ρ_{Au} are the particle volume and density of the Au NPs (19.3 g/cm^3 [39]). From Eqs. 3 and 4, the diameter (d) and particulate number (N) of the Au NPs can be written as:

$$d = \frac{6m}{S\rho_{Au}}, \tag{5}$$

$$N = \frac{S}{\pi d^2}. \tag{6}$$

The above equations suggest that S_m can be different from S when a part of the Au surface is covered by capping ligands and the available ratio of the surface can be estimated once the number and size of the Au NPs have been determined. For the

Fig. 2 Cyclic voltammograms of the bare and the Au-NP-modified GCEs in a phosphate buffer solution (pH=7.32) at a potential scan rate of 50 mV s^{-1}. The first scans were used. The *inset* displays the cathodic scan of the GCE-AuCl against the bare GCE

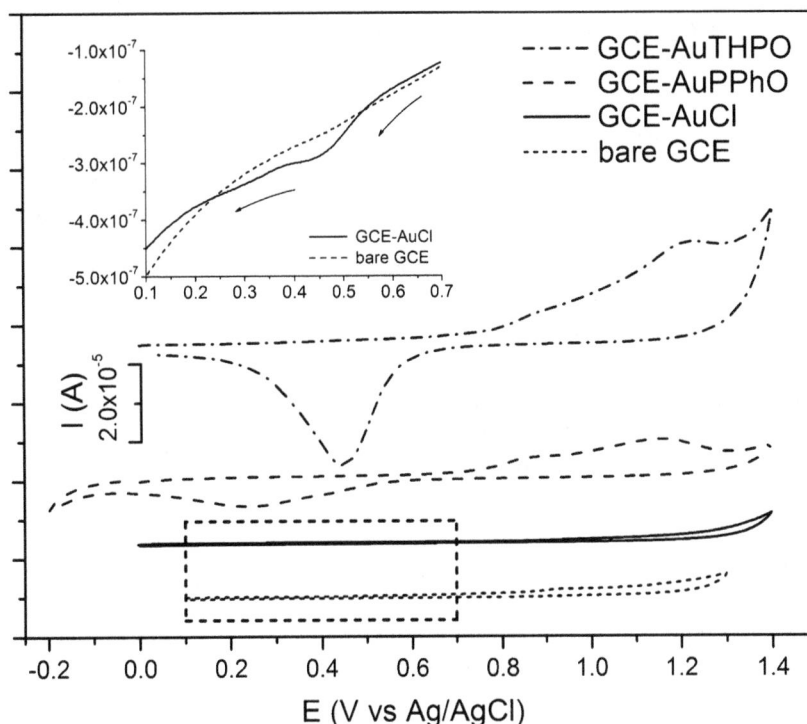

GCE-AuCl, it can be assumed that the whole surface of the Au NPs is available (ASR=100 %) or that the whole area is not affected by the possible adsorption of chloride. Therefore, the diameter of the Au NPs can be estimated directly from stripping and surface area measurements.

The total mass of the Au NPs can be calculated from the anodic stripping experiment, and the total available surface area can be measured voltammetrically in pH 7.32 phosphate buffer solution [35, 36]. During the CV experiments, the oxidation peaks for the GCE-AuTHPO and GCE-AuPPhO were found to decay with successive scans, which recovered when the experiments were restarted (not shown). This implies that the available surface of the Au-NP assemblies varied with the consumption of analyte. In contrast, such a decrease was not noticeable for GCE-AuCl possibly because of the low loading and the dispersive state of the Au NPs. Therefore, the first scans of the modified GCEs at a scan rate of 50 mV s^{-1} were used to estimate the total surface area of the Au NPs as shown in Fig. 2. GCE-AuTHPO gave the largest reduction

peak, and GCE-AuPPhO gave the second largest peak with a negative shift. The reductive shoulder for GCE-AuCl only becomes visible when the curve is compared against the unmodified GCE (shown in the inset). Based on the assumption that a monolayer of chemisorbed oxygen with a gold/oxygen ratio of 1:1 was formed, the total available surface area can be calculated from the ratio of the reduction peak area to a previously reported factor of 482 $\mu C/cm^2$, which corresponds to the reduction charge of chemisorbed oxygen according to Hoogvliet and Oesch [35, 36]. We thus obtained values of 1.8×10^{-5}, 1.2×10^{-2}, and 3.1×10^{-3} cm^2 for GCE-AuCl, GCE-AuTHPO, and GCE-AuPPhO, respectively.

Combining the results from the stripping and surface area measurements, the average diameter of the Au deposits on GCE-AuCl was calculated to be 26 nm. Using the diameter values of the Au deposits on the modified GCEs, the particle numbers can be estimated by Eq. 6, which gives 8.5×10^8, 1.2×10^{11}, and 6.5×10^{12} for GCE-AuCl, GCE-AuTHPO, and GCE-AuPPhO, respectively. The ASR values for GCE-

Table 1 Values measured and calculated from voltammetric experiments

Items	GCE-AuCl	GCE-AuTHPO	GCE-AuPPhO
Oxidation peak potential (V)	0.35	0.40	0.50
Oxidation peak current density (A/g)	2.2×10^6	1.0×10^4	1.4×10^2
Measured area (cm^2)	1.8×10^{-5}	1.2×10^{-2}	3.1×10^{-3}
Amount of Au NPs (ng)	0.15	11.2	1,150
Particle number	8.5×10^8	1.2×10^{11}	6.5×10^{12}
Available surface ratio (ASR)	100 %[a]	70.6 %	0.23 %

[a] Assuming that the absorption of chloride does not affect the available surface for the electrooxidation of formaldehyde

AuTHPO and GCE-AuPPhO were determined using Eq. 2 to be 70.6 and 0.23 %, respectively.

All the above values were measured and calculated from voltammetric experiments and the TEM images, and the data are listed in Table 1, which show that the adsorption of organophosphorus molecules obviously decreases the catalytic activity of the Au NPs. The benzene rings in the PPh_3O molecule occupy a larger space and appear more hydrophobic than THPO, which is probably responsible for the much larger obstruction to the transfer of formaldehyde to the surface of the Au NPs. This would, therefore, lead to a very small ratio of available surface area. This highlights the necessity of balancing the activity of Au-NP-based catalysts and the protection of Au NPs from aggregation.

Conclusion

The effect of capping ligands THPO and PPh_3O on the catalytic activity of Au NPs was studied voltammetrically using Au-NP-modified GCEs and compared with polycrystalline Au and another Au-NP-modified GCE without a capping molecule. The results reveal that PPh_3O causes more degradation in the catalytic activity of the Au NPs toward the electrooxidation of formaldehyde compared with THPO, probably because of the structure and hydrophobicity of the benzene rings in PPh_3O. To investigate this effect, the ASR was determined, and we found values of 70.6 % for THPO and 0.23 % for PPh_3O, indicating the significance of these values in describing the degree of obstruction generated by capping ligands toward analytes. This highlights the interplay between the functionalization and stabilization of Au-NP-based catalysts.

Acknowledgments We thank the National Natural Science Foundation of China (grant no. 21163004) and Guangxi Natural Science Foundation (grant no. 2013GXNSFAA019029). We also sincerely appreciate the academic guidance from Prof. Robert A. W. Dryfe in the University of Manchester.

References

1. Haruta M, Kobayashi T, Sano H, Yamada N (1987) Chem Lett 16: 405–408
2. Hutchings GJ (1985) J Catal 96:292–295
3. Bond GC (2001) Gold Bull 34:117–140
4. Thompson D (1999) Gold Bull 32:12–19
5. Valden M, Lai X, Goodman DW (1998) Science 281:1647–1650
6. Salisbury BE, Wallace WT, Whetten RL (2000) Chem Phys 262: 131–141
7. Prati L, Martra G (1999) Gold Bull 32:96–101
8. Cortie MB, Lingen E (2002) Materials Forum 26:1–14
9. Hughes MD, Xu Y-J, Jenkins P, McMorn P, Landon P, Enache DI, Carley AF, Attard GA, Hutchings GJ, King F, Stitt EH, Johnston P, Griffin K, Kiely CJ (2005) Nature 437:1132–1135
10. Schubert MM, Hackenberg S, van Veen AC, Muhler M, Plzak V, Behm RJ (2001) J Catal 197:113–122
11. Grisel R, Weststrate KJ, Gluhoi A, Nieuwenhuys BE (2002) Gold Bull 35:39–45
12. Thompson D (1998) Gold Bull 31:111–118
13. Bond GC, Thompson DT (2000) Gold Bull 33:41–51
14. Porta F, Prati L, Rossi M, Coluccia S, Martra G (2000) Catal Today 61:165–172
15. Yuan YZ, Kozlova AP, Asakura K, Wan HL, Tsai K, Iwasawa Y (1997) J Catal 170:191–199
16. Hernandez J, Solla-Gullon J, Herrero E (2004) J Electroanal Chem 574:185–196
17. Hernandez J, Solla-Gullon J, Herrero E, Aldaz A, Feliu JM (2006) Electrochim Acta 52:1662–1669
18. Yahikozawa K, Nishimura K, Kumazawa M, Tateishi N, Takasu Y, Yasuda K, Matsuda Y (1992) Electrochim Acta 37:453–455
19. ten Kortenaar MV, Tessont C, Kolar ZI, van der Weijde H (1999) J Electrochem Soc 146:2146–2155
20. ten Kortenaar MV, Kolar ZI, de Goeij JJM, Frens G (2002) Langmuir 18:10279–10291
21. ten Kortenaar MV, Kolar ZI, de Gopeij JJM, Frens G (2001) J Electrochem Soc 148:E327–E335
22. Bindra P, Roldan JM, Arbach GV (1984) Ibm J Res Dev 28:679–689
23. El-Deab MS, Okajima T, Ohsaka T (2003) J Electrochem Soc 150: A851–A857
24. Vaskelis A, Tarozaite R, Jagminiene A, Tamasiunaite LT, Juskenas R, Kurtinaitiene M (2007) Electrochim Acta 53:407–416
25. Binks BP, Clint JH, Fletcher PDI, Lees TJG, Taylor P (2006) Langmuir 22:4100–4103
26. Wang DY, Duan HW, Mohwald H (2005) Soft Matter 1:412–416
27. Duan HW, Wang DA, Kurth DG, Mohwald H (2004) Angewandte Chemie-Int Ed 43:5639–5642
28. Reincke F, Kegel WK, Zhang H, Nolte M, Wang DY, Vanmaekelbergh D, Mohwald H (2006) Phys Chem Chem Phys 8: 3828–3835
29. Reincke F, Hickey SG, Kegel WK, Vanmaekelbergh D (2004) Angewandte Chemie-Int Ed 43:458–462
30. Shen Z, Yamada M, Miyake M (2007) J Am Chem Soc 129:14271–14280
31. Rao CNR, Kulkarni GU, Agrawal VV, Gautam UK, Ghosh M, Tumkurkar U (2005) J Colloid Interface Sci 289:305–318
32. Rao CNR, Kulkarni GU, Thomas PJ, Agrawal VV, Saravanan P (2003) J Phys Chem B 107:7391–7395
33. Luo K, Schroeder SLM, Dryfe RAW (2009) Chem Mater 21:4172–4183
34. Duff DG, Baiker A, Edwards PP (1993) J Chem Soc Chem Comm 1: 96–98
35. Hoogvliet JC, Dijksma M, Kamp B, van Bennekom WP (2000) Anal Chem 72:2016–2021
36. Oesch U, Janata J (1983) Electrochim Acta 28:1237–1246
37. Yang H, Lu T, Xue K, Sun S, Lu G, Chen S (1999) J Mol Catal A Chem 144:315–321
38. Ragoisha GA, Jovanovic VM, Avramovivic MA, Atanasoski RT, Smyrl WH (1991) J Electroanal Chem 319:373–379
39. Hotta H, Ichikawa S, Sugihara T, Osakai T (2003) J Phys Chem B 107:9717–9725

8

UV photochemical synthesis of heparin-coated gold nanoparticles

M. del P. Rodríguez-Torres · Luis Armando Díaz-Torres ·
Pedro Salas · Claramaría Rodríguez-González ·
Martin Olmos-López

Abstract Sodium salt heparins of both reactive and pharmaceutical grade were used as the reducing and stabilizing agents for the UV photochemical synthesis of gold nanoparticles. Reduction of chloroauric acid, $HAuCl_4$, by heparin using UV irradiation as a catalyst is a simple, inexpensive method with direct biofunctionalization of the products for further use. Experiments using both heparin types were carried out separately. The varying synthesis parameters were the $HAuCl_4$ and heparin sodium concentrations. The synthesized AuNPs present spherical as well as anisotropic shapes, such as oval, triangular, hexagonal sheets, rods, and some other faceted forms, with dimensions ranging from 20 to 300 nm as well as agglomerated networks. No usage of additional surfactants or growth-inducing seeds was necessary in the syntheses.

Keywords Gold colloids · Photochemistry · Green synthesis · Glycosaminoglycans · Spectroscopy

Introduction

Colloidal metal nanoparticle synthesis techniques are used to produce different types of nanoparticle shapes for several applications, such as drug carriers, tumor detectors, photothermal agents, and so on. Chemical methods include chemical reduction of metal salts [1–3], alcohol reduction and polyol processes [4, 5], microemulsions [6], thermal decomposition [7], and electrochemical synthesis [8]. Physical methods include the exploding wire technique [9], plasma synthesis [10], chemical vapor deposition [11], microwave irradiation [12], supercritical fluids [13], sonochemical reduction [14], and gamma irradiation [15]. Photochemical synthesis is now being considered as another alternative for obtaining colloidal nanoparticles [16]. Some of its advantages are high spatial resolution, reaction selectivity and controllability as well as shape conversion. This method can be used with light sources like lasers [17], medium and low pressure Mercury [18], Xenon, LEDs and UV lamps [19], and even white light [20].

A wide range of materials and their combinations are available for light-based synthesis in solution; metal salts which act as precursors ($AgNO_3$, $HAuCl_4$, and $PtCl_6$) in a suitable solvent (ethyleneglycol, alcohols); chelate complexes, metal acetylides, polymers, and surfactants. There are also photosensitive agents that assist the reactions by means of radical formation which allow the metal reduction process leading to nanoparticle generation [17].

Heparin (also known as unfractioned heparin) is a member of the glycosaminoglycans family of carbohydrates and has a complex and heterogeneous structure. Its molecular weight can vary from 5,000 to 40,000 Da; and its sulfation degree also varies depending on the tissue it is collected from, which can be either beef lung or porcine intestinal mucosa, with 75 % of heparins made from the latter. It is formed mostly by two repeating disaccharide units (variably sulfated) shown in Fig. 1, a 2-O-sulfated iduronic acid (left ring) and a 6-O-sulfated, N-sulfated glucosamine (right ring) which can also be referred to as only iduronic acid and glucosamine, respectively. Due to the sulfate groups present, this disaccharide structure makes heparin the highest negative charge-containing of any known molecule.

Heparin is highly sensitive to light exposure (especially in the UV range), making it an excellent candidate for metal

M. del P. Rodríguez-Torres · L. A. Díaz-Torres (✉)
M. Olmos-López
Grupo de Espectroscopia de Materiales Avanzados y
Nanoestructurados (GEMANA), Centro de Investigaciones en
Optica A. C., León Gto. 37150, Mexico
e-mail: ditlacio@cio.mx

P. Salas · C. Rodríguez-González
Centro de Física Aplicada y Tecnología Avanzada, Universidad
Nacional Autónoma de México, A.P. 1-1010, Querétaro 76000,
Mexico

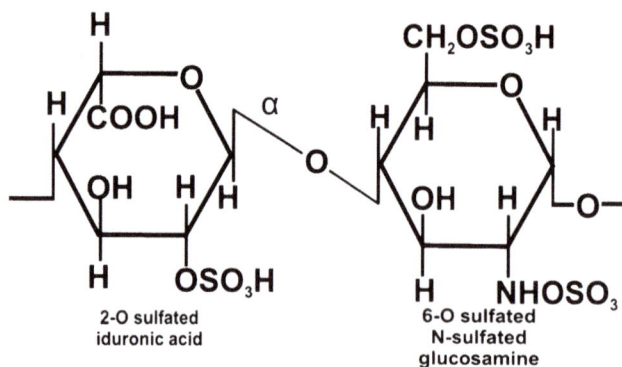

Fig. 1 Sodium heparin structure

nanoparticle synthesis. It is a very reactive material due to its composition; for example, each disaccharide repeating unit contains a carboxyl group. All repeating units contain one or more 1° or 2° hydroxyl groups and an average of 2–2.5 sulfo groups and N-sulfo groups are found in 75–85 % of all repeating units. Approximately 15–25 % of these repeating units contain a vicinal diol. Each chain contains a reducing hemiacetal end. Heparin also contains ~0.3 unsubstituted amino groups/chains and it owes its acidic nature to the covalently linked sulfate and carboxylic acid groups [21].

In this work, a photochemical method involving UV irradiation and the usage of sodium heparin, well-known biomolecule as a reductant and stabilizer in a green synthesis scheme has been developed in order to get nanowire-like, anisotropic and plate-shaped nanoparticles at room temperature. Green synthesis methods are good alternatives for avoiding the usage and generation of hazardous materials [22–24]. These forms have already been synthesized by heat-based techniques [25] at low and ambient temperatures [26] and via photochemical methods [27] not using heparin as a reagent whereas UV triggering has been employed. Heparin has been used in thermal-based methods obtaining spherical nanoparticles [28–31].

The samples obtained were characterized by UV–vis spectroscopy and TEM; the concentration of HAuCl$_4$ and heparin was varied in order to find out how this could affect the nanoparticle shape and plasmon peak. Also, heparin changes before and after UV irradiation were studied by means of UV–vis, FTIR spectroscopy and pH measurements. Finally, a formation mechanism was investigated and proposed.

Experimental

Materials

Gold (III) chloride trihydrate 99.9 % metal basis (HAuCl$_4$) and reactive-grade heparin sodium salt (195.9 USP units/mg) were purchased from Sigma-Aldrich. Pharmaceutical-grade heparin in aqueous solution (5,000 IU/mL, 10 mL, sealed container) was obtained from PISA Laboratories. Deionized water was used for preparing stock solutions as well as a solvent in all synthesis reactions. The materials for the experimental procedures were used as received without any further purification or modification.

All glassware and quartz material was cleaned with aqua regia (three parts hydrochloric acid and one part nitric acid), then rinsed with distilled water and 96 % ethanol. All washed material was dried in an oven at 70 °C before synthesis procedures take place. Gold (III) chloride trihydrate and reactive-grade heparin stock solutions were prepared using deionized water. Pharmaceutical-grade heparin was used as received.

UV irradiation of solutions

For nanoparticle synthesis, a home-made cylindrical UV reactor was used. Inside the reactor there are three vertically oriented black light lamps (peak intensity at $\lambda = 366$ nm, 4 W). Forced air cooling through top of the reactor was used to keep a reaction temperature around 30 °C. The typical synthesis solution is prepared as follows: First, deionized water is added, secondly the gold (III) chloride trihydrate, and finally the heparin, this solution is then poured in a 10-mm quartz-cell which is placed inside the reactor in the central position for it to be irradiated uniformly for seven hours as shown in Fig. 2.

In order analyze UV effects on heparin, stock solutions of heparin alone were poured in a 10-mm quartz cell and placed inside the UV reactor to be irradiated up to seven hours. The UV–vis and FTIR spectra as well as pH changes were continuously monitored.

Synthesis of gold nanoparticles with reactive-grade sodium salt heparin

Two sets of experiments were carried out, the first one keeping a constant concentration of heparin at 0.0312 mM while

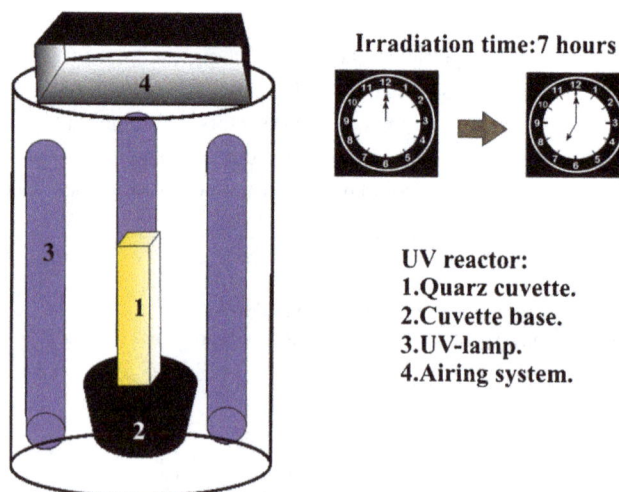

Fig. 2 Block diagram of the UV reactor used for nanoparticle synthesis

increasing the gold precursor concentration :0.0833 (sample A1), 0.333 (sample A2), and 0.4165 mM (sample A3). The second consisted of maintaining a gold concentration of 0.833 mM and varying the heparin one by 0.312, 0.624, 0.936, 1.248, and 1.56 mM.

Synthesis of gold nanoparticles with pharmaceutical-grade sodium salt heparin

The experiments carried out followed exactly the same preparation procedure for the reactive-heparin samples. Since pharmaceutical-grade heparin its vial has a 0.833-mM concentration, a maximum of 0.7 mM was used.

Characterization

UV–vis spectra were measured in a Perkin Elmer Lambda 3B UV–vis spectrophotometer with a 1-mm quartz cell; the measurements were done in air at ambient temperature (for both heparin solutions and colloidal nanoparticle samples). For infrared spectra measurement, a Perkin Elmer Spectrum BX FTIR system was used. Solid samples were prepared by

mixing heparin powder with KBr (potassium bromide), in a heparin to KBr ratio of 1:100, and milling them to form a very fine and homogeneous powder to make a pellet for obtaining the spectra. Pharmaceutical-grade heparin, as well as both UV-irradiated and non-irradiated heparin solutions, were dried at 45 °C for 6 h, and mixed with dry KBr powder. The pH changes were monitored by using a Conductronic pH120 pH meter. Transmission electron microscopy (TEM) was carried out in a JEOL model JEM-1010 operated at 80 kV using a carbon grid. TEM samples were prepared in the traditional way; an aliquot of the solution was dropped on a 3 mm diameter lacey carbon copper grid and then left to dry at room temperature for the solvent to evaporate.

Results and discussion

Reactive-grade heparin results

The synthesized samples were prepared by varying the concentrations of both, HAuCl4 and heparin. First, the HAuCl4 concentration was increased from 0.0833 mM up to 0.4165 mM,

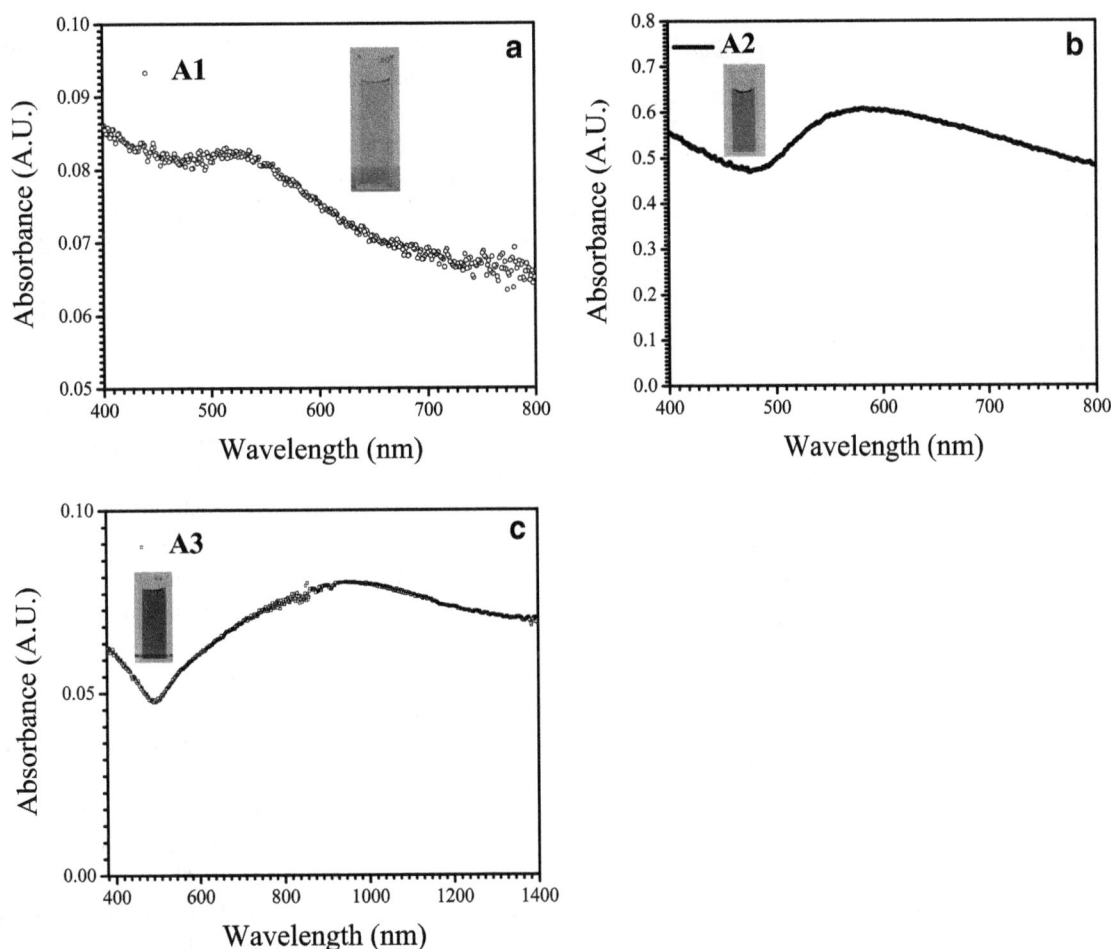

Fig. 3 a, b, c UV–vis spectra of the nanoparticles prepared with increasing gold concentration; the solution tones are also shown along

with heparin fixed at 0.0312 mM, and irradiated up to seven hours. Figure 3 shows some UV–vis spectra of samples A1, A2, and A3, for increasing HAuCl4 concentrations of 0.0833, 0.333, and 0.4165 mM, respectively. The solutions color changed from a light purple to a blue-grayish hue, this suggests a change in the metal oxidation and the type of nanoparticles formed. Additionally, the UV–vis spectra also suggest particle shape changes.

For sample A1, Fig. 3a, the spectra shows a surface plasmon resonance (SPR) peak centered at 522 nm, which is typical for nanospheres. The TEM micrographies in Fig. 4a, b show sphereroid particles of small diameters between 16 and 18 nm. It is clear that the nanoparticle yield is quite low, being in agreement with the small amplitude of the SPR peak in Fig. 3a. Besides this, there are some aggregates in a wire-like assembly, although they are still attached to much bigger clusters. This is in accordance with its UV spectrum because the shape is quite wide indicating the dispersion provoked by the aggregated assemblies and the maximum absorbance value being quite low. Samples A2 and A3 differ from A1 in their color and optical density as shown in Fig. 3b and c. These two

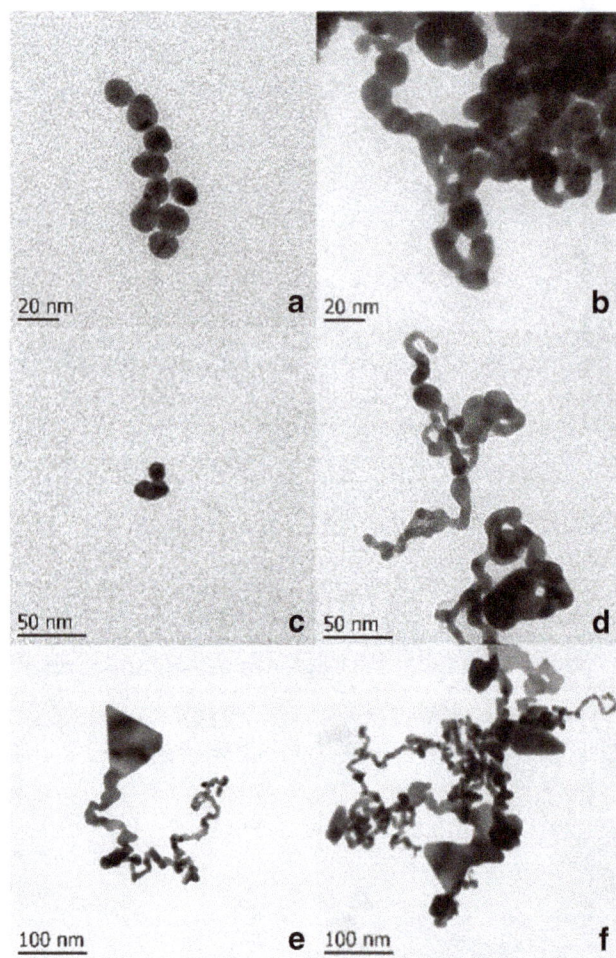

Fig. 4 TEM images of **a**, **b** sample A1, **c**, **d** sample A2 and **e**, **f** sample A3

solutions show similar tones being A3 the one with a more intense shade due to the higher gold concentration. Their plasmon peak is shifted to longer wavelengths respect to the A1 plasmon, also other weak peaks in the near IR could be observed.

Nanoparticles in A1 are spheroids, which are in agreement with their UV–vis spectrum, but it has to be taken into account that the distribution is broad probably because the particles tend to agglomerate a little and besides that they have different sizes. Sample A2 exhibits very few oval-like nanoparticles whereas there is a lot of wire-like particles that are less aggregated and more elongated, and still keeping some nanoparticles within, see Fig. 4c, d. In Fig. 4e, f, triangular plate-like nanoparticles are observed in addition to the nanowires. Sample A3 contains nanoparticles on the order of 100 ± 50 nm as well as other quasispherical nanoparticles between 10 and 30 nm, most of which are agglomerated. This can also explain why in its UV–vis spectrum a depression is found at 495 nm.

A possible explanation for the change from spheres to nanowire networks could be the fact that when the metal precursor concentration is very low, nanoparticles are formed, but the lack of stabilizer tends to form aggregated, elongated and bigger-sized structures. Taking into account that the gold concentration was being increased, the preferential growth in some directions is reached. This could explain the changes in shapes and the increase in length and decrease in diameter, indicating that there is not enough capping agent to passivate the nanoparticles surface.

The particles that have a plate-like aspect like in sample A3 could be a sign of polycrystallinity because of the varying contrasts in the TEM images. This may suggest that the crystals are oriented in different directions as the electron beam passes through the sample. Vasilev et al. [32] found similar results but using 2-mercaptosuccinic acid (MSA) as the reducing and stabilizing material, obtaining similar wire diameters and plate dimensions, but shorter wires. Other research groups have also obtained comparable results [33, 34].

When the heparin concentration was increased, the synthesized samples show narrower SPR peaks around 532 ± 4 nm that slightly shift towards longer wavelengths, show less dispersion, and increase in amplitude as heparin concentration increases, as can be observed in Fig. 5, the solutions color go from a light purple up to purplish-blue one, indicating nanoparticles' change in shape and size. When the highest heparin concentration is reached there is again a slight increase in dispersion, but still less than the one for the lowest heparin concentration.

This could be due to the fact that there are several assemblies formed by nanowires as well as some other nanoparticles of irregular shapes as shown in Fig. 6a. The high dispersion of the SPR at the lowest heparin concentration might be a consequence of the presence of several clusters formed by nanowires and some small nanoparticles within their structures; see

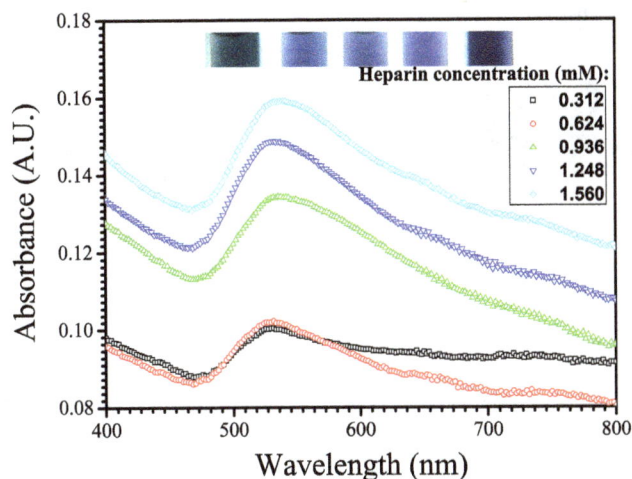

Fig. 5 UV–visible spectra of samples prepared with increasing heparin concentration. On *top*, there are the synthesis produced and their tone changes

Fig. 5. As the heparin concentration increases, the assemblies begin to connect to each other forming networks of different sizes. The reason for this may come from the fact that heparin

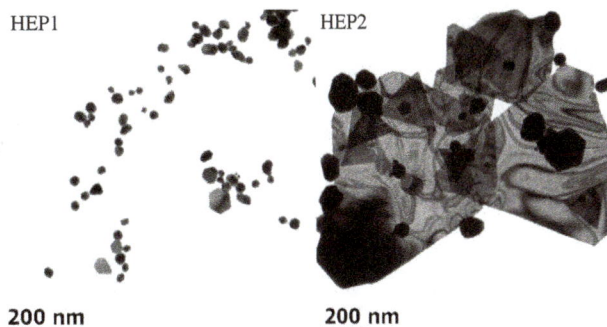

Fig. 6 TEM images when increasing heparin concentration. **a** 0.312, **b** 0.624, **c** 0.936, **d** 1.248 and **e, f** 1.56 mM

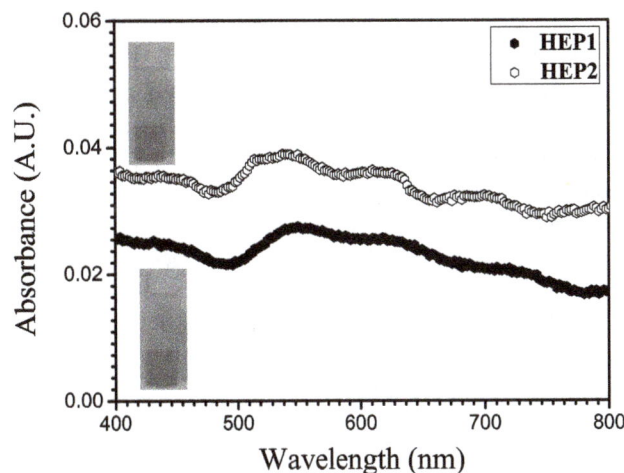

Fig. 7 HEP1 and HEP2 samples TEM images

starts to enclose nanoparticles as well as to stabilize them, so this causes the individual clusters formed to gather in networks, and at the same time it induces growth for some nanoparticles in some preferential directions. Their aggregation inside the wires is probably a result of the heparin encapsulation, mentioned before, that is not letting the nanoparticles grow outside the wire but inside it; and in consequence inducing their aggregation in some parts. There is a noticeable difference at the highest concentration, there are still formed networks but now some parts of them break apart from that main structure, in a tap-pole shaped nanoparticles (see Fig. 6f). That could make a slight dispersion contribution to the SPR peak observed in the UV–vis spectrum in Fig. 5. Similar nanoparticles shapes and networks have been obtained in previous works by means of photochemical synthesis using a mix of CTAB, NaOH and deionized water to reduce HAuCl4 with 365 nm UV light for about 9 h at room temperature [34].

Pharmaceutical-grade heparin results

The concentrations used for increasing gold concentrations are 0.0833 and 0.4165 mM (keeping heparin at 0.0312 mM) and for increasing heparin concentrations (keeping gold at 0.833 mM) are 0.312, 0.5, and 0.7 mM.

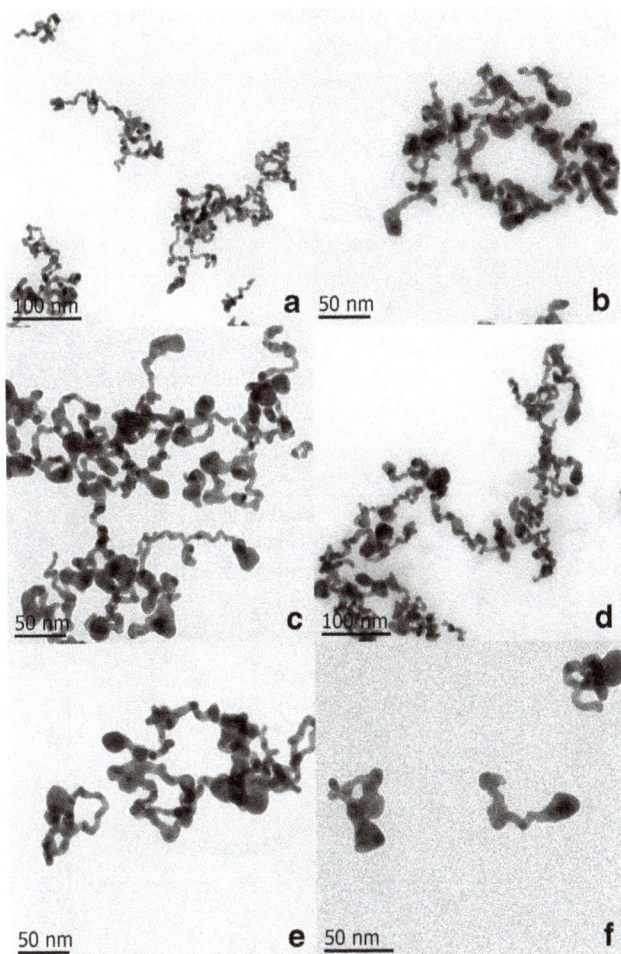

Fig. 8 UV–vis spectra of samples synthesized with decreasing gold concentration along with the prepared colloids

Fig. 9 HEP3, HEP4 and HEP5
samples TEM images

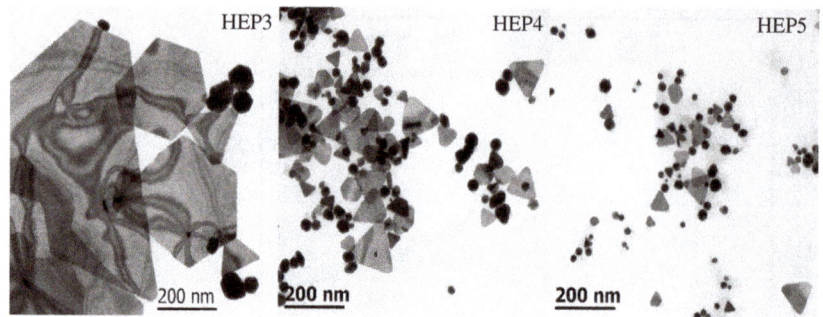

Non-spherical gold nanoparticles present multiple absorption bands correlating with their multiple axes, and they can support both propagating and localized surface plasmon resonances. The number of SPR peaks usually increases as the symmetry of nanoparticles decreases; spherical nanoparticles exhibit only one peak, whereas two and three peaks are often observed in nanorods, nanodisks, and triangular plates, respectively [30]. Nanoplates have been previously synthesized by photochemical methods [23, 35–37] with similar results. In our case, when first modifying the gold concentration, it is observed that for the HEP1 sample, few nanoparticles are generated and they are small, ranging from 13 to 20 nm. For sample HEP2 the situation is completely different because the nanoparticle sizes go up abruptly most likely due to the fact that the capping effect coming from heparin is not enough to stop their growth as shown in Fig. 7.

Also, another thing is that in contrast with reactive-grade heparin experiments, there is now a shape change. The particles show anisotropic features but instead of wire-like nanoparticles we obtained plates, faceted, pseudospherical and even some trapezoidal products that in principle do not coalesce among them to form more complex agglomerates. Other aspects indicating differences from the nanoparticles synthesized with reactive-grade heparin are the solutions tones and their UV–vis spectra, the solutions shade change from a light pink to light violet which are shown in Fig. 8.

For sample HEP1, there are three peaks: at 539, 626, and 718 nm and for HEP2 at 547, 642, and 737 nm, the appearance of three peaks can be explained due to the presence of tip-pointed nanoparticles which contribute to their plasmon resonance. Additionally, the plasmon peak red-shift indicates the nanoparticles sizes are bigger.

When pharmaceutical-grade heparin concentration was changed, the nanoparticles still show anisotropic features just like when the Au content is varied, the only difference resides in the size distribution as can be observed in Fig. 9.

As the heparin concentration rises, there is more encapsulating and size-limiting effects on the particles obtained as observed from the TEM images (Fig. 9). In the case of the HEP5 sample, it is noted that some really small nanoparticles are present due the heparin reducing power (besides its stabilizing role).

The plate sizes for these three syntheses with increasing heparin concentration vary from 100 up to 800 nm and for the other shapes from 20 to 50 nm.

Figure 10 shows the solutions tones as well as their UV–vis spectra. Unlike the previous case when decreasing heparin concentration case, for the three syntheses, the first peaks are located between 480 and 630 nm, and the second one between 700 and 900 nm , both peaks tend to blue-shift as the heparin concentration is increased indicating the nanoparticles diminish their size. Also, the FWHM for the first peak decreases from 82 to 58 nm indicating that the nanoparticle distribution is narrower, besides this peak sharpness indicates that a great contribution from the quasispherical nanoparticles is present. Similar mixed products have been obtained by Zhang et al. [36].

Structure and changes of heparins under UV irradiation

The two types of heparins used in nanoparticle synthesis exhibit different characteristics, as suggested by the UV–vis absorbance spectra in Fig. 11. Both present two peaks in the UV region due to the aldehyde groups present in their basic structure. For the pharmaceutical grade preparation, there is a peak related to the presence of benzyl alcohol [38] as its excipient. It should be noted that most pharmaceutical-grade

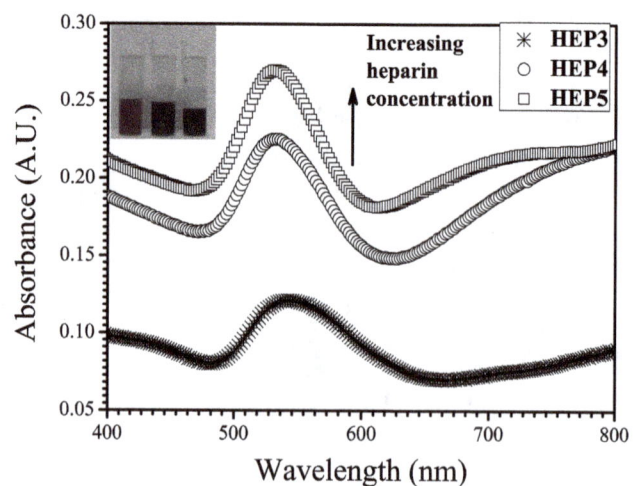

Fig. 10 UV–vis spectra of samples synthesized with increasing heparin concentration along with the colloids prepared

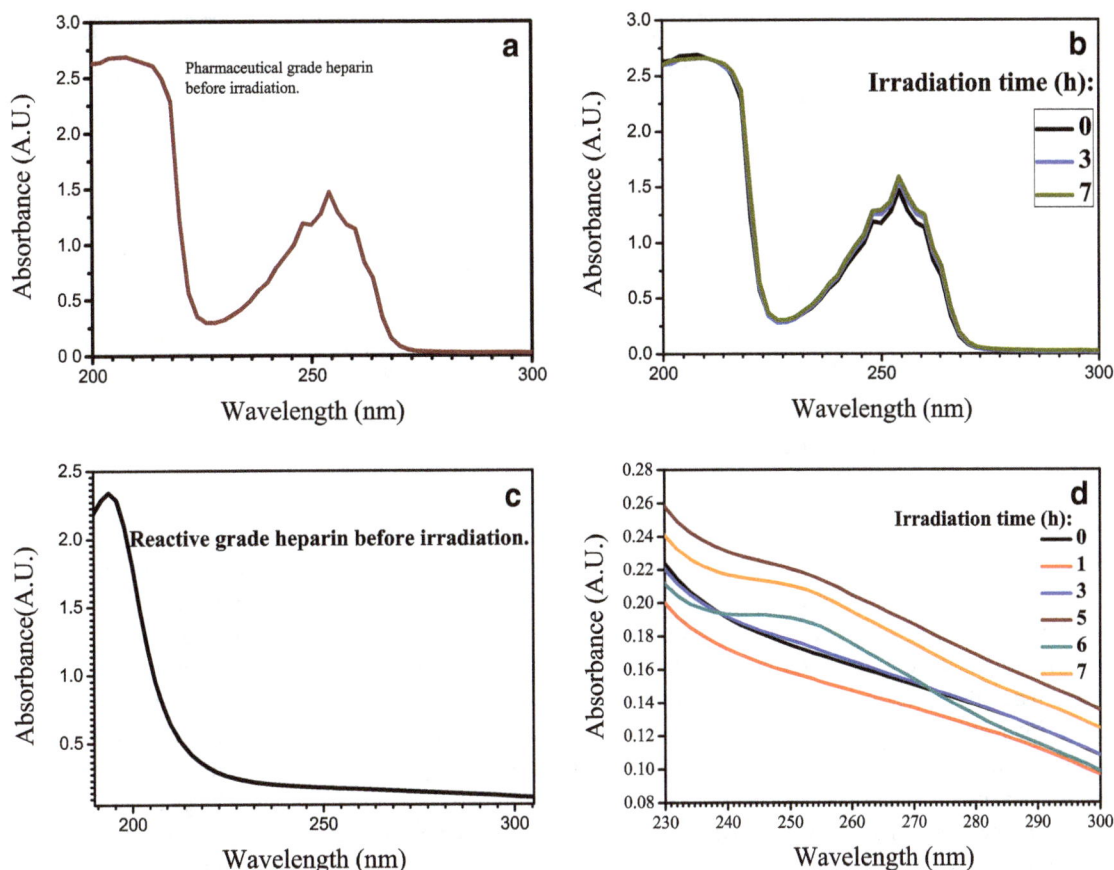

Fig. 11 UV–vis spectra of **a**, **b** pharmaceutical and **c**, **d** reactive grade heparin before and after UV irradiation

heparins usually contain different excipients, for example, distilled water and other substances like sodium chloride to render isotonic; sodium hydroxide and/or hydrochloric acid for pH adjustment, too. In our case, it was determined that pharmaceutical-grade heparin is more acidic due to benzyl alcohol presence. Balazs et al. [39] carried out several experiments using heparin, heparan sulfate and hyaluronic acid to investigate what changes occurred when exposed to UV irradiation. For heparin, they found that it is degraded by UV

which was evident in the loss of coagulation activity because the sulfate groups are affected. It was also revealed that the heparin solutions lowered their pH, meaning there is an electron loss. Heparin's reducing power (glucose monosaccharide) also increases proportionally to irradiation time indicating a breaking of glycosidic linkages and an increase of reducing end groups along with other irradiation products.

The spectra shown in Fig. 11a for pharmaceutical-grade heparin do not show abrupt changes in the structure, but a

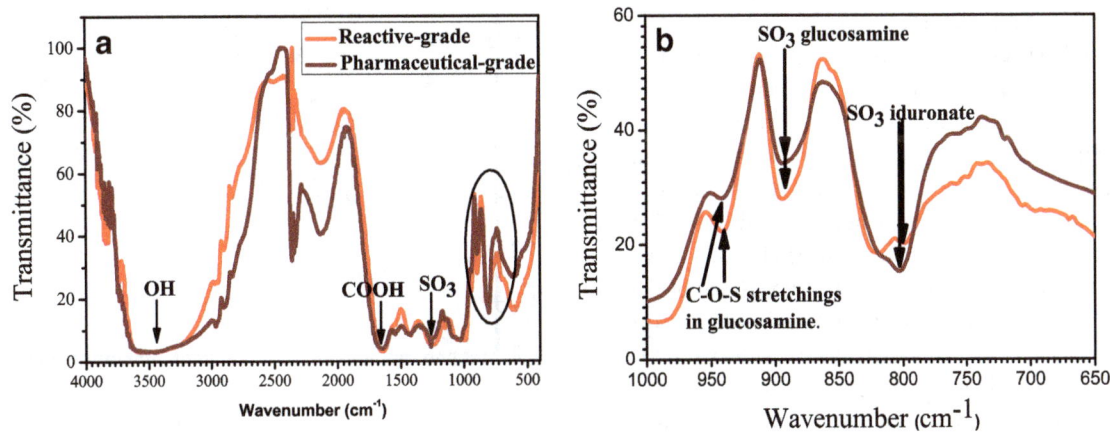

Fig. 12 FTIR spectra of heparins before they are UV-irradiated

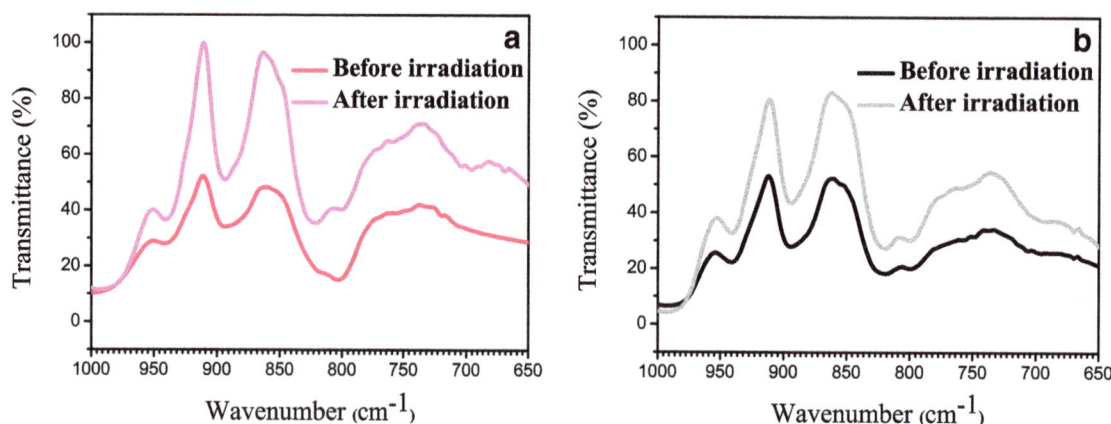

Fig. 13 FTIR spectra of **a** pharmaceutical-grade heparin and **b** reactive-grade heparin showing changes before and after UV irradiation

slight rise on optical density that may indicate the formation of radicals. However, for the spectra corresponding to reactive-grade heparin in Fig. 11b, the peaks are not as clear and intense as the observed for pharmaceutical heparin in Fig. 11a. This difference has to do with the different solvents in each case. Furthermore, the second peak in both cases tends to grow slightly in intensity and to change its shape as the irradiation time passes by. According to Bazals et al. [39], this may be due to the fact that the increase in absorbance is promoted in more alkaline media than in acidic media, indicating radical formation as well.

FTIR peak assignation

FTIR studies before UV irradiation taken for both heparin kinds show vibrational and absorption peaks, see Fig. 12. The spectra tend to depict some differences. One thing to take into account is the fact that heparin even when dried is still hydrophilic, which means that some peaks may correspond to solvent retention and even impurities, considering heparin synthesis yields heterogeneous products even when being from the same origin.

The peak at 3,467 cm^{-1} in both heparins corresponds to hydroxyl group vibrations; carboxyl and sulfate groups are responsible for the acidic character of heparin. Their location is approximately 1,645 cm^{-1} by 1,648 cm^{-1} in pharmaceutical and reactive-grade heparins, respectively. The band at approximately 1,260 cm^{-1} in pharmaceutical-grade heparin and the one at 1,236 cm^{-1} in reactive-grade heparin display different structures; these peaks and their absorption intensity are normally related to symmetric vibrations of S=O bonds [40]. The latter has a less intense and broader peak in the 800–880 range which is the sulfated sugar region [41]. It is noted that peaks corresponding to the values of 800 and 816 and 820 cm^{-1} (in pharmaceutical-grade heparin and reactive-grade heparin, respectively) belong to the iduronate residue 2-0-sulphate group and to the 6-0-sulphate group of the glucosamine residue, such bands are overlapped in pharmaceutical-grade heparin and

split into two separated contributions in reactive-grade heparin but with lower absorption. Outside the mentioned range, the peak at approximately 941 cm^{-1} indicates vibrations of components which are part of the glucosidic linkages and also belong to C–O–S stretching (in glucosamine) [42].

Figure 13 depict the changes in both heparins after UV irradiation. The peak for the sulfate groups diminishes for pharmaceutical-grade heparin. This may indicate the appearance of other absorbing materials. The peaks that correspond to the glycosidic linkages in both heparins show absorption decrease which is in agreement with the fact that they break during irradiation as expected. The peaks corresponding to S=O vibrations pose contrasting characteristics, because in the pharmaceutical-grade heparin there is less absorption after UV irradiation, contrary to what happens in reactive-grade heparin. This may be the result of major vibration in these bonds, from which it might be inferred that there are more atoms in this heparin chain which makes it bigger and heavier. The sulfate groups absorptions due to iduronate and glucosamine decrease in absorption but the peaks that are part of the D-glucosamine residue tend to be more defined, something that

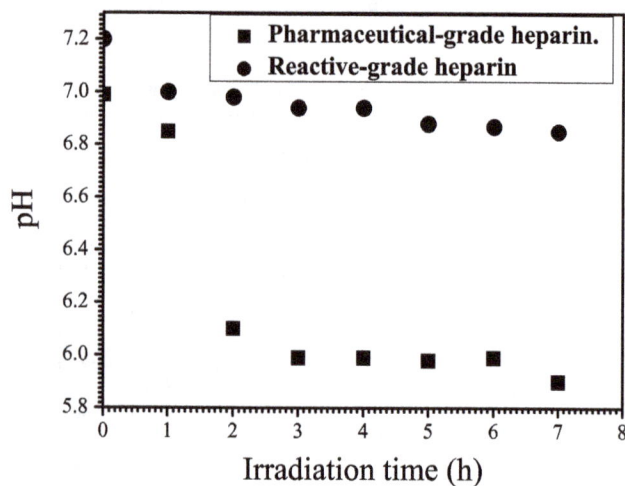

Fig. 14 pH measurement of heparins irradiated for 7 h

may be explained by suggesting an increase in the reducing end groups.

Heparin pH measurements

pH measurements were taken by preparing a reactive-grade heparin solution and a pharmaceutical grade one and irradiating them for seven hours, checking how the value changed every hour. The initial pH value for reactive-grade heparin was 7.2 which corresponds to an alkaline value, and from then on, it continued to go down reaching a final value of 6.85 while for the pharmaceutical-grade preparation it went from 6.99 to 5.9. That is the reason why there is more UV absorbing material generation as stated by Balazs et al. [32] and was verified in the the UV–vis spectra (Fig. 14).

Nanoparticle formation mechanism under UV irradiation

Heparin is a polysaccharide formed by disaccharide basic units (iduronic acid and glucosamine) which are repetitive along its structure with varying sulfation degrees and that becomes hydrolized when dissociated. This means that the glycosidic bond that joined them is broken because H and OH ions are donated by water. When this happens, these disaccharides units are separated (this occurs to the whole chain composed by these two monosaccharide units).

Once there has been hydrolysis, the glucosamine monosaccharide turns to its hemiacetal form. In this hemiacetal form, the aldehyde present, which t is reactive in the UV range, promotes its own oxidation, which at the same time allows the reduction of gold ions when present in solution, and thus this process leads to nanoparticles generation and growth as shown in Fig. 15.

Thus, from the above discussed results, it is possible that the differences between the types of nanoparticles produced by the two heparins might be due to the fact that pharmaceutical-grade heparin has a more acidic nature which on time allows more reduction of gold ions.

On the other hand, when working with reactive grade heparin, which is more alkaline, heavier, and bigger, the molecules might take some time for the reaction to happen in a certain direction. It may be slower, not allowing rapid

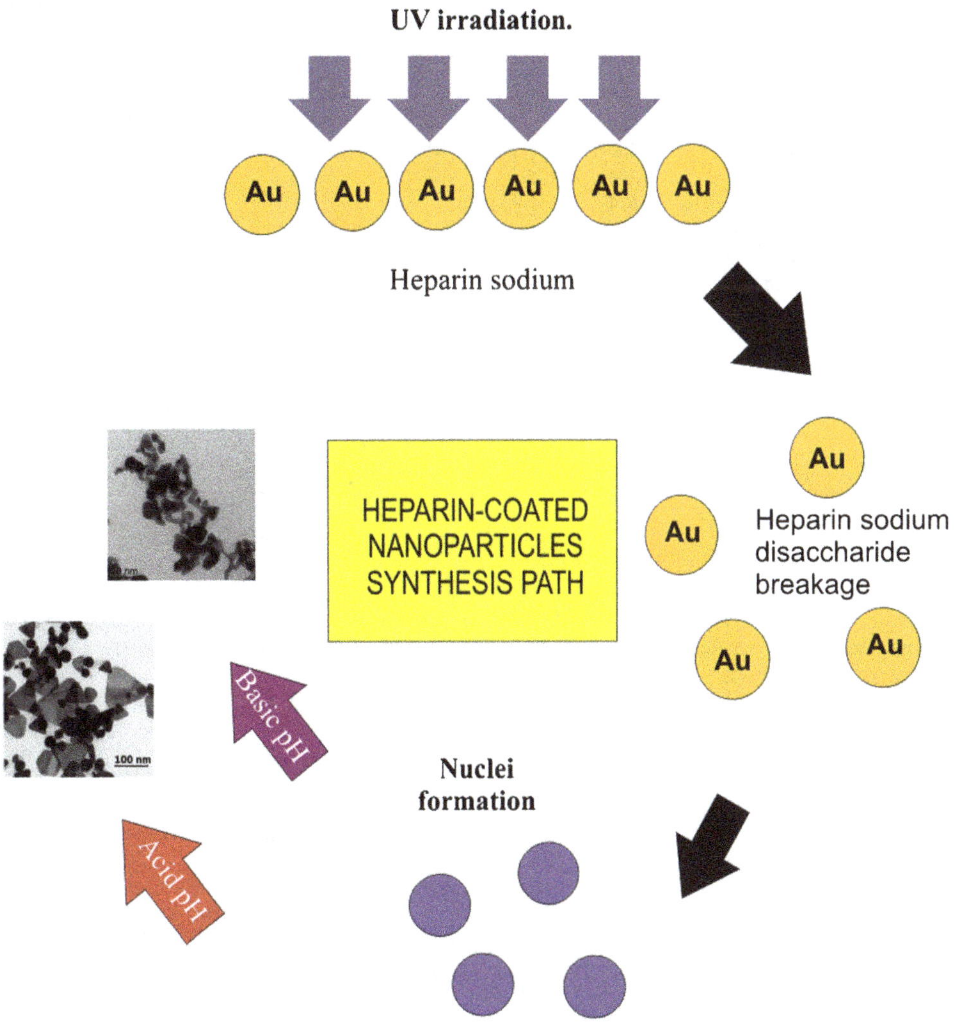

Fig. 15 Heparin-coated nanoparticle synthesis path

growth in other directions, and letting nanoparticles stick to each other or agglomerate instead of forming wire-like products. It has to be taken into account that polysaccharide hydrolysis is more favored in acidic than in alkaline media.

Conclusions

In summary, a simple while flexible strategy to synthesize colloidal gold anisotropic nanoparticles and nanoaggregated networks has been shown. Moreover, nanoparticles colloidal solutions are obtained by taking advantage of an inexpensive home-made black light UV reactor and heparin, a biomolecule used as an anticoagulant which is very reactive in the UV range.

It is found that low heparin concentrations trigger nanoparticle growth in preferential directions contrary to the case when its concentration is higher than the gold precursor one, which restrains this growth. It is suggested that pH affects the nanoparticle shapes, getting agglomerates when basic and nanoplates, faceted, pseudospherical and even some trapezoidal products when the medium is acid and agglomerated wire-like products when basic. In addition, we presume that given the biocompatibility of the capping and stabilizing agent, heparin these as-synthesized colloids could be potential candidates for plasma studies and therapy.

Acknowledgments We want to thank the following people: Ricardo Valdivia Hernandez for his technical assistance with the UV-reactor maintenance throughout this work as well as Carlos Juarez Lora, Octavio Pompa Carrera and Raul Nieto Centeno for their assistance on the UV–vis spectrophotometer and FTIR system usage. TEM measurements were aided by I.B.Q. Ma. Lourdes Plama Tirado.

References

1. Kimling J, Maier M, Okenve B, Kotaidis V, Ballot H, Plech A (2006) Turkevich method for gold nanoparticle synthesis revisited. J Phys Chem B 110:15700–15707.
2. Goulet PJG, Bruce Lennox R (2010) New insights into Brust–Schiffrin metal nanoparticle synthesis. J Am Chem Soc 132:9582–9584.
3. Perrault SD, Chan WCW (2009) Synthesis and surface modification of highly monodispersed, spherical gold nanoparticles of 50–200 nm. J Am Chem 131:17042–17043.
4. Kim K-S, Choi S, Cha J-H, Yeon S-H, Lee H (2006) Facile one-pot synthesis of gold nanoparticles using alcohol ionic liquids. J Mater Chem 16:1315–1317.
5. Dang TMD, Le TTT, Fribourg-Blanc E, Dang MC (2012) Influence of surfactant on the preparation of silver nanoparticles by polyol method. Adv Nat Sci: Nanosci Nanotechnol 3:035004.
6. Ethayaraja M, Dutta K, Muthukumaran D, Bandyopadhyaya R (2007) Nanoparticle formation in water-in-oil microemulsions: experiments, mechanism, and Monte Carlo simulation Langmuir 23:3418–3423.
7. Izaak TI, Babkina OV, Boronin AI, Drebushchak TN (2003) Constitution and properties of nanocomposites prepared by thermal decomposition of silver salts sorbed by Polyacrylate Matrix Colloid Journal 65:720–725.
8. Yu-Ying Y, Chang S-S, Chien-Liang Lee CR, Wang C (1997) Gold nanorods: electrochemical synthesis and optical properties. J Phys Chem B Lett 101:6661–6664.
9. Abdullah A, Annapoorni S (2005) Fluorescent silver nanoparticles via exploding wire technique PRAMANA. J Phys 65:815–819. doi:
10. Richmonds C, Mohan Sankaran R (2008) Plasma-liquid electrochemistry: rapid synthesis of colloidal metal nanoparticles by microplasma reduction of aqueous cations. Appl Phys Lett 93:131501–131504.
11. Kim H, Heup S (2011) Moon Chemical vapor deposition of highly dispersed Pt nanoparticles on multi-walled carbon nanotubes for use as fuel-cell electrodes. Carbon 49:1491–1501.
12. Doolittle JW Jr, Dutta PK (2006) Influence of microwave radiation on the growth of gold nanoparticles and microporous zincophosphates in a reverse micellar system. Langmuir 22:4825–4831
13. Byrappa K, Ohara S, Adschiri T (2008) Nanoparticles synthesis using supercritical fluid technology - towards biomedical applications. Adv Drug Deliver Rev 60:299–327.
14. Chen W, Cai W, Zhang L, Wang G, Zhang L (2001) Sonochemical processes and formation of gold nanoparticles within pores of mesoporous silica. J Colloid Interf Sci 238:291–295.
15. Gasaymeh SS, Radiman S, Heng LY (2010) Synthesis and characterization of silver/polyvinylpyrrolidone (Ag/PVP) nanoparticles using gamma irradiation techniques. Am J Appl Sci 7(7):892–901.
16. Sakamoto M, Fujistuka M, Majima T (2009) Light as a construction tool of metal nanoparticles: synthesis and mechanism. J Photochem Photobiol C Photochem Rev 10:33–56.
17. Amendola V, Rizzi GA, Polizzi S, Meneghetti M (2005) Synthesis of gold nanoparticles by laser ablation in toluene: quenching and recovery of the surface Plasmon absorption. J Phys Chem B Lett 109(49):23125–23128.
18. Tan S, Erol M, Attygalle A (2007) Synthesis of positively charged silver nanoparticles via photoreduction of AgNO₃ in branched polyethyleneimine/HEPES solution. Langmuir 23:9836–9843
19. Courrol LC, de Oliveira Silva FR, Gomes L (2007) A simple method to synthesize silver nanoparticles by photo-reduction. Colloid Surf A Physicochem Eng Asp 305:54–57.
20. Sato-Berrú R, Redón R, Vázquez-Olmos A, Saniger JM (2009) Silver nanoparticles synthesized by direct photoreduction of metal salts. Application in surface-enhanced Raman spectroscopy. J Raman Spectroscopy 40:376–380.
21. Murugesan S, Xie J, Linhardt RJ (2008) Immobilization of Heparin: Approaches and Applications Current Topics in Medicinal Chem 8:80–100.
22. Wu C-C, Chen D-H (2010) Facile green synthesis of gold nanoparticles with gum arabic as a stabilizing agent and reducing agent. Gold Bull 43(4):234–240.
23. Chien C-W, Chen D-H (2007) A facile and completely green route for synthesizing gold nanoparticles by the use of drink additives. Gold Bull 40(3):206–212.

24. Sivaraman SK, Kumar S, Santhanam V (2010) Room-temperature synthesis of gold nanoparticles – Size-control by slow addition. Gold Bull 43(4):275–286.

25. Fan X, Liu L, Guo Z, Ning G, Lina Y, Zhang X (2010) Facile synthesis of networked gold nanowires based on the redox characters of aniline. Mater Lett 64:2652–2654.

26. Castro L, Blázquez ML, Munoz JA, González F, García-Balboa C, Ballester A (2011) Biosynthesis of gold nanowires using sugar beet pulp. Process Biochem 46:1076–1082

27. Kim Y-J, Song JH (2006) Practical synthesis of Au nanowires via a simple photochemical route. Bull Korean Chem Soc 27:633–634

28. Guo Y, Yan H (2008) Preparation and characterization of heparin-stabilized gold nanoparticles. J Carbohydr Chem 27:309–319.

29. Park Y, Im A-R, Hong YN, Kim C-K, Kim YS (2013) Green synthesis and nanotopography of heparin-reduced gold nanoparticles with enhanced anticoagulant activity. J Nanosci Nanotechnol 11: 7570–7578.

30. Kemp MM, Kumar A, Mousa S, Park T-J, Ajayan P, Kubotera N, Mousa S, Linhardt RJ (2009) Synthesis of gold and silver nanoparticles stabilized with glycosaminoglycans having distinctive biological activities. Biomacromolecules 10(3):589–595.

31. Huang H, Yang X (2004) Synthesis of polysaccharide-stabilized gold and silver nanoparticles: a green method. Carbohydr Res 339:2627–2631.

32. Vasilev K, Zhu T, Wilms M, Gillies G, Lieberwirth I, Mittler S, Knoll W, Kreiter M (2005) Simple, one-step synthesis of gold nanowires in aqueous solution. Langmuir 21:12399–12403.

33. Wang L, Song Y, Sun L, Guo C, Sun Y, Li Z (2008) Controllable synthesis of gold nanowires. Mater Lett 62:4124–4126.

34. Kim Y-J, Song JH (2006) Practical Synthesis of Au Nanowires via a Simple Photochemical Route Bull Korean. Chem Soc 27(5):633–634

35. Yang S, Zhang T, Zhang L, Wang S, Yang Z, Ding B (2007) Continuous synthesis of gold nanoparticles and nanoplates with controlled size and shape under UV irradiation. Colloid Surf A Physicochem Eng 96:37–44.

36. Zhang G, Jasinski JB, Howell JL, Patel D, Stephens DP, Gobin AM (2012) Tunability and stability of gold nanoparticles obtained from chloroauric acid and sodium thiosulfate reaction. Nanoscale Res Lett 7:337.

37. Lee KY, Kim M, Lee YW, Choi MY, Han SW (2007) Photosynthesis of gold nanoplates at the water/oil interface. Bull Korean Chem Soc 28(12):4124–4126

38. de Micalizzi C, Pappano NB, Debattista NB (1998) First and second order derivative spectrophotometric determination of benzyl alcohol and diclofenac in pharmaceutical forms. Talanta 47:525–530

39. Balazs EA, Laurent TC, Howe AF, Varga L (1959) Irradiation of mucopolysaccharides with ultraviolet light and electrons. Radiat Res 11:149–164.

40. Sushko NI, Firsov SP, Zhbankov RG, Tsarenkov M, Marchewka M, Ratajczak C (1994) Vibrational spectra of heparins. J Appl Spectroscopy 61:5–6.

41. Oliveira GB, Carvalho LB Jr, Silva MPC (2003) Properties of carbodiimide treated heparin. Biomaterials 24:4777–4783.

42. Grant D, Long WF, Moffat CF, Williamson FB (1991) Infrared spectroscopy of heparins suggests that the region 750–950 cm-' is sensitive to changes in iduronate residue ring conformation. Biochem J 275:193–197

One-pot organometallic synthesis of well-controlled gold nanoparticles by gas reduction of Au(I) precursor: a spectroscopic NMR study

Pierre-Jean Debouttière · Yannick Coppel ·
Philippe Behra · Bruno Chaudret · Katia Fajerwerg

Abstract A stable colloidal solution of gold nanoparticles (AuNPs) has been prepared in tetrahydrofuran by gas reduction of AuCl(tetrahydrothiophene) and alkylamines (1-octylamine,1-dodecylamine, and 1-hexadecylamine) as stabilizing agents. Carbon monoxide is a better reducing agent than hydrogen. The important parameters for control of the synthesis of AuNPs are the temperature, the molar ratio of ligand/metal, and the structure of the stabilizing agent. A high concentration of long alkyl chains (10 eq.) favours the control of the growth of AuNPs of defined size and shape with a diameter of 4.7 nm and a narrow size distribution. For the first time, liquid-state combined with solid-state NMR spectroscopies were used in order to determine the role of the long chain alkylamines in the synthesis of AuNPs in CO atmosphere. This combination enables the understanding of the complex chemistry of the surface of AuNPs involved in the stabilization of the AuNPs. Indeed, carbamide species were formed during the synthesis. They were strongly coordinated to the surface of AuNPs and exchange phenomena of the alkylamines present in solution occurred, too.

Keywords Au(I) precursor · Organometallic synthesis · Gold nanoparticles · NMR spectroscopy

P.-J. DeboutTière · Y. Coppel · K. Fajerwerg
CNRS, LCC (Laboratoire de Chimie de Coordination), 205 route de Narbonne, BP 44099, 31077 Toulouse Cedex 4, France

P.-J. DeboutTière · Y. Coppel · K. Fajerwerg (✉)
UPS, INPT, Université de Toulouse, 31077 Toulouse Cedex 4, France
e-mail: katia.fajerwerg@lcc-toulouse.fr

B. Chaudret (✉)
INSA, CNRS, UPS, LPCNO, Université de Toulouse, 31077 Toulouse, France
e-mail: bruno.chaudret@insa-toulouse.fr

P. Behra
INPT, INRA, Université de Toulouse, UMR 1010, ENSIACET, 4 allée Emile Monso, 31030 Toulouse CEDEX 4, France

P. Behra
LCA (Laboratoire de Chimie Agro-industrielle), INRA, 31030 Toulouse, France

P.-J. DeboutTière
RTRA "Sciences et Technologies pour l'Aéronautique et l'Espace", 31030 Toulouse, France

Introduction

Since the 1990s, the formation of self-assembled nanostructures and the deposition of metallic nanoparticles (MNPs) have been the subject of growing interest. MNPs are now commonly used in many applications such as electronics [1, 2], catalysis [3], environmental sciences [4, 5], medicine [6, 7], and sensors [8–10]. Indeed, highly sensitive miniature sensors require advanced technology coupled with fundamental knowledge of biology, material sciences, and chemistry. Among the MNPs, gold nanoparticles (AuNPs) have been extensively studied for their optical properties [11] their absence of toxicity [12] their catalytic properties [13, 14] and their high conductivity [15–17] that make them excellent nanomaterials for various sensors such as electrochemical sensors [18, 19]. The development of simple methods to obtain AuNPs which are able to form thin films on substrates is therefore of great interest. Recently, different electrochemical methods were investigated for the deposition of AuNPs on glassy carbon electrodes for Hg(II) trace detection in water [20, 21]. Despite good performance with a detection limit of 0.40 nmol of

Hg.L^{-1}, no control of shape and dispersisty of AuNPs smaller than 15 nm were obtained by electrodeposition of HAuCl$_4$ in sodium nitrate electrolyte. As the catalysis and electrocatalysis performance depend strongly on the AuNP shape and size, it is of primary importance to control both of these parameters.

Several methods can be used to obtain well-defined AuNPs, in particular chemical reduction of molecular or ionic gold precursors. Au(III) and Au(I) are the common oxidation states of gold. Au(I), as an intermediate oxidation state, is well known to play an important role in controlling the morphology of metallic gold nanostructures synthesized from Au(III) ions [22] but only a few Au(I) precursors have been used to generate AuNPs. A recent review reports numerous advantages starting with an Au(I) precursor: (i) energy saving, (ii) easier reduction, (iii) better control of morphology, and (iv) in situ capping of nanostructures [23]. In this context, some examples of organometallic Au(I) complexes of alkylphosphines [24, 25] or (alkyl)amines [26, 27] have been reported to be efficient precursors for synthesizing gold Au$_{55}$ clusters or AuNPs with different shapes such as nanospheres, nanorods, nanorice [28], and nanowires [29].

In our group, we have developed a methodology for the synthesis of MNPs from organometallic precursors in organic solvents and their efficient stabilization by various ligands which allows us to get MNPs with tuneable surface properties [30]. The mild temperature and gas pressure applied for the precursor decomposition permit the synthesis of stable MNPs with a great versatility in particle size and shape. The reasons for this interest were (i) the possibility to control the kinetics of the decomposition of the precursor and therefore the size of the particles, (ii) the possibility to prepare novel structures through the mild conditions involved, and (iii) the control of surface state.

AuNPs synthesis by the decomposition of AuCl (tetrahydrothiophene) (AuCl(THT)) in an organic solvent under H$_2$ or CO in the presence of alkylamine of different alkyl chain length has led to well-defined NPs that can be self-assembled on a TEM grid [26]. However, good control of the size and stability of the particles could not be obtained through the mild decomposition of the AuCl-(alkylamine) complexes. In order to avoid the preparation and purification of AuCl-(alkylamine) complex, a one-pot synthesis in tetrahydrofuran (THF) using AuCl(THT) and alkylamines as stabilizing agent (1-octylamine(OA), 1-dodecylamine (DDA), and 1-hexadecylamine (HDA)) has been developed to control the growth of the AuNPs. In this work, we report the formation of a stable colloidal solution of AuNPs. To the best of our knowledge, this is the first time that the liquid-state NMR combined with the solid-state NMR has been used to determine the exact role of the alkylamine ligands in the synthesis of AuNPs based on the decomposition of AuCl-(alkylamine) complex under CO or H$_2$. This original approach is very useful in the characterization of the species coordinated at the surface and also provides the formation and

the stabilization of well-controlled AuNPs <15 nm with a narrow size distribution.

Experimental section

Reagents and general procedures

All operations concerning nanoparticle synthesis were carried out in Schlenck tubes or Fischer–Porter glassware or in a glove box in an argon atmosphere.

OA, DDA, and HDA were purchased from Acros Organics or Sigma-Aldrich-Fluka and used without further purification.

Solvents were dried and distilled before use: THF over sodium benzophenone. All reagents and solvents were degassed before use by means of three freeze–pump–thaw cycles.

Samples for TEM analyses were prepared in the glove box by slow evaporation of a drop of crude colloidal solution deposited onto porous carbon-covered copper grids under argon. The TEM analyses were performed at the "Service Commun de Microscopie Electronique de l'Université Toulouse-III Paul Sabatier" (UPS-TEMSCAN). TEM images were obtained using a JEOL 1011 electron microscope operating at 100 kV with a resolution point of 4.5 Å. The size distributions were determined through a manual analysis of 200 NPs of enlarged micrographs with the ImageTool software to obtain a statistical size distribution and a mean diameter.

All chemical shifts for ^{1}H and ^{13}C are relative to Me$_4$Si (TMS). 1D and 2D ^{1}H-^{1}H correlated spectroscopy (COSY), ^{1}H-^{13}C heteronuclear single quantum coherence (HSQC), and heteronuclear multiple bond correlation (HMBC). NMR experiments in the liquid state were recorded on a Bruker Avance 500 spectrometer equipped with a 5-mm triple resonance inverse Z-gradient probe (TBI ^{1}H, ^{31}P, BB). 2D nuclear overhauser effect spectroscopy (NOESY) measurements were done with a mixing time of 100 ms. Diffusion measurements were made using the stimulated echo pulse sequence. The recycle delay was adjusted to 3 s.

Solid-state NMR experiments were recorded on a Bruker Avance 400 spectrometer equipped with a 4-mm probe. For ^{13}C direct polarization (DPMAS) small flip angle (~30°) was used with a recycle delay of 10 s. ^{13}C cross polarization (CPMAS) spectra were recorded with a recycle delay of 5 s and contact time of 2 ms.

Synthesis of gold nanoparticles (AuNPs)

The Au(I) precursor, AuCl(THT), was synthesized according to the procedure reported by Laguna et al., [31]. This solid is stored at −22 °C and protected from UV light in a glove box and was stable for months.

For a typical synthesis, 20 mg of AuCl(THT) (0.062 mmol) stored in a glove box was introduced into a Fisher–Porter reactor, protected from UV light by aluminum foil and left in vacuum for 30 min. Solid amines were chosen as ligands (L: HDA or DDA with a molar ratio [L]/[Au]=2 corresponding to 30 and 23 mg, respectively) and were introduced into a Schlenck tube and then left in a vacuum for 30 min. 40 mL of THF, previously degassed by three freeze–pump cycles were then added onto the amines. For octylamine (liquid at room temperature), OA was introduced onto the THF with a micropipette (20 μL for 2 eq.).

The THF solution containing the alkylamine was transferred by a Teflon cannula into the reactor. The Fisher–Porter reactor was heated to 70 °C and then pressurized with 3 bars of H_2 or 1 bar of CO. The CO evacuation is done by 3 cycles of vacuum/argon. After 18 h, a homogeneous red colloidal solution was obtained and this will be referred to as "well-controlled synthesis". This ruby-red solution was evaporated in vacuum before the addition of a solution of deoxygenated pentane (2 mL). A red/violet solution was then obtained and precipitated by the addition of 50 mL of acetone. The supernatant was eliminated by filtration. The precipitate was washed a second time with 50 mL of acetone, filtered and dried under vacuum, giving rise to the AuNPs as a dark-violet powder. TEM grids are prepared from the crude colloidal solution.

The AuNPs synthesized by the typical procedure will be denoted Au@L_i for AuNPs obtained with i equivalent of the L ligand. i refers to the molar L/Au ratio (L or RNH_2/ OA ($C_8H_{17}NH_2$), DDA ($C_{12}H_{25}NH_2$), HDA($C_{16}H_{33}NH_2$)).

Results and discussion

Synthesis of gold nanoparticles from the decomposition of AuCl(THT) precursor

Scheme 1 represents the one-pot synthesis of AuNPs in organic solvent with alkylamines (RNH_2) as stabilizing agents using H_2 or CO as the reducing agent. OA, DDA, and HDA are known to be effective compounds for the stabilization of metallic nanoparticles in organic solvents [32]. In this work, neither separation nor purification of the intermediate Au(I)Cl-amine complex is needed to synthesize AuNPs in contrast with previous reports of our group [26]. Protection from the UV light was necessary as Au(I)Cl(THT) is light sensitive.

Influence of the reducing gas in the presence of octylamine

OA was considered first because of its relatively low boiling point (ca 175 °C) compared to amines with longer alkyl chain. Thus, a homogeneous deposit on the electrode could be

expected as it would be easier to remove the excess of this ligand. The first series of experiments was performed in the presence of a small excess of OA (2, 5, 10 eq./Au). The excess of OA was not higher than 10 eq. to limit the AuNP purification step. The reduction of Au(I) precursor was first studied using H_2 (3 bars) either at room temperature (r.t.) or 70 °C for 18 h (overnight). In the case of 10 eq. OA, an incomplete decomposition of the Au(I) precursor was observed. Indeed, the colour of the solution was light pink/violet regardless of the reaction temperature. TEM images only show polydisperse and polymorphous AuNPs (ESM S1). This result clearly indicates that H_2 is not a suitable reducing agent to control the shape and the size of AuNPs in the presence of an excess of OA.

In a second set of experiments, H_2 was replaced by CO. After 18 h, the solution is colourless and a black precipitate is formed whatever the temperature. This indicates that CO is an effective reducing agent for the decomposition of Au(I) precursor in the presence of OA. However, OA does not seem to be able to stabilize AuNPs under these conditions as the alkyl chain is not long enough. To minimize the formation of this black precipitate, a decrease of the reaction time appears to be an alternative. A third set of experiments were carried out under the same conditions as in the second one, but with the time of exposure to CO shortened to 20 min. After 20 min under CO at r.t., the colour of the solution is ruby red. TEM images reveal polydisperse AuNPs, i.e., small (ca 2 nm) and a few tenths of nanometer AuNPs (ESM S2). It is noteworthy that these AuNPs are stable for weeks in argon but are not stable more than 2 h in air.

Moreover, it is well known that an increase of the temperature tends to give more homogenous colloidal solutions [33]. Thus, the same synthesis (10 eq OA) was run at 70 °C in CO for 20 min. After the evacuation of CO, the solution was stirred overnight in argon at 70 °C and the final solution of AuNPs is ruby red. TEM images show polydisperse AuNPs but the general overview is better than that obtained at r.t. Indeed, the majority of the AuNPs are spherical with an average size of 9 nm whereas some polymorphous AuNPs of a few tenths of nanometers are also present (ESM S3). As in the previous case, these AuNPs are stable for months both in argon and air. From these results, CO clearly appears to be the best reducing agent and will be used for the following experiments.

Influence of the stabilizing agent

To avoid the formation of gold aggregates, primary amines with a longer alkyl chains such as DDA (C_{12}) and HDA (C_{16}) were

$$AuCl(THT) + RNH_2 \xrightarrow{THF} AuCl\text{-}(NH_2R) + THT \xrightarrow{Gas\ reduction} Au@NH_2R$$

Scheme 1 Synthesis of AuNPs by gas reduction of AuCl(THT) in the presence of alkylamine used as ligands. The first series of experiments were performed at r.t. for 18 h (overnight) with HDA as the stabilizing agent as it has a longer alkyl chain. For a molar ratio [HDA]/[Au] of 2 or 10 eq., an intense dark-blue solution was obtained after 18 h due to the presence of polymorphous AuNPs (ESM S4). These AuNPs are stable for weeks both in argon and air.

At 70 °C, the solubility of HDA in THF is enhanced, leading to a ruby-red solution after 18 h in CO. TEM analyses (Fig. 1) of this homogeneous solution show the presence of spherical AuNPs of 7.2±0.9 nm when using 2 eq. of HDA. This colloidal solution is stable for months both in argon and air.

The increase of the molar ratio [Ligand]/[Metal] is expected to favour the smaller size: the same experiment was performed with a molar ratio [HDA]/[Au] of 10. A similar ruby-red solution was obtained but containing AuNPs of 4.7±0.9 nm (ESM S5). This experiment is a strong evidence for the role of the increase in the stabilizing agent concentration (HDA) favouring the decrease in the size of spherical AuNPs to 4.7± 0.9 nm but has no morphological change. In order to compare the influence of the length of the amine alkyl chain, another set of experiments was performed with DDA under the same conditions with a molar ratio [DDA]/[Au]=2:1 or 10:1.

The colour of the solution is again ruby red for 2 or 10 eq. of DDA. These AuNPs are stable for months under argon or air. Figure 2 shows TEM images of AuNPs obtained with 2 eq. of DDA (molar ratio of [DDA]/[Au]=2:1).

The size of AuNPs is 7.4±1.0 nm and 6.1±0.7 nm for 2 and 10 eq.(ESM S6), respectively. The size of AuNPs is similar with 2 eq. DDA and HDA but is smaller with a molar ratio of 10 for HDA (4.7 nm *vs* 6.1 nm). This result indicates that a higher concentration of long alkyl chains favours the control of the growth of well-shaped and sized AuNPs.

It is noteworthy that similar AuNPs were obtained at 60 °C by Xia et al. [27] using Au(I)Cl as a precursor and oleylamine (OLA, C_{18}) or octadecylamine (ODA, $C_{18}H_{37}NH_2$) as stabilizer without any additional reducing agent. These amines have a higher steric hindrance than DDA and HDA. The

comparison between CO-assisted decomposition (our work) and thermodecomposition (Xia's work) of Au(I)precursor in the presence of amines reveals that only 2 eq. of alkylamines (DDA or HDA) are necessary to obtain AuNPs of 7.0 nm with a narrow size distribution (Figs. 1 and 2) whereas 20 eq. of OLA are needed for the synthesis of AuNPs of ca. 10 nm. In addition, the AuNPs formed in the presence of ODA showed an average size of ca. 100 nm, eight times larger than that of the AuNPs synthesized with OLA. This difference was attributed to a stronger coordination of OLA due to the presence of the olefinic C=C bond which may also coordinate to Au. In our case, the higher the concentration of HDA (or DDA), the smaller the particle size is. The AuNPs formed from the decomposition of AuCl-(NH_2R) complex in CO (1 bar) at 70 °C follow a seeding growth approach. The HDA probably plays two roles, as weak reducing agent and stabilizing agent. HDA should favour the nucleation of Au(0) atoms and control the growth of the AuNPs involving the complexation in the form of $Au_n(0)$[AuCl-HDA].

The data of the different syntheses of AuNPs obtained by the decomposition of AuCl(THT) in the THF under gas reduction in the presence of alkylamine ligands are summarized in Table S1 (ESM Table S1).

In summary, it is clearly seen that the use of DDA or HDA as stabilizing agent in THF at 70 °C in CO (1 bar) overnight leads to stable and well defined AuNPs with a narrow size distribution (<1.0 nm). NMR characterization was performed to understand the stabilization of the AuNPs.

NMR characterization of the formation of AuCl(amine) complex

The first step of the reaction between 2 eq. of amine (DDA or HDA) and the Au(I) precursor AuCl(THT) in THF at 25 °C was characterized by liquid-state NMR (ESM S7). This spectrum shows a strong evidence of the formation of AuCl-(HDA) complex (labelled # on ESM S7). Moreover,

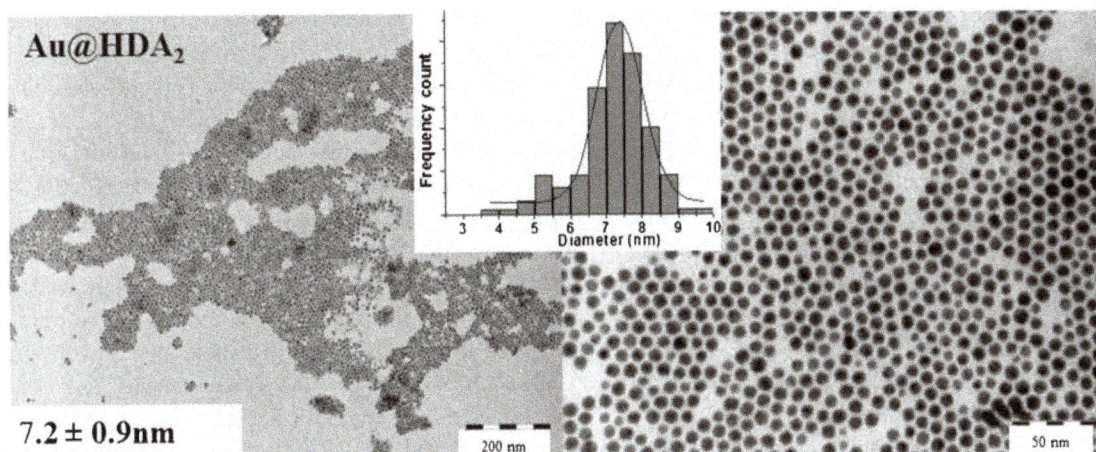

Fig. 1 Au@HDA₂ obtained from the decomposition of AuCl(THT) with HDA in THF in CO overnight at 70 °C (see Table S1)

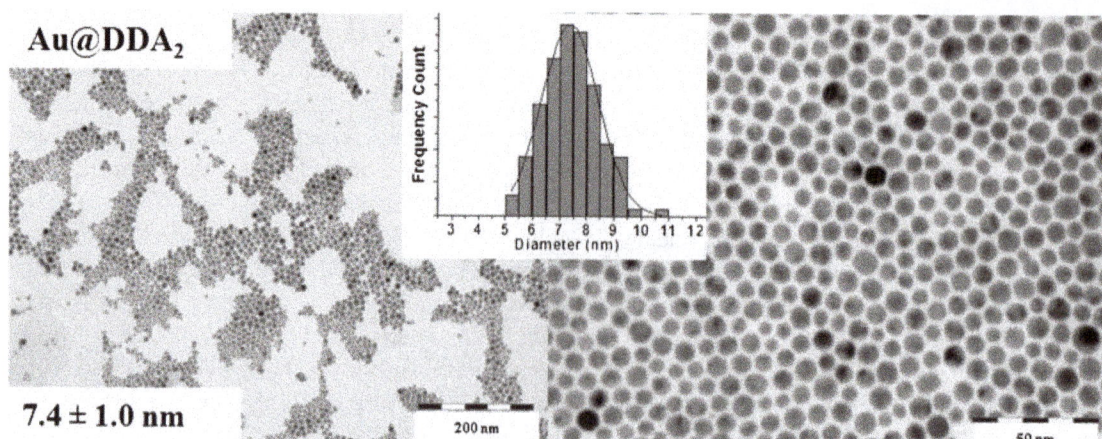

Fig. 2 Au@DDA$_2$ obtained from the decomposition of AuCl(THT) with DDA in THF in CO overnight at 70 °C (see Table S1)

the presence of 2.63 and 1.93 ppm signals confirm the formation of AuCl-(HDA) complex as they correspond to free THT also present in the solution.

NMR characterization for the understanding of the stabilization of AuNPs

As the synthesis and behaviour of AuNPs appear to be very similar in the presence of DDA or HDA, NMR characterizations were done randomly on Au@HDA or Au@DDA.

Study of the ligand stabilization of AuNPs after the decomposition of the AuCl(HDA) complex in CO was performed by liquid-state and solid-state NMR experiments. Solid-state NMR allows the characterization of all the ligands involved in the NP stabilization, especially the ones that are poorly mobile close to the NP surface. In liquid-state NMR, only species with enough mobility can be observed. Indeed, relaxation processes strongly broaden the NMR signals of rigid molecules in liquid-state NMR making them difficult or even impossible to detect. However, liquid-state NMR can give important information on the dynamics of the ligands in the NP colloidal solution. Figure 3 shows the solid-state ^{13}C MAS NMR spectra with direct polarization (DP) or cross polarization (CP) of purified Au@HDA$_2$ powder with a molar HDA/Au ratio=2. These experiments showed the presence of two species at the surface of the AuNPs. The major compound showed characteristic ^{13}C NMR signals at 159.9 ppm (carbonyl) and 39.5 ppm (CH$_2$ bonded to nitrogen) and the signals for the alkyl chain between 15 and 34 ppm. These ^{13}C chemical shifts are in good agreement with the presence of a carbamide species RNHCONHR resulting from a carbonylation of the HDA. The formation of a carbamido intermediate has already been reported for Pt NPs obtained by the decomposition of Pt(CH$_3$)$_2$(COD) in CO in the presence of oleic acid and HDA [34]. In the ^{13}C CPMAS spectrum (Fig. 3b), a second set of signals corresponding to a minor species was observed with notably a signal at 43.6 ppm. This signal at 43.6 ppm is in agreement with a CH$_2$ group

bonded to a NH$_2$ function and we assume that it comes from an amine ligand coordinated on the NP surface. The stronger amplification of the 43.6 ppm signal in the CPMAS experiment compared to the DPMAS, indicates that the amine alkyl chain has a more rigid structure than the carbamine alkyl chains, probably having fewer degrees of freedom. Indeed, local mobility reduced the dipolar coupling strength between protons and carbons and therefore the polarization transfer in the CPMAS experiment. From the ^{13}C DPMAS spectrum, we can deduce that the proportion of unreacted amine is lower than 5 % (the signal at 43.6 ppm was not detected). Furthermore, several signals showed complex line shapes (as one of the terminal methyl and carbonyl groups, ESM S8). This result could indicate the presence of heterogeneous environments for the carbamide which can result from different binding sites at the NP (such as faces, edges or apexes) or different organizations of the ligand. It is also possible that intermediate carbamido complex or urea is present at the AuNP surfaces.

Liquid-state NMR experiments were also performed on purified Au@HDA$_2$ powder (see "Experimental section") to monitor the ligand dynamics in the NP colloidal solution. This solution was characterized by ^{1}H NMR and did not reveal any trace of free HDA, undecomposed AuCl(HDA) complex or THT (Fig. 4a). Signals detected at 1.32 and 0.92 ppm correspond to the methylene or methyl alkyl chain signals of the carbamide and/or of the amine ligands.

Weak additional ^{1}H resonances at 5.03 and 3.09 ppm were also observed. From 2D ^{1}H-^{1}H COSY and ^{1}H-^{13}C HMBC NMR experiments, these signals were attributed to the NH group (5.03 ppm) and to the methylene signal connected to the nitrogen (δ ^{1}H 3.09 ppm; δ ^{13}C 39.8 ppm) of the carbamide species (δ ^{13}C of carbonyl at 157.2 ppm). They were also connected to the alkyl chain signals at 1.3 and 0.9 ppm by ^{1}H-^{1}H COSY and ^{1}H-^{13}C HMBC experiments. These signals also showed a smaller diffusion coefficient at 0.8×10^{-9} m$^2 \cdot$s^{-1} than free HDA (1.1×10^{-9} m$^2 \cdot$s^{-1}) due to a slower diffusion associated with the larger size of the carbamide species.

Fig. 3 ^{13}C DPMAS **(a)** and CPMAS **(b)** NMR of Au@HDA$_2$ powder

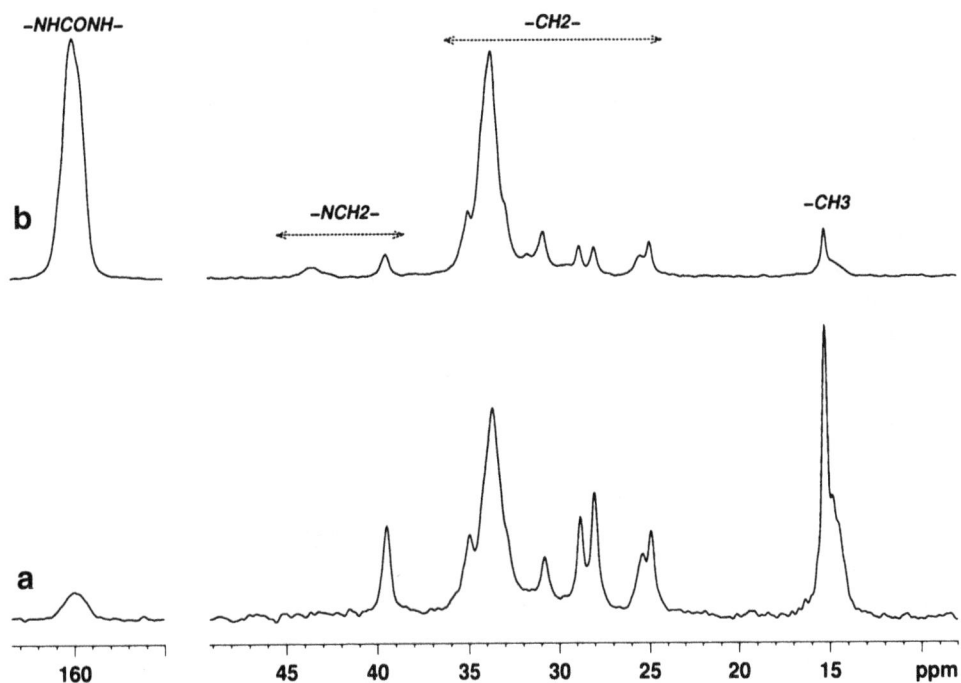

However, this diffusion coefficient is much higher than the one expected for a ligand strongly bound to NP of 7-nm diameter which should be one order of magnitude lower (less than 1×10^{-10} m$^2 \cdot$s^{-1}) [35]. These results indicated the presence of a few carbamide species in solution resulting probably from the decoordination from the NP surfaces. The amount of decoordinated carbamide seems however very small. No exchange phenomena between the NP surface and the solution could be detected for the decoordinated carbamide indicating an irreversible or slow decoordination process compared to the NMR timescales. Furthermore, integration of the methylene (1.32 ppm) and methyl alkyl chain signals (0.91 ppm) compared to that of the NH (5.03 ppm) and α-CH$_2$ signals (3.09 ppm) of the carbamide showed an excess of alkyl chain signals (for example the ratio between the α-CH$_2$ and the methyl signal is equal to 2/8

Fig. 4 ^{1}H NMR spectra at 25 °C of Au@HDA$_2$ in THF-d$_8$ **(a)** and after addition of two supplementary equivalents of HDA **(b)**. (*o*: HDA; +: carbamide; *x*: labile protons (NH$_2$, H$_2$O,..))

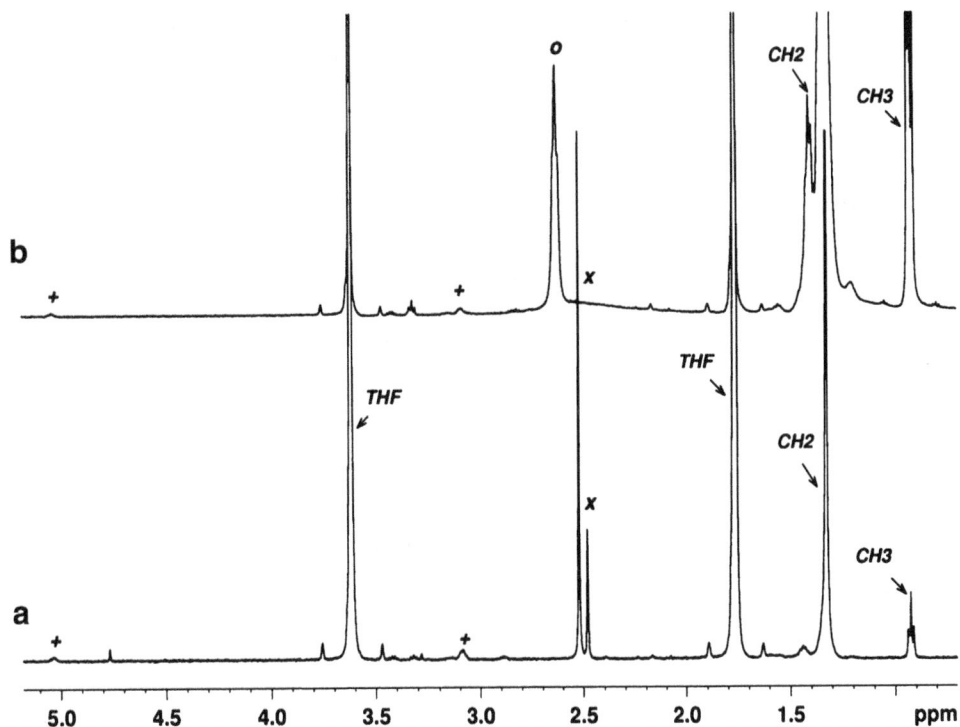

instead of the 2/3 expected for the carbamide). This showed the presence of alkyl chain signals of amine ligands superimposed with one of the free carbamide signals. However, no NMR resonance for the amine extremity, notably the α-CH$_2$ at 2.93 ppm of HDA, was observed. Disappearance of NMR signals, related to fast relaxation processes, for nuclei close to the binding site of NPs is a common observation for ligands involved in "intermediate" or slow exchange between the NP surface and the solution [36].

To study the presence of an exchange phenomenon between the NP surface and the solution for HDA, two additional equivalents of HDA were added to the solution of redispersed AuNPs@HDA$_2$ (Fig. 4b). Sharp signals of HDA were observed after this addition. The diffusion coefficient of HDA in this sample was the same as for free HDA, i.e. 1.1×10^{-9} m$^2 \cdot$s^{-1}. As the diffusion coefficient and the signal shape of HDA in excess looked very similar to one of the free HDA, a transfer NOE experiment was conducted. This experiment has been reported as a powerful technique for characterizing grafted ligands at the NP surface in rapid exchange with free ligands in the solution even when the amount of the grafted ligand is small compared to the free one [37].

Transferred NOEs were observed for the methylene signal in α-position of the amine group at 2.63 ppm (ESM S9) but not for the terminal methyl group of HDA at 0.93 ppm which instead showed zero-quantum artifact usually observed for fast tumbling molecules in NOESY experiments with short mixing times [37]. These results indicate that when an excess of HDA is present, the HDA interacts with the AuNPs and is rapidly exchanged between the NP surface and the solution at the NMR timescales. The increase of the exchange rate when the ratio of ligand/NPs increases has already been described and is related to the higher concentration of free ligands [35]. As the transferred NOE was observed on the amine extremity

and not on the methyl extremity, this also confirms that HDA interacts with AuNPs through its amine group side (Scheme 2). The fast exchange between the NP surface and the solution observed for the amine indicates the presence of weak binding modes for this ligand.

Conclusion

In this study, well-controlled syntheses of monodisperse AuNPs of small diameter have been described by the decomposition of AuCl(THT) in a CO atmosphere in the presence of alkylamine (DDA or HDA) only. Purified AuNPs are redispersible and stable in organic solvents for months. NMR characterization has evidenced the formation of a carbamide species resulting from the AuCl(amine) decomposition in CO. Moreover, the solid-state ^{13}C NMR on purified AuNPs has shown the stabilization of AuNPs by two types of ligands: a majority of carbamide-like species and a few percent of "coordinated" amine. The stabilizing carbamide ligands around the AuNPs are strongly bound to the surface of the AuNPs and do not exchange with free species in solution (in the NMR timescales). Some unreacted amine ligands might still be present and can interact with the AuNPs. The interaction of amine ligands with AuNPs is weaker than that of the carbamide as the amine ligands are able to exchange between the NP surface and the solution.

As a whole, the decomposition of the precursor through carbonylation results in the formation of a better stabilizing ligand than the original amine and explains the long-term stability of the colloidal solutions obtained in this process. The latter is attractive for the surface nanostructuration for chemical sensor applications.

Scheme 2 Weak interaction of free alkylamine with AuNP surfaces

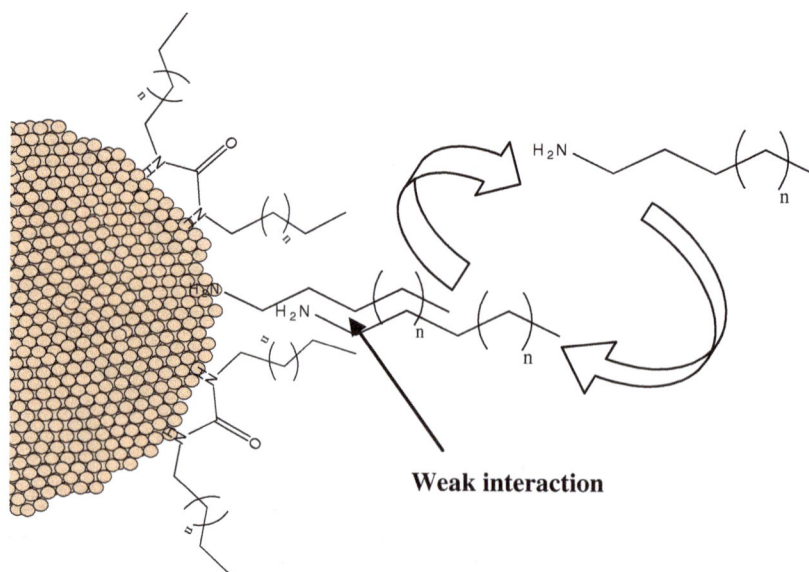

Weak interaction

Acknowledgment This work was part of a project financially supported by the Fondation STAE (Sciences et Technologies pour l'Aéronautique et l'Espace) under the acronym "MAISOE" (Microlaboratoires d'Analyses In Situ pour des Observatoires Environnementaux) and ANR (ANR project MOCANANO).

References

1. Dadosh T, Gordin Y, Krahne R, Khivrich I, Mahalu D, Frydman V, Sperling J, Yacobi A, Bar-Joseph I (2005) Nature 436:677–680
2. Pradahan S, Sun J, Deng FJ, Chen S (2006) Adv Mater 18:3279–3283
3. Somorjai GA, Frei H, Park JY (2009) J Am Chem Soc 131:16589–16605
4. Aragay G, Merkoçi A (2012) Electrochim Acta 84:49–61
5. Aragay G, Pino F, Merkoçi A (2012) Chem Rev 112:5317–5338
6. Jain KK (2007) Clin Chem 53:2002–2009
7. Yeh Y-C, Creran B, Rotello V (2012) Nanoscale 4:1871–1880
8. Wang B, Anslyn EV (2011) Chemosensors: Principles, Strategies, and Applications. John Wiley and Sons, New York, p 163
9. Hossam H (2007) J Phys D: Appl Phys 40:7173–7186
10. Zhang X, Guo Q, Cui D (2009) Sensors 9:1033–1053
11. Daniel M-C, Astruc D (2004) Chem Rev 104:293–346
12. Boissilier E, Astruc D (2009) Chem Soc Rev 38:1759–1782
13. Haruta M, Kobayashi T, Sano H, Yamada N (1987) Chem Lett 16:405–408
14. Corma A, Garcia H (2008) Chem Soc Rev 37:2096–2126
15. Bardhan R, Lal S, Joshi A, Halas NJ (2011) Acc Chem Res 44:936–946
16. Schatz GC (2007) Proc Natl Acad Sci U S A 104:6885–6892
17. Elghanian R, Storhoff JJ, Mucic RC, Letsinger RL, Mirkin CA (1997) Science 277:1078–1081
18. Lian W, Liu S, Yu J, Xing X, Li J, Cui M, Huang J (2012) Biosens bioelectrocnics 38:163–169
19. Kumar S, Kwak K, Lee D (2011) Anal Chem 9:3244–3247
20. Hezard T, Fajerwerg K, Evrard D, Collière V, Behra P, Gros P (2012) Electrochim Acta 73:15–22
21. Hezard T, Fajerwerg K, Evrard D, Collière V, Behra P, Gros P (2012) J Electroanal Chem 664:46–52
22. Li C, Shuford KL, Park QH, Cai W, Li Y, Lee EJ, Cho SO (2007) Angew Chem Int;Ed: 46:3264–3268
23. Zeng J, Ma Y, Jeong U, Xia Y (2010) J Mat Chem 20:2290–2301
24. Schmid G (1992) Chem Rev 92:1709–1727
25. Zheng N, Fan J, Stucky GD (2006) J Am Chem Soc 128:6550–6551
26. Gomez S, Philippot K, Colliere V, Chaudret B, Senocq F, Lecante P (2000) Chem Commun 1945–1946
27. Lu X, Tuan H-Y, Korgel BA, Xia Y (2008) Chem Eur J 14:1584–1591
28. Zheng Y, Tao J, Liu H, Zeng J, Yu T, Ma Y, Moran C, Wu L, Zhu Y, Liu J, Xia Y (2011) Small 167:2307–2312
29. Lu X, Yavuz MS, Tuan H-Y, Korgel BA, Xia Y (2008) J Am Chem Soc 130:8900–8901
30. Philippot K, Chaudret B (2007) In: Comprehensive Organometallic Chemistry III, RH Crabtree & MP Mingos, Elsevier, Chapter 12–03: pp71-99
31. Uson R, Laguna A (1986) Organomet Synth 3:322–342
32. Axet MR, Philippot K, Chaudret B, Cabié M, Giorgio S, Henry C (2011) Small 7:235–241
33. Debouttiere P-J, Coppel Y, Denicourt-Nowicki A, Roucoux A, Chaudret B, Philippot K (2012) Eur J of Inorg Chem 8:1229–1236
34. Latour V, Maisonnat A, Coppel Y, Colliere V, Fau P, Chaudret B (2010) Chem Comm 46:2683–2685
35. Coppel Y, Spataro G, Pagès C, Chaudret B, Maisonnat A, Kahn ML (2012) Chem Eur J 18:5384–5393
36. Hens Z, Martins JC (2013) Chem Mater 25:1211–1221
37. Fritzinger B, Moreels I, Lommens P, Koole R, Hens Z, Martins JC (2009) J Am Chem Soc 131:3024–3032

10

Experimental evidence of luminescence quenching at long coupling distances in europium (III) doped core-shell gold silica nanoparticles

Laure Bertry · Olivier Durupthy · Patrick Aschehoug ·
Bruno Viana · Corinne Chanéac

Abstract Localized surface plasmons can modify linear optical responses of materials located in their vicinity. In particular, rare earth ions luminescence can be enhanced by gold nanoparticles. Luminescence exaltation is a complex phenomenon that depends on multiple parameters, a critical one being the coupling distance between the emitting species and the plasmonic core. An original multilayer nanostructure designed to precisely control the distance between the gold cores and the luminescent ions, and to study its effect on the luminescent properties is presented here. Homogeneous silica shells with controlled thicknesses adjustable from 2 to 50 nm and rare earth ion doping rates up to 2×10^{20} Eu/cm^3 of silica were deposited onto gold nanospheres. These original nanostructures are then incorporated into densified sol–gel silica composites with high-optical quality. Luminescence properties are studied for increasing gold–europium (III) distances. Strong luminescence quenching is evidenced for coupling distances up to 28 nm.

Keywords Gold nanoparticles · Core-shell nanoparticles · Luminescence exaltation · Rare earth · Silica coating · Luminescence quenching

L. Bertry · O. Durupthy (✉) · C. Chanéac
Laboratoire de Chimie de la Matière Condensée, UPMC, CNRS, Collège de France, UMR 7574, 11 place Marcelin Berthelot, 75005 Paris, France
e-mail: olivier.durupthy@upmc.fr

P. Aschehoug · B. Viana
Laboratoire de Chimie de la Matière Condensée, UPMC, CNRS, Chimie Paristech, UMR 7574, 11 rue Pierre et Maris Curie, 75005 Paris, France

Introduction

Gold nanoparticles interact with light and exhibit strong and tunable surface plasmon resonance (SPR). These localized surface plasmons originating from gold nanoparticles in various media can enhance linear and non linear optical responses of materials located in their vicinity, mainly due to electromagnetic field enhancement. Recent studies have highlighted the possible coupling between gold nanostructures and quantum dots [1], organic fluorophores [2, 3], silicon carbide [4], or lanthanide ions [5].

In the case of lanthanide ions, the quantum yield, which is the probability for a photon to be emitted after one photon has been absorbed, is close to 1 [6]. However, the excitation process efficiency is limited by their weak absorption cross sections, due to parity-forbidden transitions. This is usually overcome by energy transfer from sensitizer ions [7, 8], ligands [7, 9], or host matrix [10], but absorption exaltation through coupling to plasmonic nanostructures could yield to even better sensibilization [5, 11].

Efficient coupling requires the surface plasmon resonance and the quantum emitter absorption band to be correctly matched in wavelength [12]. It also highly depends on the distance between the emitter and the gold core. For short distances around 1 nm, the luminescence is mainly quenched due to non-radiative decays [13, 14]. However, the radiative decay rate may be amplified by several orders of magnitude [4] for theoretical spacing distances predicted around 10 nm [15]. This optimum coupling distance has to be determined for each sensitizer/emitter couple.

Most of the exaltation studies were conducted on planar configurations [6, 16]. Pillonnet et al. [6] demonstrated limited exaltation factors around three for Eu (III) ions coupled to silver nanoparticles, with an optimum coupling distance around 15 nm. Recently, three-dimensional nanoparticles systems, coupling metallic nanoparticles with either lanthanide complexes [17] or fluorophores [18], revealed strong luminescence quenching for spacing distances up to 40 nm, showing the more complex behavior of non planar nanostructures. The nanoparticle approach, based on versatile synthetics procedures [3, 19, 20], allows for a fine control of each synthetic parameter in a wide range, and opens new applications in the field of optics. More specifically, rare earth ion incorporation into silica matrices could bring new developments for optical amplifiers [21].

In this work, Au@SiO$_2$@SiO$_2$:Eu$^{(III)}$ core-shell nanoparticles, exhibiting adequate absorption and emission spectral overlap, are synthesized via a sol–gel process allowing fine control of coupling distances and doping rates. These multilayer nanostructures are intended to the study of luminescence exaltation via plasmon-induced local field enhancement. In addition, for a better understanding of the plasmon effect, reference nanoparticles are prepared by gold core dissolution on the final multilayer nanostructures. Europium (III) photoluminescence intensities, excitation spectra, and lifetime decay profiles were analyzed to study the plasmon-induced luminescence properties.

Experimental details

Au@SiO$_2$@SiO$_2$:Eu$^{(III)}$ multilayer nanostructures, as illustrated on Fig. 1, are synthesized by the multi-step process described below. They are then incorporated into reference silica matrices for optical measurements.

Reagents

All chemicals were purchased from Sigma-Aldrich and used as received; hydrogen tetrachloroaurate (III) trihydrate (99.9 %), 3-mercaptopropyltrimethoxysilane (MPS, 95 %), sodium silicate solution (27 wt.-% SiO$_2$), europium (III) chloride hexahydrate (99.9 %), ammonia (28 wt.-%).

Au@SiO$_2$ synthesis

Spherical gold nanoparticles (NP) 50 nm in diameter were obtained by the well-known Türkevich method [22–24]. The citrate/gold molar ratio, that controls the nanospheres diameter, was set to 1. 50 mL of a 2.5 mmol L^{-1} HAuCl$_4$ aqueous

solution were added to 400 mL of boiling Milli-Q water. Once the boiling started again, 50 mL of a 2.5 mmol L^{-1} sodium citrate aqueous solution were quickly added under vigorous magnetic stirring. The heating and stirring were extended for 5 more minutes after the suspension turned violet-red.

Multiple silica shells were then successively deposited on the gold NP through optimization of the Liz-Marzan process [19, 25]. Two hundred microliters of 1 mol L^{-1} MPS solution was first added to 250 mL of the gold NP solution, corresponding to 60 % of total surface coverage based on 40-Å^2 coverage per MPS molecule. The suspension was magnetically stirred for 15 min to allow for surface functionalization. Then, 10 mL of 0.34 wt.-% (77.3 mmol L^{-1}) sodium silicate solution, previously activated at pH=10.4 with Amberlite IRN77 cation exchange resin, were added to the mixture. The suspension was maintained 5 days at 40 °C in a stove.

At this stage, the silica-coated nanoparticles are stable into water/ethanol mixtures. Homogeneous 2-nm thick silica shells are obtained, suggesting an uncompleted silica precursor heterocondensation. Indeed, upon ammonia addition in the suspension, a 25-nm homogeneous silica shell is obtained by condensation of excess unreacted silicates.

To grow silica shells with tailored intermediate thicknesses, the 2-nm silica-coated particles are first washed with same pH water (pH ~9) to remove unreacted silicates and concentrated up to 1.8×10^{14} NP/L using an ultrafiltration 30-kDa membrane. The suspension is then diluted in four volumes of ethanol. Controlled amounts of 0.02 mol L^{-1} tetraethyl orthosilicate (TEOS) and 40 µL of ammonia per milliliter of suspension are successively added under vigorous magnetic stirring. The required TEOS amount can be predicted according to the targeted shell thickness.

Au@SiO$_2$@SiO$_2$:Eu$^{(III)}$ synthesis

Lanthanide (Ln (III)) doping of the shells is obtained following the same procedure as for pure silica shells, except for the addition of a defined volume of 1 mmol L^{-1} lanthanide trichloride solution into the mixture just before ammonia. The suspension is magnetically stirred during 24 h before the next doped silica layer coating step. Typically, the addition of 420 µL of TEOS solution, 42 µL of Eu (III) solution and 800 µL of 28 wt.-% ammonia into 20 mL of suspension gives 12-nm silica shells doped with 10^{20} Eu/cm^3 of silica.

When the desired multilayer nanostructure is reached, the suspension is washed with water/ethanol mixture (1:4 v/v) by 3 cycles of 30-min centrifugation at 29,220g.

Gold core dissolution for reference nanoparticles synthesis

The gold cores were dissolved via a cyanide-etching process adapted from [26]. Typically, 1 mL of core-shell suspension was diluted into 29 mL of water/ethanol mixture and 755 µL

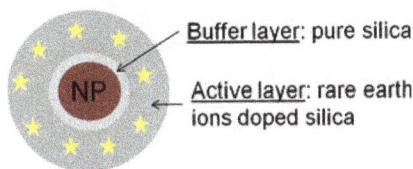

Fig. 1 Multilayer nanostructures; *NP* metallic nanoparticle

of 0.1 mol L^{-1} KCN were added (CN/Au=50). The dissolution was carried on for 5 days under magnetic stirring. The suspension was then washed with water/ethanol mixture by 3 cycles of 30-min centrifugation at 29,220g.

Incorporation into silica matrices

The nanostructures were incorporated into sol–gel silica monoliths to study the luminescence properties of the samples. The condensation process was optimized to minimize organic and silanols residuals into the final densified silica.

Equimolar TEOS to triethoxysilane (TREOS) mixtures [27] were diluted into ethanol (TEOS/TREOS/EtOH molar ratios are 1:1:8.6) and prehydrolyzed with HCl at pH=1 (25 µL for 1.82 mmol of TEOS) during 2 h under vigorous magnetic stirring. Core-shell suspension and water were then successively added and the mixture was kept stirring for 1 h to ensure sol homogeneity. Then, 0.75 mL of the sol was cast onto a 5-mL polypropylene vessel with pierced lid. After 5 days of condensation at room temperature, the gel is dried for 5 more days in a 40 °C stove. Thermal densification is then performed in a furnace up to 600 °C for 12 h.

Densified silica monoliths (1 cm^2 × 1 mm) with good optical quality are thus obtained. For the sake of reproducibility in luminescence emission measurements, the monoliths are then grinded into fine powder.

Characterizations

UV-visible absorption spectra were recorded from 400 to 800 nm with 1-nm steps on a SECOMAM UVIKON XL spectrophotometer.

Transmission electron microscopy (TEM) grids were prepared by evaporating one drop of core-shell suspension onto a carbon-coated copper TEM grid. The samples were then characterized with a Tecnai G2 Spirit apparatus operating at 120 kV. Particle size distributions were evaluated from 200 nanoparticles.

Lifetime measurements were conducted directly on the monoliths. Photoluminescence (PL) emission spectra were measured on grinded monoliths. A tunable optical parametric oscillator pumped with the third harmonic of a Nd:YAG Q-switched laser (Ekspla NT-342B-SH, 10-Hz repetition frequency, 6-ns pulse width) was used as excitation source. The luminescence was analyzed with a Jobin Yvon HR250 monochromator coupled to a Roper ICCD camera. With the ICCD camera, integrated and delay times can be adjusted to the emission ion. For europium (III), a typical set-up is 5-ms integrated time and 5-µs delay time. Lifetime measurements were performed during a total time of 7 ms with 1-µs increment in the delay time.

Materials description

Metallic core

Gold salt reduction with citrate ions in water yields a stable suspension of roughly spherical gold nanospheres with a diameter of 49.3±5.2 nm (see Fig. 2 and size distribution in Online Resource 1). The monodispersity level reached is not perfect but good enough for the targeted application. A diameter of 50 nm is estimated to be a good compromise between the enhancement of local electromagnetic field by larger nanoparticles and the problem of light diffusion when particles size is too large [15].

The gold nanosphere suspension exhibits a surface plasmon resonance extinction spectrum (Fig. 3b), characterized by a strong absorption band in the 450–600-nm spectral range and a maximum of 535 nm, as evidenced by the red-wine color of the suspension. The europium (III) excitation and emission spectra are reported in Fig. 3. The $^7F_0 \rightarrow {}^5D_1$ absorption band is perfectly matched with the SPR maximum, thus allowing coupling at absorption. The $^7F_0 \rightarrow {}^5D_2$ absorption band is shifted towards shorter wavelengths compared to the gold surface plasmon resonance, enabling non resonant excitation at 465 nm.

As for the $^5D_0 \rightarrow {}^7F_2$ europium (III) emission band, the 80-nm red-shift compared to the plasmon maximum is larger than the usual exaltation studies conducted with organic dyes [2].

Pure silica buffer layers with controlled thicknesses

The vitreophobic surface of gold nanoparticles is first functionalized with MPS to promote silica heterogeneous condensation [19]. A 60 % theoretical coverage of the NP surface is required to allow further coating, while excess MPS molecules would induce gold nanospheres aggregation, eventually leading to precipitation.

By condensation of sodium silicates, silica shells as thin as 2 nm (see Fig. 4a1, a2) are obtained in water. All the gold

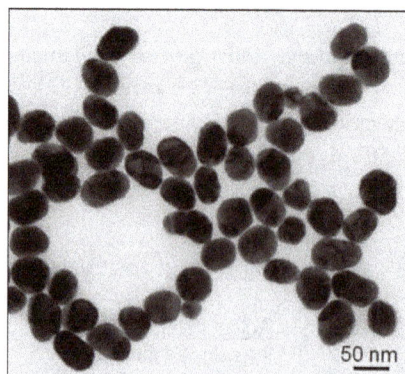

Fig. 2 Transmission electron micrograph of the 50-nm roughly spherical gold nanoparticles

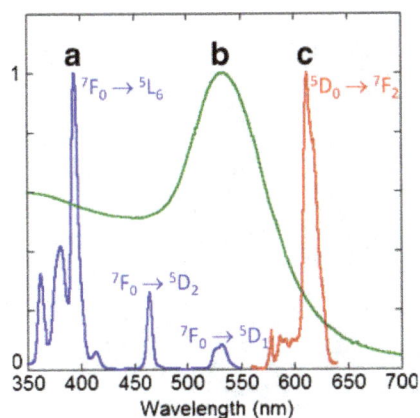

Fig. 3 Intensities of *a*) europium (III) excitation spectra recorded at 615 nm; *b*) gold nanospheres suspension extinction spectra; and *c*) europium (III) emission spectra for excitation at 532 nm. For comparison sake, intensities are normalized to 1

cores are totally coated. This coating step does not impact the UV-visible absorption spectra (not shown), indicating the formation of a very thin silica layer and a good nanoparticle dispersion. This will be further confirmed by the next coating steps, leading to only a very few multi-core nanostructures.

The 2-nm silica-coated nanoparticles are stable into water/ethanol 1:4 *v/v* mixtures. To grow silica shells with tailored intermediate thicknesses, removal of excess unreacted silicates is needed; otherwise, silica condensation preferently occurs on small gold-free pure silica nuclei. Washing was performed by ultrafiltration to avoid any aggregation that may occur during centrifugation or dialysis [19].

Pure silica shell growth is performed through ammonia-catalyzed sol–gel condensation [28]. Ammonia concentration has to be controlled to develop a coating process compatible with potential further doping with lanthanide ions. In consequence, concentrated gold suspensions were used. Ultrafiltration allows concentration of the sols up to 6×10^{14} NP/L, but the suspension stability towards aggregation is then strongly affected. The optimum concentration was determined to be 1.8×10^{14} NP/L. In these conditions, the optimum ammonia concentration was 0.569 mol/L.

With this procedure, pure silica shells with controlled thicknesses ranging from 10 to 35 nm (see Fig. 4b1–2, c1–2) are obtained in one single TEOS addition step. To grow even thicker shells, multiple TEOS addition steps can be performed. As illustrated in Fig. 4, the silica shells are very homogeneous and only a few (<1 %) multi-core nanostructures are obtained, confirming the high stability of this colloidal suspension at each step.

Despite the washing step, some core-free pure silica nanoparticles are systematically formed. This leads to a bias of about 5 nm between the predicted and observed silica shell thicknesses. These core-free particles will also contribute to the luminescence baseline, not affected by the plasmon, when doped with europium ions.

As the silica shell thickened, the SPR is slightly red-shifted (see Online Resource 2). This is correlated to an increase in the local refractive index around the gold core [19]. The maximum shift observed in our samples for 35-nm silica shells was 8 nm. This still allows good coupling in absorption with the $^7F_0 \rightarrow ^5D_1$ europium (III) absorption band presented in Fig. 3.

Silica shells doped with lanthanide ions

Once the gold nanoparticles have been coated with a first 2-nm pure silica shell, any of the subsequent shells can be doped with lanthanide (III) ions, even prior to washing of excess unreacted silicates.

Ammonia-catalyzed condensation occurring in basic conditions, electrostatic interactions can favor lanthanide (III) incorporation into the condensing negatively charged silica network. For europium (III) ions, the maximum accessible doping rate was determined to be 2×10^{20} Eu/cm^3 of silica. For higher doping rates, the silica network condensation is highly disturbed (see Online Resource 3). Bright-contrasted nuclei, attributed to europium hydroxides [29], appear and promote the formation of a high number of core-free pure silica nanoparticles, leading to silica shell thicknesses being significantly lower than the predicted ones. For rare earth concentration in the shell larger than 1×10^{17} Eu/cm^3 of silica, clustering of the Eu(III) ions may occur but its consequences on the luminescence properties is compensated by the use of the appropriate reference samples, as described below.

Due to the low-doping rates in the final nanostructures and silica matrices, no quantitative measurement by chemical analysis could confirm the luminescent ion incorporation efficiency. The doping is, however, confirmed by the luminescence studies presented below.

Gold core dissolution for reference nanoparticles

In-depth study of the possibility to enhance lanthanide ions emission through plasmonic exaltation, requires reference samples to study the gold core effect on optical properties. Two main strategies emerge from the literature: the use of very thick spacer layers to get rid of any coupling [13, 17], or the removal of the plasmonic part [3, 6]. Due to the potential long-range coupling with gold [17, 18], the latter was favored in this study and achieved by dissolution of the plasmonic core.

Gold core dissolution can be achieved by etching with either cyanide ions [26, 30] or aqua regia [31]. Aqua regia, being a highly acidic medium, might promote lanthanide (III) ions release from the silica network. Dissolution through cyanide ions complexation was therefore applied. It can be summarized by the following equation:

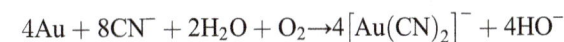

$$4Au + 8CN^- + 2H_2O + O_2 \rightarrow 4\left[Au(CN)_2\right]^- + 4HO^-$$

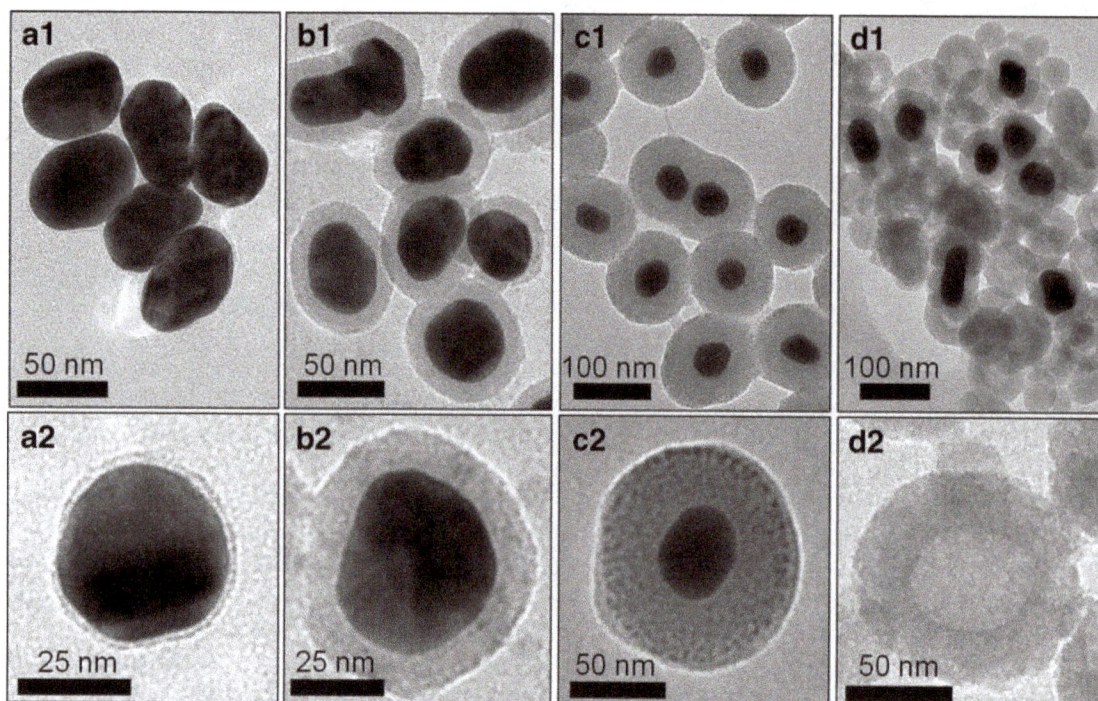

Fig. 4 Transmission electron micrographs of Au@SiO$_2$ nanostructures coated with **a** 2 nm, **b** 10 nm, and **c** 35-nm pure silica shells. **d** Nanostructures with 25-nm pure silica shells after etching with cyanide ions for 5 days

Complete gold core dissolution proceeds by inward diffusion of cyanide ions and outward diffusion of gold-cyanide complexes through the all silica shells thickness [26]. This process is diffusion-limited and requires a few reaction days. Potassium counter-ions are known as silica network modifiers. They can lead to partial silica shell dissolution. Moreover, during the dissolution reaction, released HO$^-$ anions induce a small increase in pH that may also promote silica dissolution. An optimum between the dissolution rate and the silica integrity was experimentally found for a cyanide/gold molar ratio of 50. After 5 days of etching, more than 90 % of the gold cores are dissolved, as proved by the decrease in SPR intensity (see Online Resource 4).

Typical multilayer nanostructures after etching with cyanide ions are presented in Fig. 4d1–2. Silica shells' thicknesses and homogeneity are preserved. In some cases, the external part of the shells is partially dissolved during etching [30], leading to a shrinking of the shell and a partial release of lanthanide ions. To get perfect reference nanoparticles, the cyanide to gold molar ratio should be optimized for each shell thickness.

Composites silica matrices

After incorporation of the nanostructures into optical quality-densified silica monoliths involving a heating treatment at 600 °C under air, the SPR extinction spectrum exhibits the same shape as the gold suspensions, indicating that the nanostructures are well dispersed inside the matrix.

Luminescence properties

Luminescence excitation measurements conducted for gold-loaded and dissolved nanostructures with only 2-nm pure silica buffer layer and 25 nm of Eu (III)-doped silica are presented in Fig. 5. The excitation spectra is not modified by the presence of the gold core compared to europium (III) ions dispersed in a pure silica matrix. A strong decrease in the emission intensity, by about 50 %, is observed for the gold-loaded sample. Surprisingly, the europium (III) emission is quenched for excitation wavelengths in transitions both resonant (532 nm) and non resonant (465 nm) with the gold SPR.

This gold-induced luminescence quenching is further confirmed by lifetime measurements. Luminescence decays were found to be exponential in both cases. Dissolved nanostructures systematically exhibit a longer lifetime than gold-loaded ones, arising from the disappearance of the plasmon-induced non-radiative decay channel. Lifetime values are reduced by ~10 % in these strong quenching conditions. Such weak lifetime variations have already been reported [3, 17].

This strong luminescence quenching for very short coupling distances is in good agreement with theoretical predictions [14]. Quenching may come from luminescence re-absorption by the gold nanoparticles [13, 32], especially for

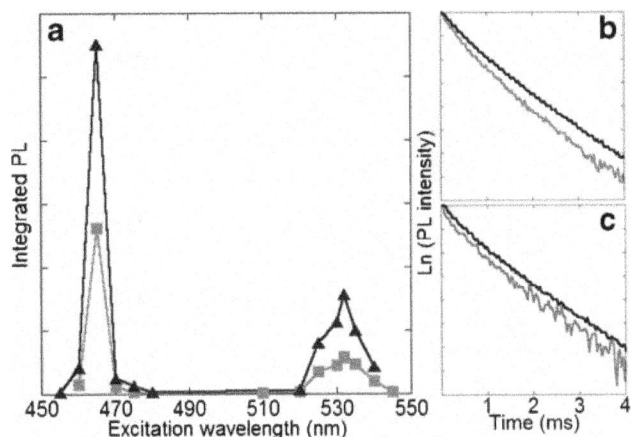

Fig. 5 **a** Excitation spectra recorded at 615 nm and **b–c** decay curves for excitation at **b** 465 nm and **c** 532 nm for either gold-loaded samples (*gray closed squares*) or gold-dissolved references (*black closed triangles*) with 2-nm pure silica buffer layer and 25 nm of Eu (III)-doped silica

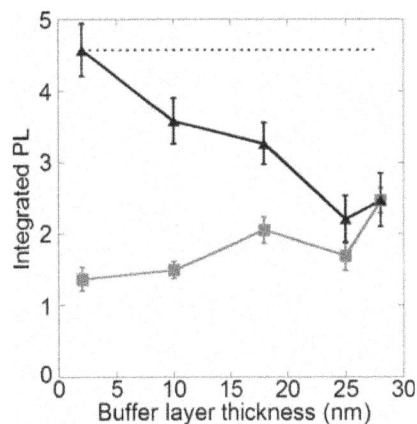

Fig. 6 Integrated PL intensities (excitation at 532 nm) for either gold-loaded samples (*gray closed squares*) or gold-dissolved references (*black closed triangles*) with increasing pure silica buffer thicknesses and identical Eu (III)-doping. The best reference, defined in the text, is identified as a *dotted line*. *Error bars* are estimated from three measurements

non resonant excitation. Figure 3 shows that the SPR is still quite intense at the europium (III) emission wavelength, and this is amplified by the difference in oscillator strengths [33, 34].

Figure 6 shows photoluminescence intensities recorded under excitation at 532 nm for samples with increasing buffer layer thicknesses. For this study, after pure silica layers condensation, the same amounts of TEOS and europium (III) salt were added to all the samples. The shells are doped with 10^{20} Eu/cm^3 of silica and have different thicknesses ranging from 20 nm for the sample with 2-nm buffer layer, to 9 nm for the sample with 28-nm buffer layer.

Reference samples gave unexpectedly different signal intensities depending on the buffer layer thickness. This may be explained by a partial dissolution of the external silica shell, releasing some luminescent ions; for similar etching conditions, as the final silica thickness increases, one can predict that a higher percentage of europium (III) ions are released, leading to photoluminescence decrease. The best reference sample would thus be the one with the thinner buffer layer and the higher luminescence intensity.

Compared to this best reference, identified as a dotted line on Fig. 6, luminescence quenching is observed for all studied buffer thicknesses. The quenching intensity diminishes as the buffer thickness increases: 70 % at 2 nm and 46 % at 28 nm. Same behavior is observed for excitation at 465 nm. Quenching at such long distances is not predicted theoretically [14] but has been sometimes observed experimentally [17, 18]. It is usually attributed to non-radiative energy transfer towards gold nanoparticles, but a possible modification of the efficiency to populate the Eu$^{(III)}$ $^5D_0$ emitting level has also been reported [17]. No clear lifetime difference was evidenced among the samples, but because of the low-doping rates, the luminescence decays (not shown) signals are very weak.

Once again, we believe that this unexpected quenching at long distance may come from luminescence re-absorption by the gold cores. Compared to oriented planar configurations, systems reported here favor interactions between gold nanoparticles and rare earth ions. Indeed, in bulk samples, europium (III) emission has to come out through the whole sample thickness. Interactions with multiple gold nanoparticles with large absorption cross sections may occur, especially as emission bands partially overlap with gold SPR.

Conclusion

Multilayer nanostructures made of a spherical 50-nm gold core coated with silica shells, either pure or europium (III)-doped, were successfully prepared. Optical reference nanoparticles were obtained by gentle gold core dissolution with cyanide ions. A systematic study of the effect of the coupling distance between the gold core and the luminescent ions was conducted. Strong quenching is evidenced for coupling distances up to 28 nm, the quenching intensity increasing as the pure silica buffer layer diminishes.

This quenching is attributed to a re-absorption of the emitted luminescence by the gold cores. Emitting species exhibiting larger red-shifts compared to the SPR wavelength could probably help overcome this problem [35]. In this particular study, no efficient coupling is achieved between the metallic core and the luminescent ions. This may arise from the thickness of the doped shells; only a very small portion of the luminescent ions is actually in an optimal coupling configuration. Anisotropic nanoparticles, such as gold nanorods [36–38], could also yield more efficient coupling [35].

Acknowledgments This work is supported by the French Agence Nationale de la Recherche (ANR) under the Fenoptic project (ANR-09-NANO-23), part of the Nanosciences, Nanotechnologies and Nanosystems (P3N2009) program. Authors thank Draka – Prysmian Group, the Laboratoire de Physico-Chimie des Matériaux Luminescents and the Institut Carnot de Bourgogne for their collaboration.

References

1. Naiki H, Masuhara A, Masuo S et al (2013) Highly controlled plasmonic emission enhancement from metal–semiconductor quantum dot complex nanostructures. J Phys Chem C 117:2455.

2. Chen H, Ming T, Zhao L et al (2010) Plasmon–molecule interactions. Nano Today 5:494.

3. Schneider G, Decher G, Nerambourg N et al (2006) Distance-dependent fluorescence quenching on gold nanoparticles ensheathed with layer-by-layer assembled polyelectrolytes. Nano Lett 6:530. doi:

4. Sui N, Monnier V, Zakharko Y et al (2012) Plasmon-controlled narrower and blue-shifted fluorescence emission in (Au@SiO₂) SiC nanohybrids. J Nanopart Res 14:1004.

5. Fukushima M, Managaki N, Fujii M et al (2005) Enhancement of 1.54-μm emission from Er-doped sol–gel SiO₂ films by Au nanoparticles doping. J Appl Phys 98:024316

6. Pillonnet A, Berthelot A, Pereira A et al (2012) Coupling distance between Eu³⁺ emitters and Ag nanoparticles. Appl Phys Lett 100:153115.

7. Binnemans K (2009) Lanthanide-based luminescent hybrid materials. Chem Rev 109:4283.

8. Eliseeva SV, Bünzli J-CG (2011) Rare earths: jewels for functional materials of the future. New J Chem 35:1165.

9. Dossing A (2005) Luminescence from lanthanide(3+) ions in solution. Eur J Inorg Chem 8:1425.

10. Huignard A, Gacoin T, Boilot J-P (2000) Synthesis and luminescence properties of colloidal YVO4:Eu phosphors. Chem Mater 12:1090.

11. Chiasera A, Ferrari M, Mattarelli M et al (2005) Assessment of spectroscopic properties of erbium ions in a soda-lime silicate glass after silver–sodium exchange. Opt Mater 27:1743.

12. Thomas M, Greffet J-J, Carminati R, Arias-Gonzalez JR (2004) Single-molecule spontaneous emission close to absorbing nanostructures. Appl Phys Lett 85:3863.

13. Dulkeith E, Morteani A, Niedereichholz T et al (2002) Fluorescence quenching of dye molecules near gold nanoparticles: radiative and nonradiative effects. Phys Rev Lett 89:12.

14. Anger P, Bharadwaj P, Novotny L (2006) Enhancement and quenching of single-molecule fluorescence. Phys Rev Lett 96:3.

15. Colas des Francs G, Bouhelier A, Finot E et al (2008) Fluorescence relaxation in the near-field of a mesoscopic metallic particle: distance dependence and role of plasmon modes. Opt Express 16:17654

16. Kalkman J, Kuipers L, Polman A, Gersen H (2005) Coupling of Er ions to surface plasmons on Ag. Appl Phys Lett 86:041113.

17. Lin C, Berry MT, Stanley May P (2010) Influence of colloidal-gold films on the luminescence of Eu(TTFA)₃ in PMMA. J Lumin 130:1907.

18. Huang Y-F, Ma K-H, Kang K-B et al (2013) Core–shell plasmonic nanostructures to fine-tune long "Au nanoparticle-fluorophore" distance and radiative dynamics. Colloids Surf, A 421:101.

19. Liz-Marzán LM, Giersig M, Mulvaney P (1996) Synthesis of nanosized gold-silica core-shell particles. Langmuir 12:4329.

20. Bahadur NM, Furusawa T, Sato M et al (2011) Fast and facile synthesis of silica coated silver nanoparticles by microwave irradiation. J Colloid Interface Sci 355:312.

21. Pastouret A, Gonnet C, Collet C, et al. (2009) Nanoparticle doping process for improved fibre amplifiers and lasers. Proc of Spie.

22. Frens G (1973) Controlled nucleation for the regulation of the particle size in monodisperse gold suspensions. Nature Phys Sci 241:20

23. Turkevich J, Stevenson C, Hillier J (1953) The formation of colloidal gold. J Phys Chem 57:670

24. Ji X, Song X, Li J et al (2007) Size control of gold nanocrystals in citrate reduction: the third role of citrate. J Am Chem Soc 129:13939.

25. Obare SO, Jana NR, Murphy CJ (2001) Preparation of polystyrene- and silica-coated gold nanorods and their use as templates for the synthesis of hollow nanotubes. Nano Lett 1:601.

26. Ung T, Liz-Marzán LM, Mulvaney P (1998) Controlled method for silica coating of silver colloids. Influence of coating on the rate of chemical reactions. Langmuir 14:3740

27. Soraru GD, D'Andrea G, Campostrini R, Babonneau F (1995) Characterization of methyl-substituted silica gels with Si-H functionalities. J Mater Chem 5:1363.

28. Stöber W, Fink A, Bohn E (1968) Controlled growth of monodisperse silica spheres in the micron size range. J Colloid Interface Sci 26:62.

29. Zhao D, Qin W, Zhang J et al (2005) Modified spontaneous emission of europium complex nanoclusters embedded in colloidal silica spheres. Chem Phys Lett 403:129.

30. Giersig M, Ung T, Liz-marzan LM, Mulvaney P (1997) Direct observation of chemical reactions in silica-coated gold and silver nanoparticles. Adv Mater 9:570.

31. Allabashi R, Stach W, Escosura-Muñiz A et al (2008) ICP-MS: a powerful technique for quantitative determination of gold nanoparticles without previous dissolving. J Nanopart Res 11:2003. doi:10.

32. Jian Z (2005) SPR induced quenching of the ⁵D₁ → ⁷F₁ emission of Eu³⁺ doped gold colloids. Phys Lett A 341:212.

33. Mahato KK, Rai SB, Rai A (2004) Optical studies of Eu³⁺ doped oxyfluoroborate glass. Spectrochim Acta, Part A 60:979. doi:10.

34. Haiss W, Thanh NTK, Aveyard J, Fernig DG (2007) Determination of size and concentration of gold nanoparticles from UV–vis spectra. Anal Chem 79:4215.

35. Liaw J-W, Tsai H-Y (2012) Theoretical investigation of plasmonic enhancement of silica-coated gold nanorod on molecular fluorescence. J Quant Spectrosc Radiat Transf 113:470.

36. Tréguer-Delapierre M, Majimel J, Mornet S et al (2008) Synthesis of non-spherical gold nanoparticles. Gold Bull 41:195

37. Sau TK, Murphy CJ (2004) Seeded high yield synthesis of short Au nanorods in aqueous solution. Langmuir 20:6414.

38. Nikoobakht B, El-Sayed MA (2003) Preparation and growth mechanism of gold nanorods (NRs) using seed-mediated growth method. Chem Mater 15:1957.

11

One-pot synthesis of Au nanoparticles/reduced graphene oxide nanocomposites and their application for electrochemical H_2O_2, glucose, and hydrazine sensing

Xiaoyun Qin · Qingzhen Li · Abdullah M. Asiri · Abdulrahman O. Al-Youbi · Xuping Sun

Abstract In this paper, Au nanoparticles/reduced graphene oxide (AuNPs/rGO) nanocomposites were prepared through a one-pot strategy, carried out by heating the mixture of $HAuCl_4$ and graphene oxide solution at 90 °C under alkaline condition. The resultant AuNPs/rGO nanocomposites were found to exhibit good catalytic performance toward H_2O_2 reduction and oxidation as well as hydrazine oxidation. The electrochemical sensing application of the nanocomposites for H_2O_2, glucose, and hydrazine was also demonstrated successfully.

Keywords Au nanoparticles · Reduced graphene oxide · Electrochemical detection · H_2O_2 · Glucose · Hydrazine

Introduction

Graphene, a two-dimensional aromatic sheets composed of sp^2-bonded carbon atoms, has received enormous interest in various areas of research owing to its large specific surface area, excellent thermal and electrical conductivity, strong mechanical strength, good biocompatibility, and low manufacturing cost [1–3]. On the other hand, noble metal nanostructures are a class of functional materials with unique physical and chemical properties [4, 5]. Furthermore, the integration of two-dimensional graphene with zero-dimensional noble metal nanoparticles (NPs) into hybrid structures has received increased attention in the past few years [6–12]. These noble metal NPs/graphene nanocomposites not only combine the merits of each component, but possess interesting structural, electrochemical, electromagnetic, and other properties that are not available in their respective components. We have fabricated noble metal NPs/graphene nanocomposites via chemical reduction and photocatalytic strategies [13–23]. Zhou et al. demonstrated the one-step synthesis of AgNPs on graphene oxide (GO) and reduced graphene oxide (rGO) surfaces absorbed on 3-aminopropyltriethoxysilane-modified Si/SiO_x substrates without using any surfactant or reducing agent [24]. Similarly, our group has also successfully reduced GO to rGO in liquid phase and decorated AgNPs onto thus obtained rGO under strong alkaline conditions without any reducing agent [25, 26]. In this process, AgNPs are deposited onto rGO by chemical reduction of silver ions by hydroxyl group of GO accompanied with the conversion of GO into rGO under strong alkaline conditions as well as heat treatment process [27–30]. However, high temperature or multistep reactions are required. Accordingly, from a point view of material science, synthesis of noble metal NPs/graphene nanocomposites by a more facile method and exploiting their catalytic applications are still highly desirable. Herein, we prepared AuNPs/rGO nanocomposites through a one-pot route by heating the mixture of $HAuCl_4$ and GO solution under alkaline condition at 90 °C for 30 min. It suggests that the resultant AuNPs/rGO nanocomposites show good catalytic performance toward H_2O_2 reduction and oxidation as well as hydrazine oxidation. We further demonstrate the electrochemical sensing application of the nanocomposites for H_2O_2, glucose, and hydrazine.

X. Qin · Q. Li · X. Sun
Chemical Synthesis and Pollution Control Key Laboratory of Sichuan Province, School of Chemistry and Chemical Industry, China West Normal University, Nanchong 637002 Sichuan, China

A. M. Asiri · A. O. Al-Youbi · X. Sun
Chemistry Department, Faculty of Science, King Abdulaziz University, Jeddah 21589, Saudi Arabia

A. M. Asiri · A. O. Al-Youbi · X. Sun (✉)
Center of Excellence for Advanced Materials Research, King Abdulaziz University, Jeddah 21589, Saudi Arabia
e-mail: sun.xuping@hotmail.com

Material and methods

Reagents and materials

Graphite powder, NaCl, NaH$_2$PO$_4$, Na$_2$HPO$_4$, HAuCl$_4$, and H$_2$O$_2$ (30 %) were from Aladin Ltd. (Shanghai, China). Glucose, NaNO$_3$, H$_2$SO$_4$ (98 %), and KMnO$_4$ were purchased from Beijing Chemical Corp. glucose oxidase (GOD) was purchased from Aldrich Chemical Inc. All chemicals were used as received without further purification. The water used throughout all experiments was purified through a Millipore system and a fresh solution of H$_2$O$_2$ was prepared daily. Phosphate-buffered saline (PBS) was prepared by mixing stock solutions of NaH$_2$PO$_4$, Na$_2$HPO$_4$, and NaCl.

Preparation of GO

GO was prepared from natural graphite powder through a modified Hummers method [31]. In a typical synthesis, 1 g of graphite was added into 23 mL of 98 % H$_2$SO$_4$, followed by stirring at room temperature over a 24-h period. After that, 100 mg of NaNO$_3$ was introduced into the mixture and stirred for 30 min. Subsequently, the mixture was kept below 5 °C by ice bath, and 3 g of KMnO$_4$ was slowly added into the mixture. After being heated to 35–40 °C, the mixture was stirred for another 30 min. After that, 46 mL of water was added into above mixture during a period of 25 min. Finally, 140 mL of water and 10 mL of 30 % H$_2$O$_2$ were added into the mixture to stop the reaction. After the unexploited graphite in the resultant mixture was removed by centrifugation, as-synthesized GO was dispersed into individual sheets in distilled water at a concentration of 0.5 mg/mL with the aid of ultrasound for further use.

Preparation of AuNPs/rGO nanocomposites

In a typical experiment, 155 μL of HAuCl$_4$ aqueous solution (24.3 mM) was mixed with the 4,650-μL 0.25 mg/mL of GO and 195 μL 8 M NaOH. Then, the mixture was heated to 90 °C for 30 min in a hot bath. The products were collected by centrifugation and washed with water three times. Finally, the resulting precipitates were redispersed in water for characterization and further use.

Electrochemical measurements

The electrochemical measurements were performed with a CHI 660D electrochemical analyzer (CH Instruments, Inc., Shanghai). A conventional three-electrode cell was used, including a glassy carbon electrode (GCE) (geometric area= 0.07 cm^2) as the working electrode, a Ag/AgCl (3 M KCl) electrode as the reference electrode, and platinum foil as the

counter electrode. All potentials given in this work were referred to the Ag/AgCl electrode. All the experiments were carried out at ambient temperature. The modified electrode was prepared via a simple casting method. Prior to the surface coating, the GCE was polished with 1.0 and 0.3 μm alumina powder, respectively. After that, the GCE was rinsed with distilled water, followed by sonication in ethanol and distilled water, respectively. Then, the electrode was allowed to dry in a stream of nitrogen. For the determination of H$_2$O$_2$, 3 μL of AuNPs/rGO nanocomposites was dropped on the surface of pretreated GCE and left to dry at room temperature. Then, 4 μL of 38 mg/mL GOD aqueous solution was dropped on the resulting AuNPs/rGO/GCE to dry at 4 °C for 3 h. For current time experiment, 2 μL of 1 wt% chitosan solution was used as a fixative and additionally casted on the surface of the above materials modified GCE and dried at 4 °C for 2 h before electrochemical experiments.

Instruments

UV–vis spectra were obtained on a UV5800 spectrophotometer. Raman spectra were obtained on J-Y T64000 Raman spectrometer with 514.5 nm wavelength incident laser light. Powder X-ray diffraction (XRD) data were recorded on a Rigaku D/MAX 2550 diffractometer with Cu Kα radiation (λ=1.5418 Å). Transmission electron microscopy (TEM) measurements were made on a Hitachi H-8100 EM (Hitachi, Tokyo, Japan) with an accelerating applied potential of 200 kV. The sample for TEM characterization was prepared by placing a drop of the dispersion on carbon-coated copper grid and drying at room temperature.

Results and discussion

Characterization of AuNPs/rGO nanocomposites

Figure 1 shows the Raman spectra of aqueous dispersion of GO (curve a) and the products (curve b). It is seen that GO exhibits a D band at 1,361 cm^{-1} and a G band at 1,608 cm^{-1}, while the corresponding bands of the products are 1,354 and 1,579 cm^{-1}, respectively. The G band of the products redshifts from 1,608 to 1,579 cm^{-1}, which is attributed to the high ability for recovery of the hexagonal network of carbon atom [32]. It is also found that the products show relative higher intensity of D to G band (0.97) than that of GO (0.79), further confirming the formation of new graphitic domains [33]. These observations confirm the successful conversion of GO to rGO after the heat treatment process under alkali conditions.

Figure 2 shows the UV–vis spectra of aqueous dispersion of GO and resulting products. As expected, GO exhibits strong bands centered at 230 and 290 nm, corresponding to

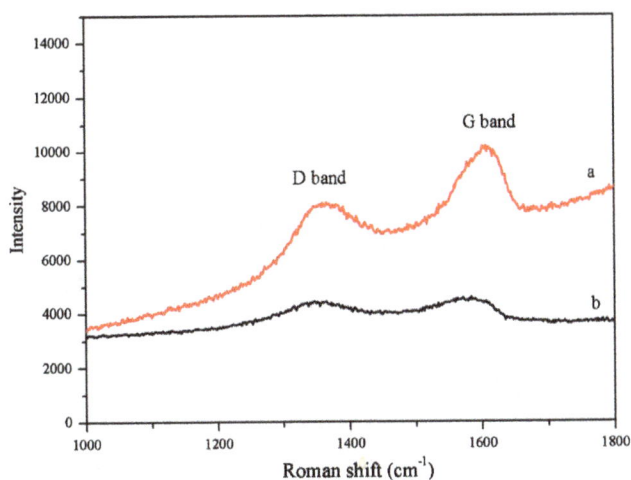

Fig. 1 Raman spectra of (*curve a*) GO and (*curve b*) the products obtained

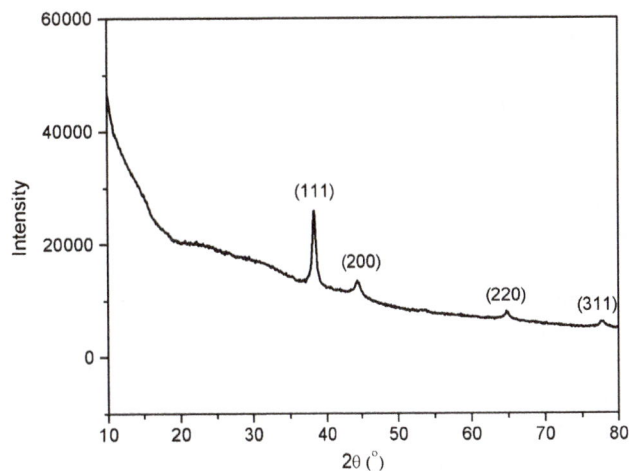

Fig. 3 XRD pattern of products obtained

$\pi-\pi^*$ transitions of aromatic C=C band and $n-\pi^*$ transitions of C=O band in GO, respectively (curve a) [34]. It is clearly seen that the adsorption peak of the obtained composites gradually red-shifts from 230 to 260 nm, and the absorbance in the whole spectral region increases after heat treatment (curve b), suggesting the successful reduction of GO [13, 35]. It is worthwhile mentioning the obvious color change from pale yellow to black after heat treatment of the mixture in alkaline conditions, revealing another piece of evidence to support the formation of rGO. Additionally, a new absorption band appears at 528 nm ascribing to the characteristic of the colloidal Au surface plasmon resonance band, indicating the formation of AuNPs [36].

The XRD pattern of the products obtained is shown in Fig. 3. The four peaks located at 38.2, 44.5, 64.8, and 77.6° are assigned to 111, 200, 220, and 311 faces of a Au crystal, respectively, demonstrating the formation of metallic Au

(JCPDS 04-0784). The broad peak at $2\theta=20-30°$ appears, indicating the disordered stacking of rGO sheets in the composites [37]. All of these observations confirm the formation of AuNPs/rGO nanocomposites after the heat treatment of the mixture of HAuCl$_4$ and GO solution under alkali conditions.

The formation of AuNPs/rGO nanocomposites was further confirmed by TEM observations. Figure 4 shows typical TEM images and the corresponding energy-dispersed spectrum (EDS). Low magnification image (Fig. 4a) indicates that numerous nanoparticles are attached onto the surface of rGO. A higher magnification image reveals the AuNPs are about 40 nm in diameter and the shape is mostly spherical (Fig. 4b). The chemical composition of the nanocomposites was determined by EDS (Fig. 4c), further confirming the existence of C, O, and Au elements. Other peaks originate from the ITO-coated glass substrate.

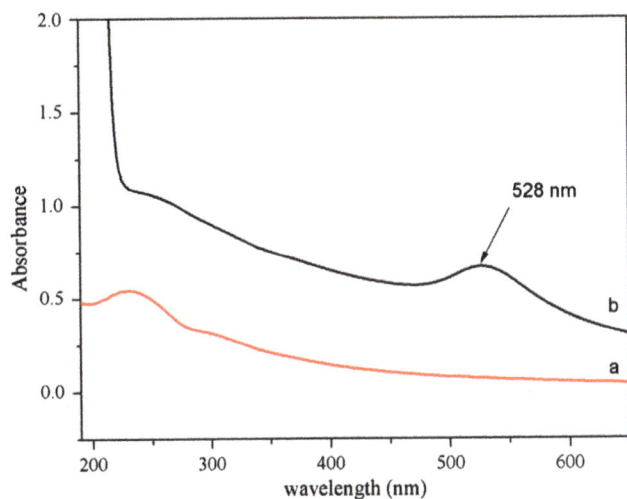

Fig. 2 UV–vis absorption spectra of aqueous dispersions of (*curve a*) GO and (*curve b*) the products obtained

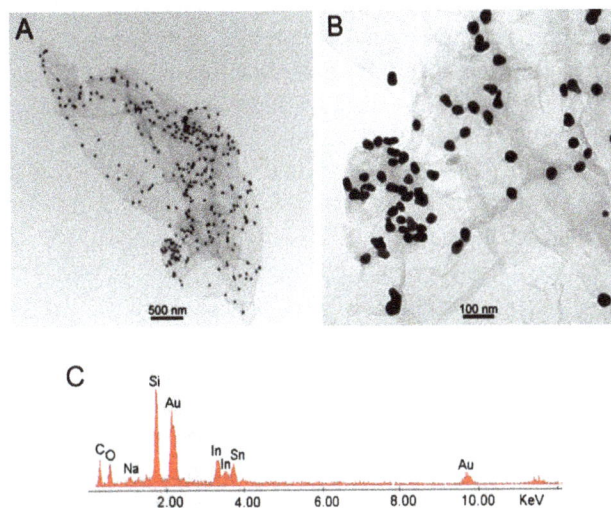

Fig. 4 Typical TEM images at (**a**) low and (**b**) high magnifications and (**c**) corresponding EDS of the obtained AuNPs/rGO nanocomposites

The reducing nature of GO under alkaline conditions has been discussed by Kannan and Zhou et al. The hydroxyl groups of the molecules attached to the hexagonal basal plane make GO a proper agent to reduce $AuCl_4^-$ under alkaline conditions [27]. Because the electrons in the negatively charged rGO can participate in the reduction of metal complex, $AuCl_4^-$ could obtain the electrons in the negatively charged rGO surface to form AuNPs. The big difference between the reduction potential of rGO and $AuCl_4^-$ also help to the spontaneous reduction process [24]. Meanwhile, the alkaline conditions can also accelerate the formation of rGO and AuNPs [28–30]. The strong alkali, NaOH, plays a dual role in the conversion of GO and the formation of AuNPs in this system. In this way, AuNPs are deposited onto rGO by chemical reduction of $AuCl_4^-$ by hydroxyl group of GO accompanied with the conversion of GO into rGO under strong alkaline conditions as well as heat treatment process.

Electrocatalytic effect toward H_2O_2 of AuNPs/rGO/GCE

To demonstrate the sensing application of AuNPs/rGO nanocomposites, we first constructed an enzymeless H_2O_2 sensor by immobilizing AuNPs/rGO nanocomposites with chitosan as a fixative onto a GCE surface. Figure 5a shows cyclic voltammograms (CVs) of bare GCE and the AuNPs/rGO/GCE in 0.2 M PBS at pH 7.4 in the presence of 1 mM H_2O_2. It is seen that the response of the bare GCE toward H_2O_2 is quite weak. In contrast, the AuNPs/rGO/GCE exhibits notable catalytic current in the process of both reduction and oxidation of H_2O_2. It is also important to note that the AuNPs/rGO/GCE exhibits no electrochemical response in the absence of H_2O_2. All these observations indicate that such AuNPs/rGO nanocomposites exhibit notable electrocatalytic activity toward both the reduction and oxidation of H_2O_2. The typical current–time curve of the AuNPs/rGO/GCE was shown in Fig. 5b. The amperometric response of the sensor was studied by successively dropping the H_2O_2 solution with different concentrations into the PBS under optimized conditions at an applied potential of −0.3 V vs. Ag/AgCl electrode. The linear range of the H_2O_2 detection was from 0.1 to 9 mM ($r=0.999$), and the detection limit was estimated to be 1.5 μM based on the criterion of a signal-to-noise ratio of 3.

Determination of glucose at GOD/AuNPs/rGO/GCE

Based on the high electrocatalytical activity of AuNPs/rGO/GCE toward H_2O_2, a glucose sensor was further developed by immobilizing GOD onto the surface of AuNPs/rGO/GCE. The sensing mechanism is that GOD can selectively catalyze the oxidation of glucose in the presence of oxygen to form H_2O_2, which can be electrochemically detected [38]. Differential pulse voltammogram (DPV) has

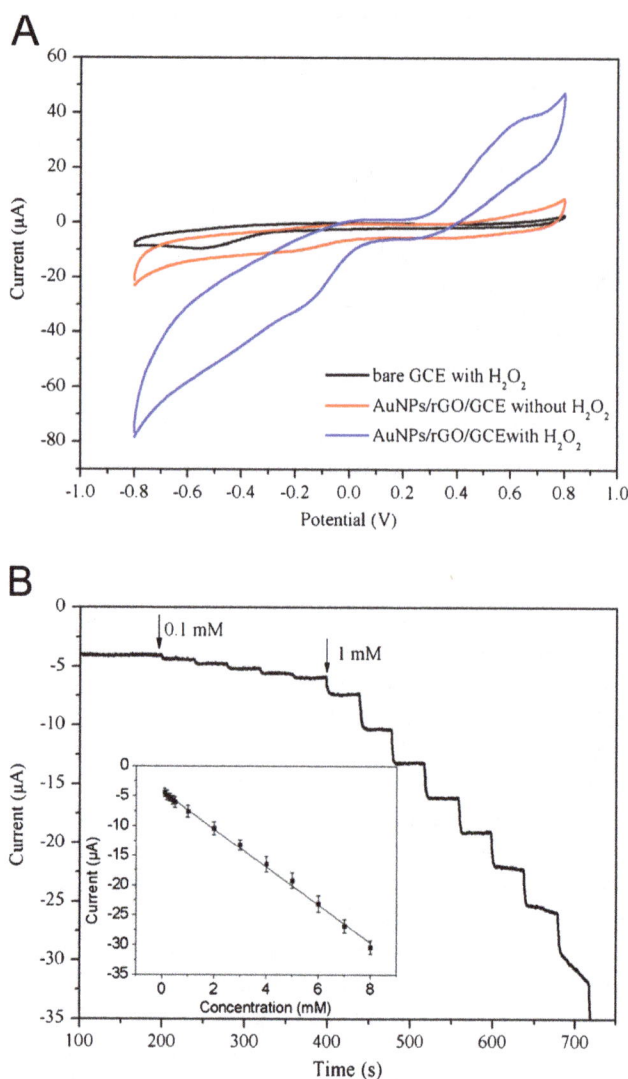

Fig. 5 **a** CVs of different electrodes in N_2-saturated 0.2 M PBS at pH 7.4 in the presence and absence of 1 mM H_2O_2 (scan rate, 50 mV s^{-1}). **b** Typical steady-state response of the AuNPs/rGO/GCE to successive injection of H_2O_2 into the stirred N_2-saturated 0.2 M PBS at pH 7.4 (applied potential, −0.3 V). *Inset*: the fitting of the experimental data by the regression line

higher sensitivity than CV and is used for quantitative measurements in the current study. Figure 6 shows the typical DPVs of the GOD/AuNPs/rGO/GCE in 0.2 M PBS solution at pH 7.4 with various concentrations of glucose in saturated O_2. It is seen that well-defined anodic peaks at −0.07 V are observed, which can be attributed to the electrochemical oxidation of H_2O_2. It is also found that the oxidation current increases with the increased amount of glucose in saturated O_2. The inset in Fig. 6 shows the calibration curves to corresponding amperometric responses. Good linear relationships are observed between the catalytic current and glucose concentration at ranges from 0 to 1 mM ($r=0.998$) and from 3 to 21 mM ($r=0.996$), respectively. The detection limit is estimated to be 20 μM with a signal-to-noise ratio of

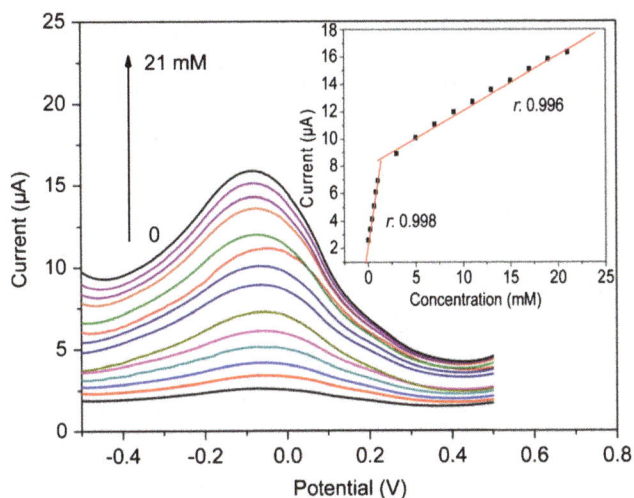

Fig. 6 DPVs of GOD/AuNPs/rGO/GCE in O_2-saturated 0.2 M PBS at pH 7.4 with various concentrations of glucose (from down to top 0, 0.2, 0.4, 0.6, 0.8, 1, 3, 5, 7, 9, 11, 13, 15, 17, 19, 21 mM). *Inset*: the calibration curves corresponding to the responses at −0.07 V

3. Note that our present sensing system gives lower detection limit than that of GOD/AuNPs/rGO/chitosan-based system (180 μM) [39], GOD/AuNPs/rGO/ionic liquid (IL) biosensor (130 μM) [40] and GOD-chemically modified graphene-IL (376 μM) [41], etc. The relative standard deviation of the response to 1 mM of glucose is 1.9 % for ten successive measurements, indicating the good reproducibility of GOD/AuNPs/rGO/GCE. To evaluate the long-term stability of the glucose biosensor, the GOD/AuNPs/rGO/GCE was stored at 4 °C when not in use. The response to glucose at the GOD/AuNPs/rGO/GCE decreased to about 95 % of its initial response current on the third day and about 91 % after 10 days. The loss of the response current may be ascribed to the decrease of the enzyme activity during these days.

Electrocatalytic effect toward hydrazine of AuNPs/rGO/GCE

It is found that the resultant AuNPs/rGO composites exhibit good catalytic performance toward hydrazine oxidation. Figure 7a shows the electrocatalytic responses of bare GCE and AuNPs/rGO/GCE toward the oxidation of hydrazine in 0.2 M PBS at pH 7.4 in the presence of 10 mM hydrazine. The response of the bare GCE toward the oxidation of hydrazine is quite weak. In contrast, the AuNPs/rGO/GCE shows a notable current peak about 75 μA in intensity centered at 0.3 V vs. Ag/AgCl; however, it exhibits no electrochemical response in the absence of hydrazine. These observations indicate that the observed current peak originates from hydrazine oxidation and the nanocomposites can effectively catalyze the electrochemical oxidation of hydrazine. The amperometric response of the hydrazine sensor was studied by successively dropping the hydrazine aqueous solution into the PBS at an applied

potential of 0.3 V vs. Ag/AgCl electrode (Fig. 7b). When an aliquot hydrazine was dropped into the stirred PBS solution, the oxidation current rose steeply to reach a stable value. The sensor could accomplish 96 % of the steady state current within 3 s, indicating a fast amperometric response behavior. The inset in Fig. 7b shows the calibration curve of the sensor. The linear detection range is estimated to be from 5 to 900 μM ($r=0.999$), and the detection limit is estimated to be 0.08 μM at a signal-to-noise ratio of 3. Note that our present sensing system gives lower detection limit than that of platinum screen-printed electrodes (0.12 μM) [42], AuNPs on thiolated single-stranded DNA-modified Au electrode

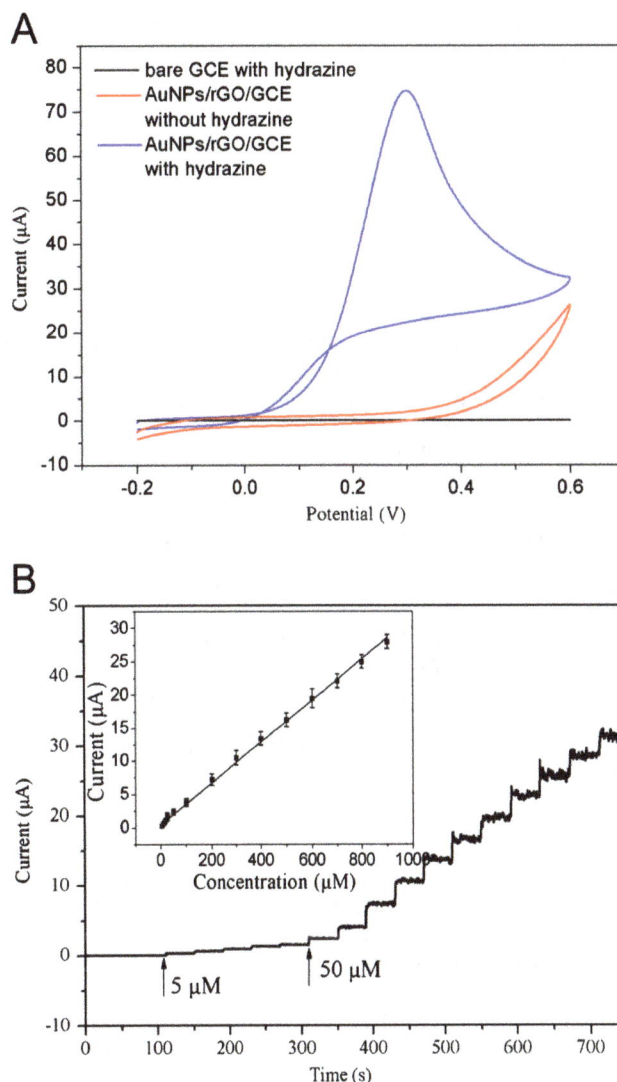

Fig. 7 **a** CVs of different electrodes in 0.2 M PBS at pH 7.4 in the presence and absence of 10 mM hydrazine (scan rate, 50 mV s^{-1}). **b** Typical steady-state response of the AuNPs/rGO/GCE to successive injection of hydrazine into the stirred 0.2 M PBS at pH 7.4 (applied potential, 0.3 V). *Inset*: the fitting of the experimental data by the regression line

(0.56 μM) [43], and hybrid nickel hexacyanoferrate-functionalized multiwalled carbon nanotube/GCE [44].

Conclusions

In summary, heating treatment of $HAuCl_4$ and preformed GO solution under alkaline condition has been proven to be an effective strategy to one-pot preparation of AuNPs/rGO nanocomposites without an extra reducing agent. Such nanocomposites exhibit good electrocatalytic activity toward H_2O_2 reduction and oxidation as well as hydrazine oxidation. Electrochemical detection of H_2O_2, glucose, and hydrazine is also demonstrated successfully. Our present study is important because it provides us a simple method for preparing noble metal nanoparticles/rGO composites for sensing and other applications.

References

1. Novoselov KS, Geim AK, Morozov SV, Jiang D, Zhang Y, Dubonos SV, Grigorieva IV, Firsov AA (2004) Science 306:666–669
2. Kim K, Park HJ, Woo BC, Kim KJ, Kim GT, Yun WS (2008) Nano Lett 8:3092–3096
3. Lee C, Wei X, Kysar JW, Hone J (2008) Science 321:385–388
4. Huang X, Li S, Huang Y, Wu S, Zhou X, Li S, Gan CL, Boey F, Mirkin CA, Zhang H (2011) Nat Commun 2:292
5. Huang X, Li H, Li S, Wu S, Boey F, Ma J, Zhang H (2011) Angew Chem Int Ed 50:12245–12248
6. Yan L, Zheng Y, Zhao F, Li S, Gao X, Xu B, Weiss PS, Zhao Y (2012) Chem Soc Rev 41:97–114
7. Huang X, Zhou X, Wu S, Wei Y, Qi X, Zhang J, Boey F, Zhang H (2010) Small 6:513–516
8. Huang X, Li S, Wu S, Huang Y, Boey F, Gan CL, Zhang H (2012) Adv Mater 24:979–983
9. Wu S, He Q, Zhou C, Qi X, Huang X, Yin Z, Yang Y, Zhang H (2012) Nanoscale 4:2478–2483
10. Guo S, Dong S (2011) Chem Soc Rev 40:2644–2672
11. Huang X, Qi X, Boey F, Zhang H (2012) Chem Soc Rev 41:666–686
12. Tan C, Huang X, Zhang H (2013) Mater Today 16(16):29–36
13. Liu S, Tian J, Wang L, Sun X (2011) Carbon 49:3158–3164
14. Liu S, Tian J, Wang L, Sun X (2011) J Nanopart Res 13:4539–4548
15. Zhang Y, Chang G, Liu S, Tian J, Wang L, Lu W, Qin X, Sun X (2011) Catal Sci & Technol 1:1636–1640
16. Zhang Y, Liu S, Lu W, Wang L, Tian J, Sun X (2011) Catal Sci & Technol 1:1142–1144
17. Zhang Y, Liu S, Wang L, Qin X, Tian J, Lu W, Chang G, Sun X (2012) RSC Adv 2:538–545
18. Lu W, Ning R, Qin X, Zhang Y, Chang G, Liu S, Luo Y, Sun X (2011) J Hazard Mater 197:320–326
19. Liu S, Wang L, Tian J, Lu W, Zhang Y, Wang L, Sun X (2011) J Nanopart Res 13:4731–4737
20. Wu T, Liu S, Luo Y, Lu W, Wang L, Sun X (2011) Nanoscale 3:2140–2142
21. Li H, Liu S, Tian J, Wang L, Lu W, Luo Y, Asiri AM, Al-Youbi AO, Sun X (2012) Chem Cat Chem 4:1079–1083
22. Li H, Chang G, Zhang Y, Tian J, Liu S, Luo Y, Asiri AM, Al-Youbi AO, Sun X (2012) Catal Sci & Technol 2:1153–1156
23. Li H, Zhang Y, Chang G, Liu S, Tian J, Luo Y, Asiri AM, Al-Youbi AO, Sun X (2012) Chem Plus Chem 77:545–550
24. Zhou X, Huang X, Qi X, Wu S, Xue C, Boey F, Yan Q, Chen P, Zhang H, Phys J (2009) Chem C 113:10842–10846
25. Tian J, Liu S, Zhang Y, Li H, Wang L, Luo Y, Asiri AM, Al-Youbi AO, Sun X (2012) Inorg Chem 51:4742–4746
26. Qin X, Luo Y, Lu W, Chang G, Asiri AM, Al-Youbi AO, Sun X (2012) Electrochim Acta 79:46–51
27. Kannan PR, Swami A, Srisathiyanarayanan D, Shirude PS, Pasricha R, Mandale AB, Sastry M (2004) Langmuir 20:7825–7836
28. Fan X, Peng W, Li Y, Li X, Wang S, Zhang G, Zhang F (2008) Adv Mater 20:4490–4493
29. Wang X, Wu H, Kuang Q, Huang R, Xie Z, Zheng L (2010) Langmuir 26:2774–2778
30. Nishimure S, Mott D, Takagaki A, Maenosono S, Ebitani K (2011) Phys Chem Chem Phys 13:9335–9343
31. Hummers WS Jr, Offeman R (1958) J Am Chem Soc 80:1339–1339
32. Li Z, Yao Y, Lin Z, Moon KS, Lin W, Wong C (2010) J Mater Chem 20:4781–4783
33. Zhang Y, Guo L, Wei S, He Y, Xia H, Chen Q, Sun H, Xiao F (2010) Nano Today 5:15–20
34. Guo Y, Guo S, Ren J, Zhai Y, Dong S, Wang E (2010) ACS Nano 4:4001–4010
35. Sediri F, Gharbi N (2007) Mater Sci Eng B 139:114–1177
36. Daniel MC, Astruc A (2004) Chem Rev 104:293–346
37. Chandra V, Park J, Chun Y, Lee JW, Hwang I, Kim KS (2010) ACS Nano 4:3979–3986
38. Tsai M, Tsai Y (2009) Sens Actuators B 141:592–598
39. Shan C, Yang H, Han D, Zhang Q, Ivaska A, Niu L (2010) Biosens Bioelectron 25:1070–1074
40. Yang M, Choi BG, Park TJ, Hong WH, Lee SY (2011) Electroanal 23:850–857
41. Yang MH, Choi BG, Park H, Hong WH, Lee SY, Park TJ (2010) Electroanal 22:1223–1228
42. Metters JP, Tan F, Kadara RO, Banks CE (2012) Anal Methods 4:1272–1277
43. Chang G, Luo Y, Lu W, Hu J, Liao F, Sun X (2011) Thin Solid Films 519:6130–6134
44. Kumar AS, Barathi P, Pillai KC (2011) J Electroanal Chem 654:85–95

12

Facile one-pot synthesis of amoxicillin-coated gold nanoparticles and their antimicrobial activity

Marco Demurtas · Carole C. Perry

Abstract Nanomaterials have been the object of intense study due to promising applications in a number of different disciplines. In particular, medicine and biology have seen the potential of these novel materials with their nanoscale properties for use in diverse areas such as imaging, sensing and drug vectorisation. Gold nanoparticles (GNPs) are considered a very useful platform to create a valid and efficient drug delivery/carrier system due to their facile and well-studied synthesis, easy surface functionalization and biocompatibility. In the present study, stable antibiotic conjugated GNPs were synthesised by a one-step reaction using a poorly water soluble antibiotic, amoxicillin. Amoxicillin, a member of the penicillin family, reduces the chloroauric acid to form nanoparticles and at the same time coats them to afford the functionalised nanomaterial. A range of techniques including UV–vis spectroscopy, dynamic light scattering (DLS), transmission electron microscopy (TEM) and thermogravimetric analysis (TGA) were used to ascertain the gold/drug molar ratio and the optimum temperature for synthesis of uniform monodisperse particles in the ca. 30–40 nm size range. Amoxicillin-conjugated gold showed an enhancement of antibacterial activity against *Escherichia coli* compared to the antibiotic alone.

Keywords Gold · Nanoparticles · Amoxicillin · *Escherichia coli* · Antibacterial activity

Introduction

Since the discovery of the first antibiotic [1], the scientific community has spent a great amount of effort in studying, developing and synthesising new types of antibiotic to fight bacteria. A large number of antibiotics, divided into various families having structural or functional similarity, have been discovered as a result of this intensive research. Unfortunately, the extensive use of these antibiotics has caused bacteria to develop new resistance mechanisms including prevention of drug interaction with the target, efflux of the antibiotic from the cell and direct destruction or modification of the compound [2, 3]. To win the battle against new strains of bacteria, an alternative approach may be to use the already well-studied drug and conjugate it with other entities in order to enhance anti-microbial activity. In the last decade, nanomaterials have been widely used in biotechnology [4]. In particular, drug-conjugated gold nanoparticles (GNPs) have been intensively investigated as delivery/carrier systems [5–7]. GNPs are characterised by properties that are potentially suitable for use in this area. First, gold nanoparticles can be synthesised in a controlled fashion affording a wide range of sizes (1–150 nm) [5]. Also, the surface of the gold nanoparticles can be functionalised through covalent or non-covalent interaction without structural modification of the drug [8]. The conjugation of a drug with the inorganic particles could increase the bio-stability, the bio-distribution and also suppress some side effects [5–7]. In addition, the high surface area typical of nanoparticles guarantees a huge drug load; for example, Zubarev et al. were able to couple ≈70 molecules of paclitaxel to a GNP of 2 nm [5]. Several examples are available in the literature of the anti-microbial activity of antibiotics conjugated with gold nanoparticles [9–11]. Usually, the synthetic pathway starts with the reduction of a chloroauric acid solution via a range of methods [12–15]. Historically, Turkevich et al. reported [12] the production of a gold colloid solution using sodium citrate as reducing and stabilising agent in 1951. In 1973, Frens et al. showed [13] the possibility of controlling the gold nanoparticle size by changing the molar ratio of the reagents. In 1994, Brust et al. proposed [14] a novel method to produce stable gold nanoparticles using a two-phase technique with $NaBH_4$ as the

M. Demurtas · C. C. Perry (✉)
Interdisciplinary Biomedical Research Centre, School of Science and Technology, Nottingham Trent University, Clifton Lane, Nottingham NG11 8NS, UK
e-mail: carole.perry@ntu.ac.uk

reducing agent. In this method, chloroauric acid is transferred to toluene using tetraoctylammonium bromide as the phase-transfer reagent and the gold reduced by $NaBH_4$ in the presence of dodecanethiol. Following the formation of the gold nanoparticles, their functionalization involves one or more reaction steps in order to attach the drug to the particle surface. Various approaches of functionalization have been proposed in the literature [5–7].

Recently, we reported a simpler water-based approach to antibiotic-coated gold particles [16] where cefaclor, a second generation antibiotic from the cephalosporin family, was shown to act as a reducing agent to produce the gold nanoparticles and, at the same time, coat them affording the desired anti-microbial nanomaterial, all in a single synthesis step. We were able to control the particle size, the amount of drug loaded and the production time just by changing reaction temperature. Gold nanoparticles alone did not show any antibacterial property, and we were able to show that conjugation between the particles and the drug was necessary to enhance antibacterial activity against common gram-positive and gram-negative bacteria compared to cefaclor alone [16]. Moreover, we were able to coat a functionalised glass slide with cefaclor-reduced gold nanoparticles creating an antimicrobial film which was very robust and inhibited the growth of *Escherichia coli* on the surface over a range of pH conditions designed to mimic common cleaning scenarios [16].

Starting from these results, we wanted to investigate the possibility of using this one-pot synthesis with other drugs, including antibiotics. The idea was to demonstrate that the one-step synthesis with cefaclor was not an isolated case, and that this approach, where the desired drug acts as reducing and capping agent, can be used to create a wide range of drug-coated GNPs. We identified various possible drug candidates that share some features with cefaclor, such as chemical structure and more importantly, the presence of a primary amine, which we demonstrated [16] to be at least partly responsible for the reducing/capping actions in the reaction. Different molecules from the cephalosporin and penicillin families, both β-lactam drugs, were investigated, and among them, amoxicillin, an antibiotic having poor water solubility, was chosen. Amoxicillin (Fig. 1) is a broad-spectrum β-lactam antibiotic used to treat bacterial infections of the chest, urine or ear [17]. Amoxicillin is effective against a range of bacteria inhibiting the synthesis of the bacterial wall cell, and it is widely used because it is better adsorbed compared with other similar antibiotics [17]. Herein, we describe the single-step synthesis of amoxicillin-coated gold nanoparticles, optimization of the reaction and its antimicrobial activity against *E. coli*. Also, we prove that an antibiotic, with very limited water solubility such as amoxicillin, is suitable for the one-pot synthesis without decreasing the antibacterial properties and bio-distribution.

Experimental

Materials

$HAuCl_4$ and amoxicillin were purchased from Sigma-Aldrich and used as received. Cellulose dialysis tubing (10 kDa) was purchased from Spectrum Laboratory Inc. and the membrane prepared for use by removing any glycerine and preservative according to the manufacturers procedure. *E. coli* culture (K12) was used for the antibacterial study.

Synthesis and characterization of amoxicillin-coated gold nanoparticles

Aqueous solutions of amoxicillin (0.01 (not fully solubilised), 0.001 M) and $HAuCl_4$ (0.01 M) were prepared for use in the one-pot synthesis according to our previous work [16]. Materials were prepared at a range of gold/antibiotic molar ratios and temperatures (20–70 °C). An example preparation (as for the material used for antimicrobial activity testing) is as follows: 9.65 mL of dd. H_2O, 100 μL of $HAuCl_4$ (0.01 M) and 250 μL of amoxicillin (0.001 M) were mixed and the resulting solution heated in a water bath at 40 °C for 4 h. Freshly synthesised amoxicillin-coated gold nanoparticles were dialysed using a 10 kDa cut-off cellulose membrane against double-distilled water for 24 h to remove any unreacted chemicals. Once dialysed, the samples were freeze-dried at 223 K using a VirTis freeze dryer.

The shape and size of the gold nanoparticles was determined by transmission electron microscopy (TEM) (JEOL 2010; accelerating voltage of 200 kV). TEM samples were prepared by depositing a 10 μl drop of drug-coated gold nanoparticle solution on a carbon-coated grid (Quantifoil grids: Cu 200 mesh). Samples were air-dried for 5–10 min and excess solution removed using a tissue paper before being left to dry at room temperature overnight. One hundred particles were measured and their average size computed. The size and the charge carried by the gold and gold/antibiotic nanoparticles in solution was also determined by dynamic light scattering/zeta potential measurement (DLS) using a Malvern Instrument Nano-S.

The antibiotic 'load' of the nanoparticles was determined using thermogravimetric analysis (TGA) using a Mettler–Toledo (TGA/SDTA851) instrument. Samples were heated in air over a temperature range of 30–900 °C, at 10 °C/min.

Antimicrobial activity testing

E. coli was grown in lysogeny broth (LB) for 24 h at 37 °C and cell counts were quantified by OD_{590} measurement. Amox–GNPs solutions were added to *E. coli* suspensions, incubated and shaken at 37 °C prior to dilution, plating and incubation on LB Agar plates. Antibacterial activity was

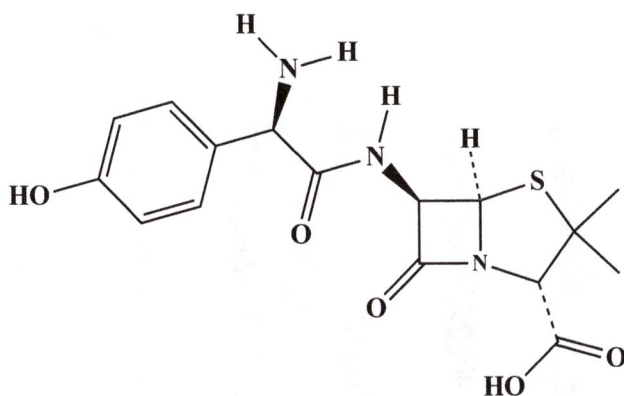

Fig. 1 Chemical structure of amoxicillin

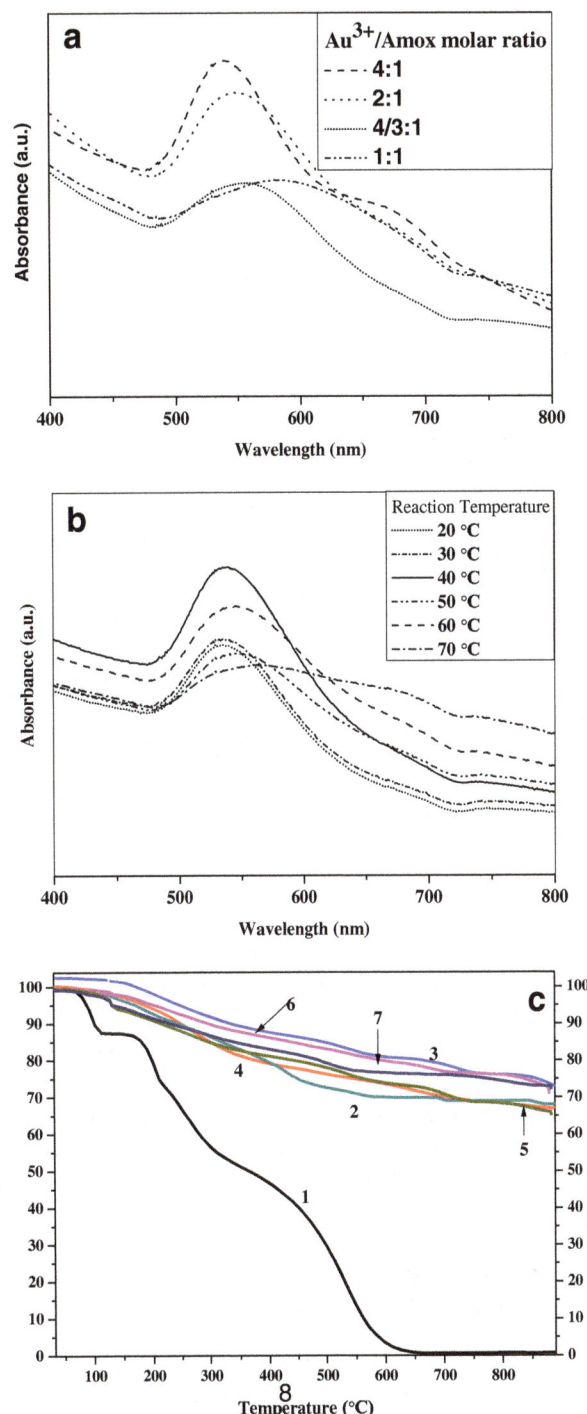

determined by plotting colony-forming units against incubation times 0, 2 and 4 h. For each experiment, three different Amox–GNPs solutions were added to a suspension of bacteria (100, 200 and 300 µg/mL), which corresponded to 33, 66 and 99 µg/mL of amoxicillin present on the GNPs, respectively. Also, a solution containing amoxicillin alone (150 µg/mL) was prepared as a control to compare the antibacterial behaviour. The microbiology experiments were performed in triplicate.

Colony counting was performed using a Synbiosis aCOLyte (7510/SYN) colony counter. The minimum inhibition concentration was calculated according to the method used by Sambhy et al. [18]

Results and discussion

Amoxicillin-coated gold nanoparticles were prepared according to the one-pot synthesis of cefaclor–GNPs presented in our previous work [16]. In this case, the molar ratio of the reagents was varied, in part due to the lower water solubility of amoxicillin. A complete UV–vis study was carried out in order to establish the best reaction conditions to achieve the coated gold nanoparticles. Three properties (peak position, peak breadth and peak intensity) of the UV–vis spectrum were used. The surface plasmon resonance (SPR) band typical of the gold nanoparticles lies in the visible region of the electromagnetic spectrum with any changes to the surroundings of these particles, such as surface modification and aggregation, leading to a colorimetric change of the dispersion. Thus, the SPR peak position provides information about the particle size and the broadness of the SPR absorption peak information on particle size distribution, a broad band signifying a broad particle size distribution and aggregation. Finally, the absorption value is proportional to the amount of gold nanoparticles produced [19, 20]. Figure 2a shows the UV–vis spectra of amoxicillin–GNPs samples produced using four different molar ratios of Au(III)/amoxicillin. The best result was obtained with a molar ratio of 4:1 as confirmed by a sharp peak in

Fig. 2 a UV–vis spectra of amoxicillin-coated gold nanoparticles obtained at different molar ratios of Au(III) and amoxicillin. b UV–vis spectra of amoxicillin-coated gold nanoparticles synthesised at different temperatures (gold: antibiotic molar ratio of 4:1). c TGA analyses of amoxicillin-coated gold nanoparticles synthesised at different temperatures with the same molar ratio as in 1b. Amoxicillin (1), amoxicillin–GNPs at 20 °C (2), 30 °C (3), 40 °C (4), 50 °C (5), 60 °C (6) and 70 °C (7)

Fig. 2a. Our previous work [16] showed the importance of temperature on the one-pot synthesis, where for cefaclor increasing reaction temperature led to a decrease in reaction

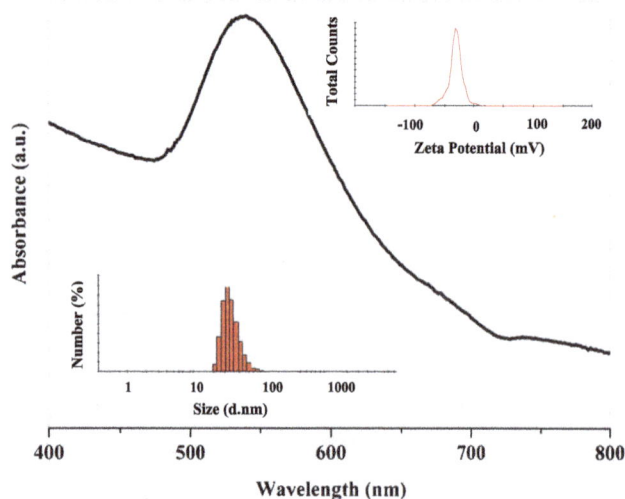

Fig. 3 UV–vis spectrum of amoxicillin-conjugated gold nanoparticles synthesised at 40 °C. Histogram of size distribution (inset *bottom left*) and zeta-potential measurement (inset *top right*) of the amoxicillin-conjugated gold nanoparticles obtained at 40 °C for the 4:1 *gold*: antibiotic sample

Fig. 5 Histogram plot showing antimicrobial activity of amoxicillin-conjugated gold nanoparticles (synthesis as per Fig. 2) and amoxicillin on *E. coli* after different incubation times (0, 2, and 4 h). All concentrations are in units of µg/ml. The 'blank' without addition of amoxicillin or amoxicillin-conjugated gold nanoparticles is shown for $t = 0$ only where it was possible to measure the colony forming units

time, a linear decrease in particle size and decrease in amount of drug loaded on the nanoparticles [16]. In the current study, a linear dependence of particle size with temperature was not seen; rather, the UV–vis spectra of the resulting gold colloid solutions presented in Fig. 2b show that temperature largely affected the growth and aggregation of the particles with the best (least aggregated/more uniform in size) particles being obtained at 40 °C.

With the optimum synthesis conditions, an aqueous solution of amoxicillin and chloroauric acid (1:4) heated at 40 °C in a water bath for 4 h gave the typical red wine colour of a gold colloid solution. The UV–vis absorption spectrum in Fig. 3 shows the characteristic SPR band with the maximum peak at 540 nm. The sharp absorption band is evidence of a fine nanoparticle dispersion without aggregation and precipitation (flocculation). Zeta potential measurements performed

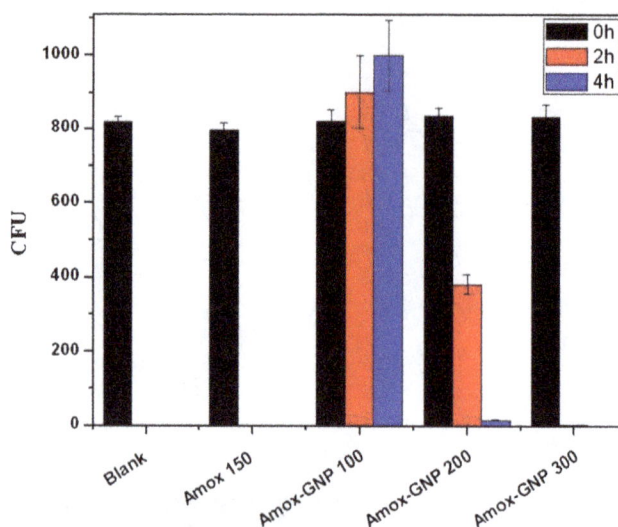

on the same sample confirmed the stability of the negatively charged amoxicillin-coated gold nanoparticles with a value of −31.3 mV. TEM analysis (Fig. 4) confirmed the uniform size and particle distribution with an average spherical diameter of 32.61 ± 4.78 nm. The lighter areas surrounding the ca. 30 nm gold–antibiotic nanoparticles also contain gold in the form of much smaller nanoparticles of the order of 1–2 nm in diameter, also associated with organic material (antibiotic) that are not discernible separately in the images shown. DLS analysis corroborated this finding with a measured particle size of 37.85 ± 13.20 nm, Fig. 3 inset. The amount of antibiotic bound to the particles as measured by TGA, Fig. 2c, was around 33 % by weight.

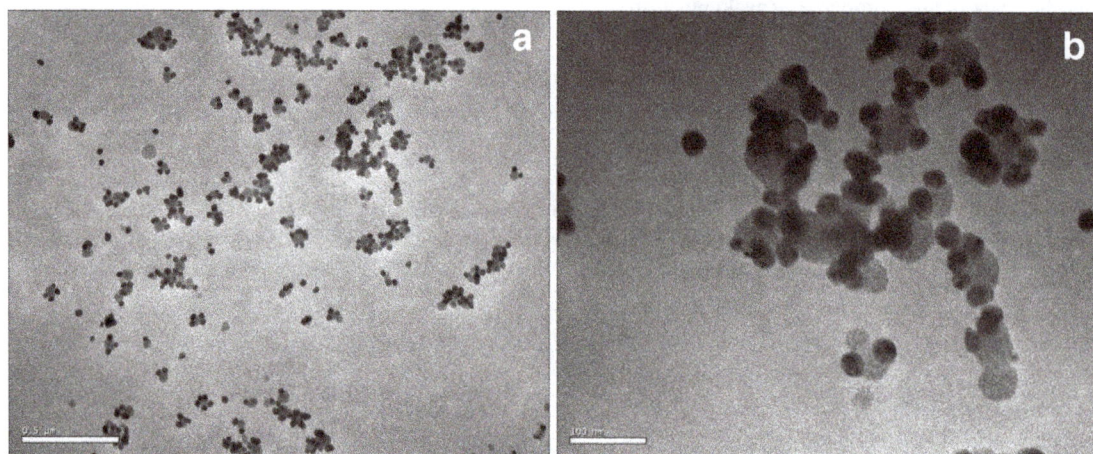

Fig. 4 TEM images of the coated gold nanoparticles synthesised at a 4:1 M ratio of *gold*: antibiotic at 40 °C recorded at different magnifications. *Scale bar* in TEM images is 500 (**a**) and 100 nm (**b**)

The antimicrobial activity of the antibiotic-coated nanoparticles was determined against a common gram-negative bacteria *E. coli*. Heating to 40 °C during the formation of the antibiotic loaded particles did not affect the antibacterial capability of the amoxicillin present on the gold nanoparticles.

Figure 5 showed that the lowest concentration sample Amox–GNPs 100 μg/mL did not inhibit the growth of *E. coli* colonies over 4 h of incubation at 37 °C. The presence of 33 μg/mL of amoxicillin associated with the gold particles was enough to slow down the growth of the gram-negative bacteria. The Amox–GNPs 200 μg/mL sample killed half of the *E. coli* colonies in 2 h of incubation, and after 4 h, only 2 % of the number was observed. The 300 μg/mL Amox–GNPs (99 μg/mL of amoxicillin) sample killed 99 % of the bacteria in 2 h, showing the same antibacterial activity compared to amoxicillin alone when the antibiotic alone was used at a higher concentration (150 μg/mL). The conjugation of the drug with gold nanoparticles enhanced the antimicrobial activity of the antibiotic. The minimum time required to inhibit the complete growth of *E. coli* was 2 h, and the minimum inhibition concentration of gram-negative bacteria obtained from the amoxicillin-reduced gold nanoparticles was determined according to the method of Sambhy et al. [18] as 300 μg/mL (99 μg/mL of amoxicillin).

Conclusions

In summary, we describe an easy way to produce antibiotic-coated gold nanoparticles using a single reaction step where the amoxicillin reduced the gold ions and capped the resulting gold nanoparticles. We show that this particular synthetic pathway can be used with antibiotics that are poorly soluble in aqueous solution and that preparation of the gold-coated particles at temperatures above body temperature for extended times does not destroy the activity of the antibiotic. The conjugation of amoxicillin with gold nanoparticles enhances antibacterial activity as the same microbiological activity is achieved with ca. 33 % less amoxicillin. Currently, we are investigating the possibility to use this one-pot synthesis to coat gold nanoparticles with other families of antibiotics and/or drugs, which must display a primary amine group, but also have enough reducing power to generate gold nanoparticles.

Acknowledgments We gratefully acknowledge support for this research from the World Gold Council.

References

1. Fleming A (1929) On the antibacterial action of cultures of a *Penicillium*, with special reference to their use in the isolation of *B. influenzae*. Br J Exp Pathol 10(3):226–236
2. Kumar A, Schweizer HP (2005) Bacterial resistance to antibiotics: active efflux and reduced uptake. Adv Drug Deliv Rev 57:1486–1513
3. Neu HC (1992) The crisis in antibiotic resistance. Science 257:1064–1073
4. De M, Ghosh PS, Rotello VM (2008) Applications of nanoparticles in biology. Adv Mater 20(22):4225–4241
5. Ghosh P, Han G, De M, Kim CK, Rotello VM (2008) Gold nanoparticles in delivery applications. Adv Drug Deliv Rev 60(11):1307–1315
6. Duncan B, Kim CK, Rotello VM (2010) Gold nanoparticle platforms as drug and biomacromolecule delivery systems. J Control Release 148(1):122–128
7. Pissuwan D, Niidome T, Cortie MB (2011) The forthcoming applications of gold nanoparticles in drug and gene delivery systems. J Control Release 149(1):65–71
8. Sperling RA, Gil PR, Zhang F, Zanella M, Parak WJ (2008) Biological applications of gold nanoparticles. Chem Soc Rev 37:1896–1908
9. Kim CK, Ghosh P, Rotello VM (2009) Multimodal drug delivery using gold nanoparticles. Nanoscale 1:61–67
10. Gu H, Ho PL, Tong E, Wang L, Xu B (2003) Presenting vancomycin on nanoparticles to enhance antimicrobial activities. Nano Lett 3(9):1261–1263
11. Gibson JD, Khanal BP, Zubarev ER (2007) Paclitaxel-functionalized gold nanoparticles. J Am Chem Soc 129(37):11653–11663
12. Turkevich J, Stevenson PC, Hillier J (1951) A study of the nucleation and growth processes in the synthesis of colloidal gold. Faraday Soc 11:55–75
13. Frens G (1973) Controlled nucleation for the regulation on particle size in monodisperse gold suspension. Nature Phys Sci 241(105):20–22
14. Brust M, Walker M, Bethell D, Schiffrin DJ, Whyman R (1994) Synthesis of thiol-derivatised gold nanoparticles in a two-phase Liquid–Liquid system. J Chem Soc Chem Commun 7:801–802
15. Martin MN, Basham JI, Chando P, Eah SK (2010) Charged gold nanoparticles in non-polar solvents: 10-min synthesis and 2D self-assembly. Langmuir 26(10):7410–7417
16. Rai A, Prabhune A, Perry CC (2010) Antibiotic-mediated synthesis of gold nanoparticles with potent antimicrobial activity and their application in antimicrobial coatings. J Mater Chem 20:6789–6798
17. Holten KB, Onusko EM (2000) Appropriate prescribing of oral beta-lactam antibiotics. Am Fam Physician 62(3):611–620
18. Sambhy V, MacBride MM, Peterson BR, Sen A (2006) Silver bromide nanoparticle/polymer composites: dual action tunable antimicrobial materials. J Am Chem Soc 128:9798–9808
19. Daniel MC, Astruc D (2004) Gold nanoparticles: assembly, supramolecular chemistry, quantum-size-related properties, and applications toward biology, catalysis, and nanotechnology. Chem Rev 104:293–346
20. Boisselier E, Astruc D (2009) Gold nanoparticles in nanomedicine: preparations, imaging, diagnostics, therapies and toxicity. Chem Soc Rev 28:1759–1782

13

Hollow γ-Al₂O₃ microspheres as highly "active" supports for Au nanoparticle catalysts in CO oxidation

Jie Wang · Zhen-Hao Hu · Yu-Xin Miao · Wen-Cui Li

Abstract Consisted of closely packed nanoflakes, γ-Al$_2$O$_3$ hollow microspheres with ca. 4–6 μm in diameter, and 500–700 nm in shell thickness have been hydrothermally synthesized through utilizing Al(NO$_3$)$_3$·9H$_2$O as precursor, urea as precipitant agent and sulfate K$_2$SO$_4$, (NH$_4$)$_2$SO$_4$, or KAl(SO$_4$)$_2$·12H$_2$O as additive, followed by a calcination step. The samples were further characterized by thermogravimetric analysis, scanning electron microscope, x-ray powder diffraction, nitrogen adsorption, and in situ diffuse reflectance infrared Fourier transform spectroscopy (DRIFTS) of adsorbed CO etc. The morphology of alumina products was strongly dependent on the presence of SO$_4{}^{2-}$. Then via a deposition–precipitation method, 3 wt.% Au nanoparticles supported on γ-Al$_2$O$_3$ hollow microspheres exhibit excellent performance with a complete CO conversion at 0 °C ($T_{100\%}$=0 °C) and 50 % conversion at −25 °C ($T_{50\%}$=−25 °C). The good catalytic activity is associated with the special hollow microsphere structures assembled by nanoflakes of γ-Al$_2$O$_3$ support. The DRIFTS confirms the presence of Au$^{\delta+}$ and Au^0 on the surface of γ-Al$_2$O$_3$ hollow microspheres. As a contrast, Au catalyst prepared using alumina support with undefined morphology shows low activity under the same catalytic test conditions ($T_{100\%}$=190 °C, $T_{50\%}$=80 °C).

Keywords Hollow microspheres · γ-Al₂O₃ · Au catalyst · CO oxidation

J. Wang · Z.-H. Hu · Y.-X. Miao · W.-C. Li (✉)
State Key Laboratory of Fine Chemicals, School of Chemical
Engineering, Dalian University of Technology, Dalian 116024,
People's Republic of China
e-mail: wencuili@dlut.edu.cn

Introduction

In the past decades, considerable attention has been paid to the supported gold catalysts for CO oxidation at low temperatures [1]. Various oxides, such as TiO$_2$ [2], Fe$_2$O$_3$ [3], Al$_2$O$_3$ [4–6], CeO$_2$ [7], MnO$_x$ [8] etc., have been employed to disperse and stabilize Au nanoparticles. Thereinto, gold catalysts supported on reducible metal oxides, in particular on CeO$_2$, TiO$_2$, Fe$_2$O$_3$, MnO$_x$, are well known for their high activity for CO oxidation at low temperatures.

Among those nonreducible oxide supporting gold nanoparticles, Au/γ-Al$_2$O$_3$ is still a very interesting system for both practical application and academic study due to the extraordinary advantages of γ-Al$_2$O$_3$ such as high surface area and good mechanical and especially thermal, chemical stability [9]. However, the less active of Au/Al$_2$O$_3$ catalyst for CO oxidation has been observed experimentally. For example, gold nanoparticles supported on mesoporous γ-Al$_2$O$_3$ give a complete CO conversion when the reaction is performed at 150 °C [10]. When using commercial γ-Al$_2$O$_3$ support, the CO conversion of Au catalyst was only ca. 50 % at 65 °C ($T_{50\%}$) and reached to 90 % at 100 °C. Even after Au nanoparticles supported on γ-Al$_2$O$_3$nanofibers, the enhanced catalytic activity can only achieve the complete CO conversion at 40 °C ($T_{100\%}$=40 °C) [11]. Thus, it remains a grand challenge to obtain highly active Au/Al$_2$O$_3$ catalyst for CO oxidation. Recently, we reported the γ-Al$_2$O$_3$with thin sheets and rough surface acted as an extraordinary catalyst support for the stabilization of gold nanoparticles from sintering [4]. It reveals that, apart from its intrinsic quality, the morphology of support material plays a significant role in the reactive activity of Au catalysts.

Herein, assembled from closely packed nanoflakes, γ-Al$_2$O$_3$ hollow microsphere structures with ca. 4–6 μm in diameter and 500–700 nm in shell thickness have been hydrothermally prepared through adjusting the molar ratio of

Al^{3+} and SO_4^{2-}, followed by a calcination step. As such unique structures of γ-Al_2O_3 support could efficiently stabilize Au nanoparticles, the obtained Au catalyst exhibits extraordinarily high activity towards CO oxidation, where a complete CO conversion at 0 °C was achieved ($T_{50\%}=$ -25 °C). This work could be distinguished by special hollow microspheres structures assembled by nanoflakes of γ-Al_2O_3 support and excellent catalytic activity for CO removal.

Experimental

Synthesis of Au/Al₂O₃ catalysts

Typically, aluminum nitrate, sulfates (K_2SO_4, $(NH_4)_2SO_4$ or $KAl(SO_4)_2 \cdot 12H_2O$) and urea were dissolved in 100-mL deionized water. The obtained mixture was transferred into a 150-mL Teflon-lined stainless steel autoclave and heated at 180 °C for 3 h. Then, the white precipitates were washed with deionized water and dried at 80 °C. The hydrothermally synthesized and dried samples were denoted as Al_2O_3-x-$hydro$ (where $x=1$, 2, and 3, corresponding to different sulfate). After a calcination step at 500 °C for 2 h, the final products were obtained and named as Al_2O_3-x. The concentration of Al^{3+} was fixed as 0.05 M. The detail synthesis conditions and textural parameters were listed in Table 1. For comparison, alumina (denoted as Al_2O_3-4) was synthesized by a common precipitation method using aluminum nitrate as precursor and ammonium carbonate as precipitant.

Preparation of Au/Al₂O₃ catalyst and catalytic test

Au nanoparticles were deposited on the surface of the sample Al_2O_3-x ($x=1$, 2, 3, and 4) by a deposition–precipitation method with $HAuCl_4$ solution (7.9 g L^{-1}) at pH 8–9 in 60 °C for 2 h. After washing and drying, the precipitants were thermal treated at 250 °C for 2 h in the air to generate the Au/Al_2O_3 catalysts, denoted as Au-$(Al_2O_3$-$x)$. The Au content was theoretically estimated as 3 wt.%. The activity of Au catalysts for CO oxidation was evaluated in a fixed bed quartz reactor using 50 mg of catalyst (20–40 mesh) with a composition of 1 vol.% CO, 20 vol.% O_2, and 79 vol.% N_2 and the total rate of feed gas was 67 mL min^{-1} (80,000 mL h^{-1} g_{cat}^{-1}). The products were analyzed using a GC-7890 gas chromatograph equipped with a thermal conductivity detector.

Characterization

Thermogravimetric and differential scanning calorimetry analysis (TG-DSC) were conducted on a thermogravimetric analyzer STA 449 F3 (NETZSCH) under an air atmosphere with a heating rate of 10 °C min^{-1}. X-ray diffraction patterns (XRD) were obtained with a D/MAX-2400 diffractometer using Cu K_α radiation (40 kV, 100 mA, $\lambda=1.54056$ Å). Nitrogen adsorption/desorption isotherms were measured with a TriStar 3000 adsorption analyzer (Micromeritics) at liquid nitrogen temperature. The samples were degassed at 200 °C for 4 h prior to analysis. The Brunauer–Emmett–Teller (BET) method was used to calculate the specific surface areas (S_{BET}). Pore size distributions (PSDs) were derived from the desorption branches of the isotherms using the Barrett–Joyner–Halenda model. Scanning electron microscope (SEM) images were obtained with a Hitachi S-4800 instrument. Transmission electron microscope (TEM) images were obtained with a Tecnai G220 S-Twin microscope with accelerative voltage of 200 kV. The remained K and S contents were analyzed using an inductively coupled plasma atomic emission spectrometer (ICP-AES) on the Optima 2000 DV. Infrared Fourier transform spectra (FT-IR) were recorded using a Nicolet 6700 FT-IR spectrometer at a resolution of 4 cm^{-1} and scale at 4,000–640 cm^{-1}. Diffuse reflectance infrared Fourier transform spectra were recorded using a Nicolet 6700 FT-IR spectrometer at a resolution of 4 cm^{-1} and scale at 4,000–640 cm^{-1}. Self-supporting disks were prepared from the sample powders and treated directly in the IR cell. The catalysts were connected to a vacuum-adsorption apparatus with a residual pressure below 10^{-3} Pa. Prior to the CO adsorption (25 °C, 5 vol.% CO and N_2 in balance), the catalysts were evacuated for 30 min for 1 h at 100 °C.

Table 1 Synthesis conditions and structure parameters of γ-Al_2O_3 samples

Sample	Al^{3+} precursor	Sulfate	Precipitant agent	S_{BET} (m^2 g^{-1})	V_{total} (cm^3 g^{-1})	D_{peak} (nm)
Al_2O_3-1	$Al(NO_3)_3 \cdot 9H_2O$	K_2SO_4	$CO(NH_2)_2$	186	0.74	4/14
Al_2O_3-2	$Al(NO_3)_3 \cdot 9H_2O$	$(NH_4)_2SO_4$	$CO(NH_2)_2$	187	0.74	4/17
Al_2O_3-3	$KAl(SO_4)_2 \cdot 12H_2O$	$KAl(SO_4)_2 \cdot 12H_2O$	$CO(NH_2)_2$	209	0.66	3
Al_2O_3-4	$Al(NO_3)_3 \cdot 9H_2O$	–	$(NH_4)_2CO_3$	268	0.92	19

S_{BET} specific surface area calculated by the Brunauer–Emmett–Teller (BET) method, V_{total} total pore volumes at $P/P_0=0.997$, D_{peak} pore sizes at maxima of the pore size distributions (PSDs)

Fig. 1 **a** TG and **b** DSC curves of hydrothermal products (without calcination), **c** XRD patterns and **d** FT-IR curves of alumina supports (calcined at 500 °C for 2 h)

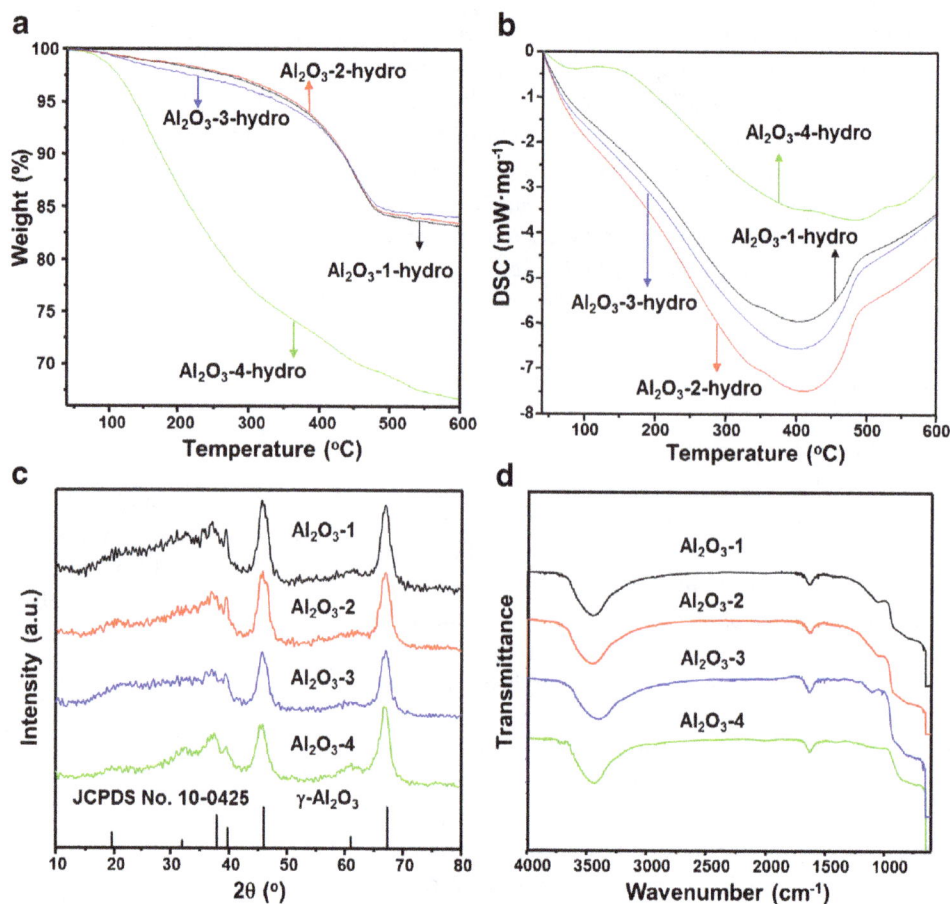

Results and discussion

Alumina hollow microspheres

To explore the thermal decomposition behavior, hydrothermal products were characterized by TG-DSC (Fig. 1a, b) under air, with a heating rate of 10 °C min^{-1}. In the case of samples Al_2O_3-x-hydro (x = 1, 2, and 3), the TG curves show a sharp weight loss before 470 °C with the actual loss is ~18 %, mainly corresponding the release of adsorbed water and the decomposition of AlO(OH) (2AlO(OH) → Al_2O_3 + H_2O, theory weight loss value is ~15 %) [12]. The DSC curve displays a broad exothermic peak at 420 °C, which indicates the phase conversion. We thus selected 500 °C as the calcination temperature. However, Al_2O_3-4-hydro sample shows the weight loss of ~34 % below 600 °C, which is corresponding to the intermediate product Al(OH)$_3$ (2Al(OH)$_3$ → Al_2O_3 + 3H_2O, theory weight loss value is ~34.6 %) [13].

Fig. 2 **a** N$_2$ sorption isotherms and **b** pore size distributions of Al_2O_3-1, Al_2O_3-2, Al_2O_3-3 and Al_2O_3-4 aluminas. The isotherms of Al_2O_3-4, Al_2O_3-1 and Al_2O_3-2 were offset up by 400, 300, and 100 cm^3 g^{-1}

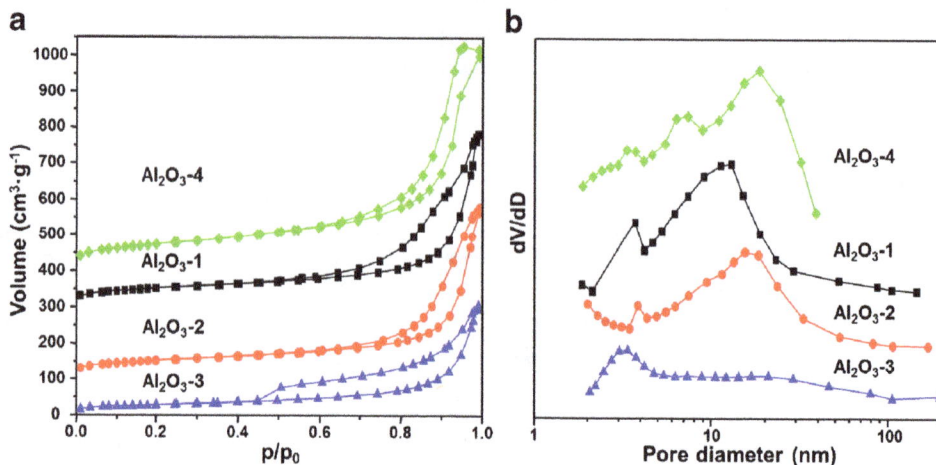

Fig. 3 SEM images of alumina supports (**a** Al_2O_3-1, **b** Al_2O_3-2, **c** Al_2O_3-3, **d** Al_2O_3-4)

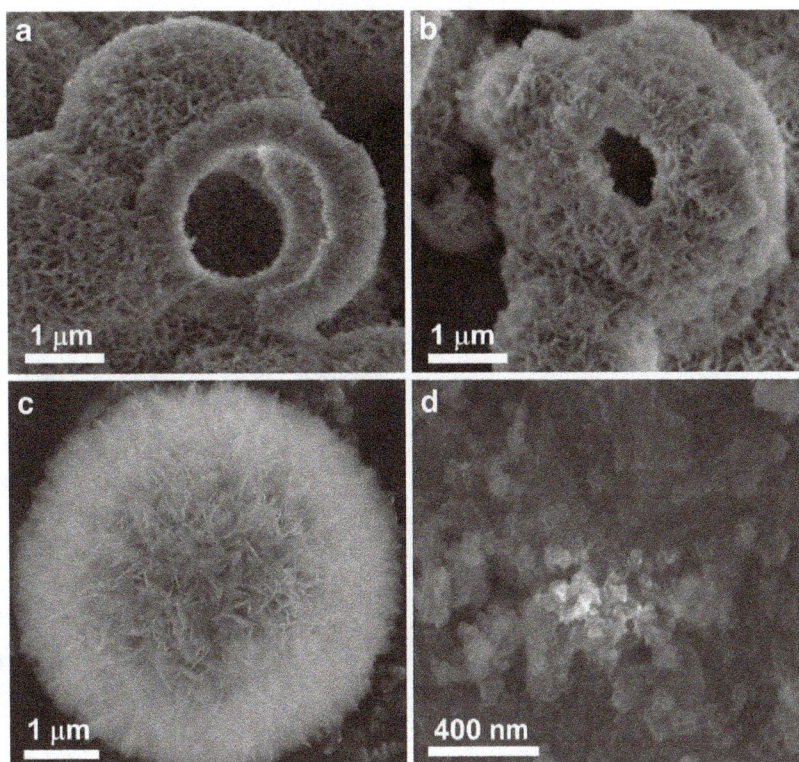

Based on the XRD analysis results in Fig. 1c, all the diffraction peaks of calcined samples Al_2O_3-x (x=1, 2, 3, and 4) can be assigned to γ-Al_2O_3 (JCPDS No. 10-0425). According to the Scherrer equation, the crystallite size of nanoflake composed hollow microsphere is ca. 13, 10, 12, and 10 nm for Al_2O_3-1, Al_2O_3-2, Al_2O_3-3, and Al_2O_3-4, respectively. More characteristics were detected in the FT-IR spectra (Fig. 1d). Basically, four samples show similar IR signals. The intensive bands at 3,447–3,449 and 1,637–1,638 cm^{-1} belong to the stretching and bending vibrations of the O–H bonds adjoining the Al atoms and a functional group of water [14–16]. The intensive peaks at 1,160 and 1,070 cm^{-1} are due to the δ_{as} Al-O-H and δ_s Al-O-H modes,

and the bands observed at 746 cm^{-1} represent the stretching modes of AlO_6 [17].

After annealing at 500 °C for 2 h, the alumina materials were further characterized by N_2 sorption measurements. As seen in Fig. 2, the nitrogen sorption isotherms of Al_2O_3-1, Al_2O_3-2, Al_2O_3-3, and Al_2O_3-4 are essentially of type IV [18], reflecting a mesoporous characteristic. Determined from desorption branches, the pore size distributions of Al_2O_3-1, Al_2O_3-2, and Al_2O_3-3 are centered at 4/17, 4/17, and 3 nm, whereas Al_2O_3-4 shows a rather broad distribution with mesopore size concentrated at 19 nm. Al_2O_3-4 exhibits a slightly higher BET surface area (268 m^2 g^{-1}) and total pore volume (0.92 m^3 g^{-1}) than those of Al_2O_3-3 (209 and 0.66 m^3 g^{-1},

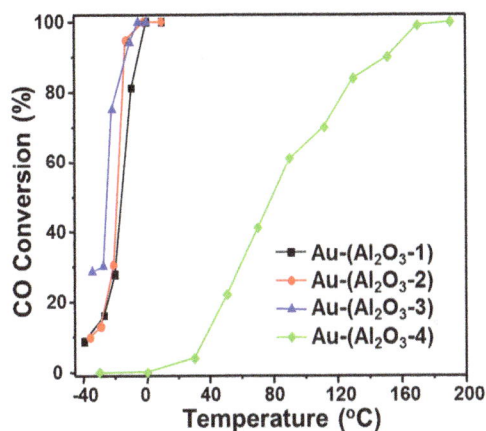

Fig. 4 CO conversion curves of gold catalysts Au-(Al_2O_3-1), Au-(Al_2O_3-2), Au-(Al_2O_3-3) and Au-(Al_2O_3-4)

Table 2 Catalytic activity of gold nanoparticles supported on different alumina

Gold catalyst	Au content (%)[a]	$T_{100\%}$ (°C)[b]	$T_{50\%}$ (°C)[b]	Rate (mol h^{-1} g$_{Au}$$^{-1}$)[c]
Au-(Al_2O_3-1)	2.8	0	−15	1.267
Au-(Al_2O_3-2)	2.6	−1	−18	1.352
Au-(Al_2O_3-3)	2.6	0	−25	1.413
Au-(Al_2O_3-4)	2.7	190	80	–

[a] The actual Au content was detected by ICP technique

[b] $T_{50\%}$ and $T_{100\%}$ represent the temperatures for 50 and 100 % CO conversion, respectively

[c] The corresponding reactive rate of Au catalysts at 0 °C

Fig. 5 TEM images of gold catalyst Au-$(Al_2O_3$-$3)$ and size distribution of gold nanoparticles

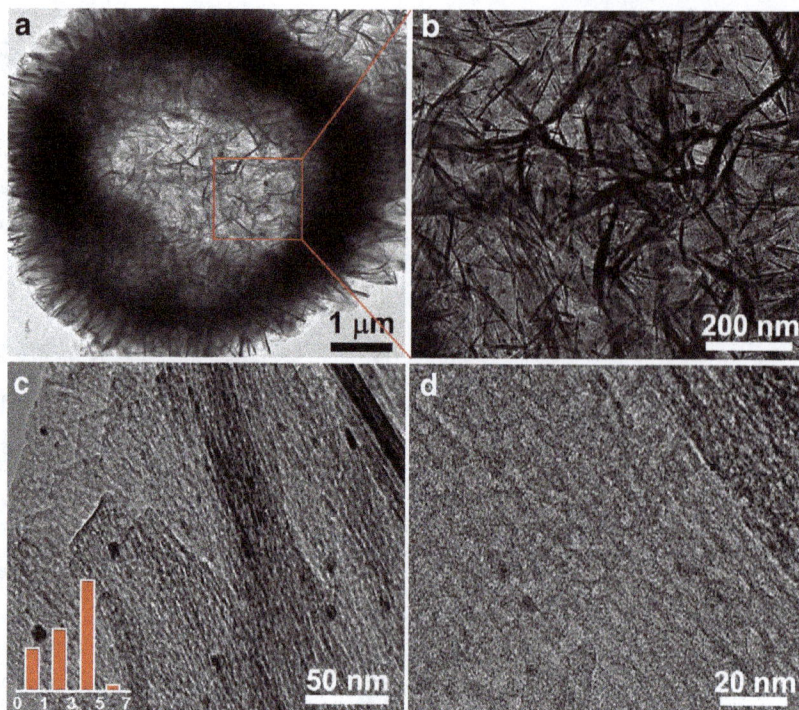

accordingly). The detail structure parameters of γ-Al_2O_3 materials are listed in Table 1.

In addition, the morphologies of alumina were characterized by SEM technique. It can be clearly seen in Fig. 3a, b that Al_2O_3-1 and Al_2O_3-2 samples display hollow microsphere structures (the diameter ~4 μm and shell thickness 500–600 nm) with open mouths assembled from closely packed nanoflakes. When $KAl(SO_4)_2$·$12H_2O$ was used as alumina precursor and sulfate additive (keeping the molar ratio of Al:SO_4^{2-}=1:2), then, the perfect γ-Al_2O_3 hollow microspheres with ca. 6 μm in diameter, 600–700 nm in shell thickness can be obtained (Fig. 3c). Whereas, Al_2O_3-4 has an undefined morphology (Fig. 3d), which is totally different from alumina hollow microspheres prepared with the addition of sulfates. As the promotion of SO_4^{2-}, Al^{3+} and urea can alternatively hydrolyze and polycondense, which leads to the precipitation of amorphous aluminum oxyhydroxide spheres [18–23]. It can be deduced that the increase in concentration of SO_4^{2-} facilitates formation of larger microspheres through secondary nucleation and growth of AlO(OH) crystal nuclei until the system reaches equilibrium with the surrounding solution. It confirms that the sulfate anion plays a significant role in the formation of hollow microspheres morphology.

Au/Al$_2$O$_3$ catalyst for CO oxidation

As a simple and typical probe reaction, CO oxidation was selected to identify the promotion of γ-Al_2O_3 with unique hollow microspheres morphology assembled by closely packed nanoflakes for the catalytic performance of Au

nanoparticle catalysts. Supported on obtained γ-Al_2O_3 hollow microspheres, the catalytic activity of 3 wt.% Au/Al_2O_3 catalysts were shown in Fig. 4 and Table 2.

Remarkably, the corresponding Au catalysts exhibited excellent catalytic performance with a complete CO conversion at ~0 °C. Particularly in the case of sample Au-$(Al_2O_3$-$3)$, the 50 % CO conversion can be achieved at −25 °C, and the specific rate at 0 °C was calculated as 1.413 mol h^{-1} g_{Au}^{-1}. For comparison, the regular Al_2O_3 support (Al_2O_3-4) with undefined morphology was synthesized by a common precipitation method using $Al(NO_3)_3$·$9H_2O$ and $(NH_3)_4CO_3$. The obtained Au-$(Al_2O_3$-$4)$ catalyst tested under the same conditions is generally inactive at 0 °C. The results in turn indicate

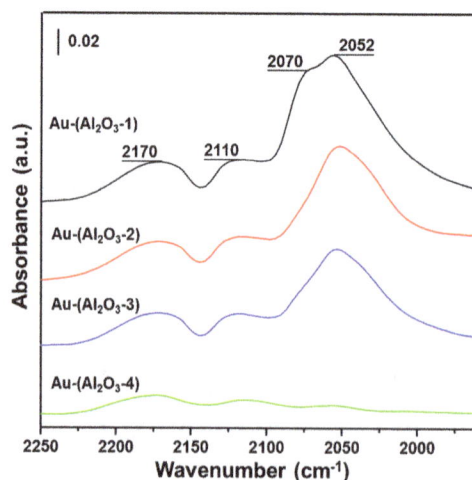

Fig. 6 In situ DRIFTS spectra of gold catalysts Au-$(Al_2O_3$-$1)$, Au-$(Al_2O_3$-$2)$, Au-$(Al_2O_3$-$3)$ and Au-$(Al_2O_3$-$4)$

that such γ-Al_2O_3 hollow microspheres can efficiently stabilize Au nanoparticles, which exhibit higher catalytic performance for CO oxidation. In addition, despite the different catalytic test conditions, such as the composition of feed gas, the loading content of gold particles, and the temperature of calcination, the rate values in this work are comparable and even higher to the results in literatures [10, 13, 24–27].

Presently, from SEM images (Fig. 3) we know Al_2O_3-1, Al_2O_3-2, and Al_2O_3-3 have quite similar morphologies while Al_2O_3-4, as controlled sample, is in another case. On the other hand, the activity of Au/Al_2O_3-1, Au/Al_2O_3-2, and Au/Al_2O_3-3 samples are also quite similar in CO oxidation, while Au/Al_2O_3-4 shows a very low activity (Fig. 4). The Au contents of all samples were maintained nearly same (Table 2). Therefore, one representative sample Au/Al_2O_3-3 was characterized by TEM. As shown in Fig. 5a, b, gold catalysts supported on the surface of γ-Al_2O_3 hollow microspheres consisted of closely packed nanoflakes. Besides, Au nanoparticles are highly dispersed with an average size of 3.0 ± 0.5 nm estimated from the TEM images in Fig. 5c. In addition, Fig. 5d shows a rough surface of γ-Al_2O_3 hollow microspheres, which are beneficial to stabilize gold nanoparticles efficiently. From our previous work [13], the controlled sample Au/Al_2O_3-4 has a gold particle size of 2.5 ± 0.5 nm based on the TEM observation, which is more or less similar to that of sample Au/Al_2O_3-3. All these information demonstrates the advantages of the rough surface of γ-Al_2O_3 hollow microspheres. Furthermore, concerning the influence of the possible remained heteroatoms (for example, sulfur and potassium) on the activity [28–30], sulfur contents and potassium contents of samples Al_2O_3-1, Al_2O_3-2 and Al_2O_3-3 were analyzed by ICP-AES technique, correspondingly. All three samples show ignorable sulfur and potassium content. Therefore, one can basically eliminate the influence of sulfate and K^+ on the catalytic performance. It further suggests that the shape and surface property of γ-Al_2O_3 support have strong influence on the activity of the Au nanoparticles.

To further explore the surface chemical property of Au nanoparticles on γ-Al_2O_3 hollow microsphere supports, in situ DRIFTS was applied to measure the adsorbed CO. As shown in Fig. 6, the strong absorption bands of CO positioned around 2,110 and 2,070 cm^{-1} appeared on the Au/Al_2O_3 samples dispersed on γ-Al_2O_3 hollow microspheres which belong to two linear CO species adsorbed on the metallic Au sites [31]. The peak at 2,170 cm^{-1} was attributed to CO adsorbed on cationic gold (Au(I) or Au(III)) [32]. The DRIFTS measurements suggest that the presence of $Au^{\delta+}$ and Au^0 on the surface of Au/Al_2O_3 catalysts, although weaker absorption bands were observed for Au nanoparticles supported on alumina Al_2O_3-4. Meanwhile, the absorption band appeared at 2,052 cm^{-1}, which can be assigned to CO adsorbed on a negatively charged gold surface [33]. Apparently, Al_2O_3-4 synthesized by the common precipitation method displays an inferior surface for CO adsorption and thus results in a catalyst with a low catalytic activity.

Conclusions

It has been demonstrated that AlO(OH) hollow microspheres consist of closely packed nanoflakes were synthesized through a hydrothermal process by using $Al(NO_3)_3\cdot9H_2O$ as a precursor, urea as precipitant agent and sulfate K_2SO_4, $(NH_4)_2SO_4$, and $KAl(SO_4)_2\cdot12H_2O$ as additive. When adjusting the molar ratio of Al^{3+} and SO_4^{2-} to 1:2, γ-Al_2O_3 hollow microspheres with ca. 4–6 μm in diameter, 500–700 nm in shell thickness can be obtained after a calcination step. As such, γ-Al_2O_3 supports with special morphology and rough surface can stabilize gold nanoparticles efficiently; an excellent catalytic performance for CO oxidation can be achieved with a complete CO conversion at 0 °C and 50 % conversion at −25 °C. The DRIFTS confirms the presence of $Au^{\delta+}$ and Au^0 on the surface of Au/Al_2O_3 catalysts. In contrast, Au catalyst on alumina support with undefined morphology shows low activity at low temperature under the same catalytic test conditions.

Acknowledgments The project was supported by the National Natural Science Foundation of China (No.20973031) and the Ph.D. Programs Foundation (20100041110017) of Ministry of Education of China.

References

1. Haruta M, Kobayashi T, Sano H, Yamada N (1987) Novel gold catalysts for the oxidation of carbon monoxide at a temperature far below 0 °C. Chem Lett 16:405–408
2. Li WC, Comotti M, Schüth F (2006) Highly reproducible syntheses of active Au/TiO2 catalysts for CO oxidation by deposition-precipitation or impregnation. J Catal 237:190–196
3. Herzing AA, Kiely CJ, Carley AF, Landon P, Hutchings GJ (2008) Identification of active gold nanoclusters on iron oxide supports for CO oxidation. Science 321:1331–1335
4. Wang J, Lu AH, Li MR, Zhang WP, Chen YS, Tian DX, Li WC (2013) Thin porous alumina sheets as supports for stabilizing gold nanoparticles. ACS Nano 7:4902–4910
5. Qi CX, Su HJ, Guan RG, Xu XF (2012) An investigation into phosphate-doped Au/alumina for low temperature CO oxidation. J Phys Chem C 116:17492–17500
6. Costello CK, Kung MC, Oh HS, Wang Y, Kung HH (2002) Nature of the active site for CO oxidation on highly active Au/γ-Al2O3. Appl Catal A Gen 232:159–168
7. Yang J, Wang FF, Ma XM, Tang QH, Wang K, Guo YM, Yang L (2013) Facile one-step synthesis of porous ceria hollow nanospheres for low temperature CO oxidation. Micropor Mesopor Mat 176:1–7

8. Frey K, Iablokov V, Sáfrán G, Osán J, Sajó I, Szukiewicz R, Chenakin S, Kruse N (2012) Nanostructured MnO$_x$ as highly active catalyst for CO oxidation. J Catal 287:30–36

9. Comotti M, Li WC, Spliethoff B, Schüth F (2006) Support effect in high activity gold catalysts for CO oxidation. J Am Chem Soc 128:917–924

10. Yuan Q, Duan HH, Li LL, Li ZX, Duan WT, Zhang LS, Song WG, Yan CH (2010) Homogeneously dispersed ceria nanocatalyst stabilized with ordered mesoporous alumina. Adv Mater 22:1475–1478

11. Han YF, Zhong Z, Ramesh K, Chen F, Chen L (2007) Effects of different types of γ-Al$_2$O$_3$ on the activity of gold nanoparticles for CO oxidation at low-temperatures. J Phys Chem C 111:3163–3170

12. Cai WQ, Yu JG, Jaroniec M (2010) Template-free synthesis of hierarchical spindle-like γ-Al$_2$O$_3$ materials and their adsorption affinity towards organic and inorganic pollutants in water. J Mater Chem 20:4587–4594

13. An AF, Lu AH, Sun Q, Wang J, Li WC (2011) Gold nanoparticles stabilized by a flake-like Al$_2$O$_3$ support. Gold Bull 44:217–222

14. Ahmad AL, Mustafa NNN (2007) Sol-gel synthesized of nanocomposite palladium–alumina ceramic membrane for H$_2$ permeability: preparation and characterization. Int J Hydrogen Energ 32:2010–2021

15. Liu F, Asakura K, He H, Shan W, Shi X, Zhang C (2011) Influence of sulfation on iron titanate catalyst for the selective catalytic reduction of NO$_x$ with NH$_3$. Appl Catal B Environ 103:369–377

16. Romero-Pascual E, Larrea A, Monzón A, González RD (2002) Thermal stability of Pt/Al$_2$O$_3$ catalysts prepared by sol-gel. J Solid State Chem 168:343–353

17. Feng YL, Lu WC, Zhang LM, Bao XH, Yue BH, Lv Y, Shang XF (2008) One-step synthesis of hierarchical cantaloupe-like AlOOH superstructures via a hydrothermal route. Cryst Growth Des 8:1426–1429

18. Yu JG, Liu W, Yu HG (2008) A one-pot approach to hierarchically nanoporous titania hollow microspheres with high photocatalytic activity. Cryst Growth Des 8:930–934

19. Yu HG, Yu JG, Liu SW, Mann S (2007) Template-free hydrothermal synthesis of CuO/Cu$_2$O composite hollow microspheres. Chem Mater 19:4327–4334

20. Matijevic E (1981) Monodispersed metal (hydrous) oxides—a fascinating field of colloid science. Acc Chem Res 14:22–29

21. Castellano M, Matijevic E (1989) Uniform colloidal zinc compounds of various morphologies. Chem Mater 1:78–82

22. Ramanathan S, Roy SK, Bhat R, Upadhyaya DD, Biswas AR (1997) Alumina powders from aluminium nitrate-urea and aluminium sulphate-urea reactions—the role of the precursor anion and process conditions on characteristics. Ceram Int 23:45–53

23. Zhou JB, Wang L, Zhang Z, Yu JG (2013) Facile synthesis of alumina hollow microspheres via trisodium citrate-mediated hydrothermal process and their adsorption performances for p-nitrophenol from aqueous solutions. J Colloid Interface Sci 394:509–514

24. Wang GH, Li WC, Jia KM, Spliethoff B, Schüth F, Lu AH (2009) Shape and size controlled α-Fe$_2$O$_3$ nanoparticles as supports for gold-catalysts: synthesis and influence of support shape and size on catalytic performance. Appl Catal A Gen 364:42–47

25. Kung HH, Kung MC, Costello CK (2003) Supported Au catalysts for low temperature CO oxidation. J Catal 216:425–432

26. Okumura M, Nakamura S, Tsubota S, Nakamura T, Azuma M, Haruta M (1998) Chemical vapor deposition of gold on Al$_2$O$_3$, SiO$_2$, and TiO$_2$ for the oxidation of CO and of H$_2$. Catal Lett 51:53–58

27. Calla JT, Bore MT, Datye AK, Davis RJ (2006) Effect of alumina and titania on the oxidation of CO over Au nanoparticles evaluated by ^{13}C isotopic transient analysis. J Catal 238:458–467

28. Moma JA, Scurrell MS, Jordaan WA (2007) Effects of incorporation of ions into Au/TiO$_2$ catalysts for carbon monoxide oxidation. Top Catal 44:167–172

29. Solsona B, Conte M, Cong Y, Carley A, Hutchings G (2005) Unexpected promotion of Au/TiO$_2$ by nitrate for CO oxidation. Chem Commun 2351–2353

30. Mohapatra P, Moma J, Parida KM, Jordaan WA, Scurrell MS (2007) Dramatic promotion of gold/titania for CO oxidation by sulfate ions. Chem Commun 1044–1046

31. Somodi F, Borbáth I, Hegedűs M, Lázár K, Sajó IE, Geszti O, Rojas S, Fierro JLG, Margitfalvi JL (2009) Promoting effect of tin oxides on alumina-supported gold catalysts used in CO oxidation. Appl Surf Sci 256:726–736

32. Venkov T, Klimev H, Centeno MA, Odriozola JA, Hadjiivanov K (2006) State of gold on an Au/Al$_2$O$_3$ catalyst subjected to different pre-treatments: an FTIR study. Catal Commun 7:308–313

33. Liu X, Liu MH, Luo YC, Mou CY, Lin SD, Cheng H, Chen JM, Lee JF, Lin TS (2012) Strong metal-Support interactions between gold nanoparticles and ZnO nanorods in CO oxidation. J Am Chem Soc 134:10251–10258

Demonstrative experiments about gold nanoparticles and nanofilms: an introduction to nanoscience

Olivier Pluchery · Hynd Remita · Delphine Schaming

Abstract An important task of the scientific community is to provide non-specialized audience with explanations about what is nanoscience. Such explanations can be given during public conferences, seminars in high schools or lab work organized with teachers. And very often, the use of an experimental illustration greatly helps to raise the interest and the curiosity of the public. The present article will describe how the authors have used five simple and visual experiments in chemistry and physics to progressively introduce different audiences into the fascination of nanoscience. One experiment is the synthesis of gold nanoparticles with the Turkevich method and shows the progressive appearance of the ruby-red colour of the nanometric gold particles. The second and third experiments describe the way for modulating their colour and how to include them into a polymer and form a ruby-red coloured plastic film. The fourth experiment shows that starting from these nanoparticles, it is possible to turn them back into a yellow golden film. The last experiment is based on the optical properties of ultra-thin gold films. Using the plasmon resonance, it is possible to demonstrate that gold change colours from yellow to orange and green when a white light beam is shone on the gold interface. These visual experiments cannot be fully interpreted in front of a large audience but serve for rising curiosity.

O. Pluchery
Institut des NanoSciences de Paris, UMR 7588 CNRS, Université Pierre et Marie Curie-UPMC, 4 place Jussieu, 75252 Paris Cedex 05, France

H. Remita
Laboratoire de Chimie Physique, UMR 8000 CNRS, Université Paris-Sud, 91405 Orsay Cedex, France

D. Schaming (✉)
ITODYS, UMR 7086 CNRS, Sorbonne Paris Cité, Université Paris Diderot, 15 rue Jean-Antoine de Baïf, 75205 Paris Cedex 13, France
e-mail: delphine.schaming@univ-paris-diderot.fr

Keywords Nanoscience · Gold nanoparticles · Plasmon resonance · Colloids · Gold nanofilms · Teaching · Education

Introduction

Nanoscience awareness

Many scientists who work in the field of nanoscience have experienced that there is a gap between their daily practice and the comprehension that the common citizen can have of their research. Scientists have been trained to drive their reflections at the nanoscale, and the cutting-edge microscopes they routinely use have made them familiar with the visualization of nano-objects. Transmission electron microscopes were invented as early as 1931, and they are now commonly used to visualize the atomic structure of nanoparticles. More recently, since 1984, the scanning tunnel microscope (STM) makes possible to track individual atoms on a conductive surface and the atomic force microscope (AFM) is able to detect self-organized molecules on even non-conductive substrates. These are just examples of the way a scientist has a kind of privileged access to the "nanoworld". All these images are truly fascinating, and researchers are probably the first to be mesmerized by the beauty of the world they discover through their various microscopes. Many of these images have been used to draw the attention of the public, of the politics towards our research. Our society is puzzled to discover the amazing nanoworld that is surrounding us, or even within us. As a consequence, the nanoworld has made its way into the public subconscient.

However, once these images have made a first impression in the people's mind, they need to be explained and commented. Uncontrolled release of images or scientific information can lead to the birth of fears and mistrust towards our scientific work [1]. It is the responsibility of us, scientists, to make sure that the images we release convey a true message and limit the possible distortion. The situation is all the more important that the present status of nanoscience is still being questioned, by the population in most of the European countries at least. Nanoscience is still benefiting a positive image, but this is not sure it will be always so.

Our responsibility for explaining nanoscience

A recent poll issued by a well-known polling agency in France shows that in 2006, only 44 % of the French population has heard about nanoscience or nanotechnology [2]. In 2010, they were 59 % which is a sharp increase within only 4 years. But when one looks closer at the numbers, it turns out that among this proportion of "informed people", 22 % declare that they know precisely what nanoscience is, whereas 37 % have only a very rough idea. These 37 % should be viewed as "good-willing" citizens who are "ill-informed people" but eager to learn more about science. As scientists, we have a moral obligation to provide them with key concepts and explanation, so that they will turn into the group of "nanoscience-educated people". An efficient way to do so is probably that some scientists from our communities address the public through general conferences, vulgarization action in high schools, or scientific movies. Probably the most receptive public could be found in general-audience conferences and in high schools. The present article offers some support for colleagues who are ready to dedicate part of their time to this task by giving conferences and lectures. Since the impact of a scientific conference is largely magnified with the help of one or two live experiments, this paper will describe five experiments that can be presented in almost any conference hall without special equipment.

How to use the content of this article?

The aim of the present article is not to draw a typical conference guideline but rather to provide lecturers with experiments that illustrate selected features of nanoscience. The content of this paper will be most useful for teachers who want to introduce nanoscience experiments in their courses or lab teaching. It will also be useful for scientists who are invited to give general-audience conferences. Most of these experiments can be carried out in almost any conference hall, although it is useful to use a webcam and a video projector to magnify the observations. Of course, the most important in a conference is to set up a scenario where the experimental part will be just one aspect of the demonstration. Experimental evidences bring strong arguments in a demonstration. Usually, it is sufficient to present one or two experiments during a talk to definitively capture the public's attention. To maximize the impact of a conference, it is useful to focus on a few concepts of nanoscience and demonstrate their relevance. This is why,

in this article, each experiment is introduced with a few sentences that can give hints towards key concept conveyed by the experiment itself. Finally, these experiments have been tested several times in real conferences [3], and the details needed to reproduce them are given in the "Experimental" part. No toxic products are used; therefore, no particular security cautions have to be taken. The cost of the experiments is low, especially for the first four ones.

Experimental

All glassware and magnetic stirrings used for gold nanoparticle (Au NP) syntheses must be cleaned with freshly prepared aqua regia solution (three parts of conc. HCl and one part of conc. HNO_3). The aqua regia can be stored in a glass bottle and reused as long as the orange colour remains. The washing with aqua regia must be handled with care, wearing gloves as well as safety glasses and protective clothing. After that, items must be washed in tap water and then rinsed with distilled water (18.2 $M\Omega$ cm resistivity) and dried in air or in an oven.

The same ultrapure distilled water must be used for the preparation of the aqueous solutions. All chemical reagents are purchased from Sigma Aldrich. $KAuCl_4$ can be replaced by $HAuCl_4$. Heptane and ethanol are analytical or reagent grade and are used without purification.

Experiment 1. Synthesis of gold nanoparticles

An aqueous solution of $KAuCl_4$ is heated to the boiling point in an Erlenmeyer, and then, a sodium citrate aqueous solution is added with vigorous magnetic stirring. This solution is heated during about 10 min, and the obtained red solution is finally left cooling. In Table 1, the experimental conditions to obtain Au NPs with two well-defined sizes are given.

Experiment 2. Change of colour by aggregation with salt

At least 0.5 g of NaCl in 10 mL of distilled water is dissolved, or a saturated solution (>360 g/L, i.e. 6.1 mol/L) is used. A small amount (about 3–4 mL) of the Au NPs solution is put in two test tubes. One tube is used as a colour reference. To the other tube, 5–10 drops of the saturated NaCl solution are added. The colour of the solution changes in the second tube. This experiment could be done in a cuvet with the UV–visible spectrum recorded after each addition.

Experiment 3. Polymeric film containing gold nanoparticles

A concentrated solution of poly(vinyl alcohol) (PVA) (1 M in monomer) is prepared by dissolving 440 mg in 10 mL of distilled water. The beaker containing PVA in water is put in a water bath, and the solution is heated to boiling under

Table 1 Experimental conditions to obtain Au NPs with 15 and 30 nm of diameter

NP diameter (nm)	Gold salt solution	Sodium citrate solution	Ratio of citrate to gold
15	20 mL at 0.25 mM (1.9 mg)	1 mL at 34 mM (1 wt%)	6.8
30	20 mL at 1 mM (7.6 mg)	0.8 mL at 34 mM (1 wt%)	1.4

stirring. The water bath prevents the PVA from sticking on the bottom of the beaker. A homogeneous solution of PVA is obtained after a few tens of minutes. Then, 2 mL of the colloidal Au NPs solution is added to 3 mL of the concentrated PVA solution. After stirring, the mixture is put in a crystallizing dish (of about 3–4 cm in diameter) and heated at 70–80 °C in order to evaporate water. Care should be taken to avoid overheating, since this can lead to non-homogeneous distribution of the Au NPs in the film. A thin film of PVA containing Au NPs is obtained after a few hours of drying.

Au NPs can also be induced directly in PVA solution containing KAuCl$_4$ [4]: PVA acts as reducing and stabilizing agent [5]. The Au NPs are prepared as follows: in a typical experiment, 4 mL of 2.4 mM KAuCl$_4$ aqueous solution is added into 80 mL of 0.24 M (repeating unit) PVA aqueous solution. Then, the mixture is heated at 80 °C for several hours, yielding a purple solution. The film is then prepared in a crystallizing dish as described before.

Experiment 4. Formation of a metallic gold film from colloidal gold nanoparticles

Forty millilitres of an aqueous Au NP solution is centrifugated (around 3,000 rpm, until the supernatant becomes pale pink). The supernatant is then removed, and the concentrated aliquot of Au NPs is redissolved in 0.8 mL of ethanol. A heptane/water biphasic system is prepared in small glassware with ca. 3–4 cm^2 of section. In agreement with the densities of these two solvents, the water is located below and the heptane above. The concentrated ethanol solution of Au NPs is then added by small injections in the organic phase near the organic/aqueous interface until a golden film appears at the interface.

Experiment 5. Surface plasmon resonance and the colour of gold

The surface plasmon wave is an electromagnetic wave that arises at the interface between a metal and an insulating medium (dielectric) and whose structure is that of an evanescent wave. It can be seen as a wave confined very close to the metallic surface [6]. In the case of the gold/air interface and at a wavelength of 633 nm, this wave is confined within a 30-nm-thick layer close to the interface and crawls along the gold surface over a distance of 10 μm. Its excitation is possible only with geometries capable of generating large wave vectors parallel to the interface. The most common geometry is the Kretschmann geometry shown in Fig. 1 where the impinging wave is generated by total internal reflection on a glass prism as depicted in Fig. 1b. The gold film should be deposited on this face of the prism. The experiment consists in measuring the reflected intensity as a function of incidence angle when the light is p-polarized. If one measures this intensity when a helium–neon laser in used (λ=633 nm), the reflected intensity exhibits a sharp minimum for an angle of ca. 30° as plotted in the graph of Fig. 1c [7]. From this plot, it is easy to understand that if the angle is set at 30°, and if a white light beam from a halogen lamp is used instead of a monochromatic laser beam, the red component of light will not be reflected. Therefore, the observed resulting beam will appear green. And by adjusting the angle, it is possible to turn off various colours, and the colour of the observed beam will also vary. This is a typical effect arising from the confinement of an optical wave into a nanometer-thin gold film. A setup that exhibits this effect is described in details in a publication [7] and is available from a small company as a lab work for master students [8].

Experiment 1. Synthesis of gold nanoparticles

While gold is known since a long time for its golden colour so appealing in jewellery, gold is also used since antiquity as red dye. For instance, Au NPs have been found in several red glasses or on porcelains decorated with red or pink enamels dating from ancient times, for instance, in Roman glassworks (such as the famous Lycurgus Cup dating from the fourth century CE) or in the Chinese "Famille Rose" porcelain dating from the eighteenth century [9–11]. Indeed, bulk gold reflects the yellow colour and appears golden-coloured, whereas at the nanoscale gold appears generally red. Michael Faraday, fascinated by this ruby colour of colloidal gold solutions, was the first scientist who has discovered that the optical properties of gold colloids differ from those of the corresponding bulk metal. He reported on the first synthesis of Au NPs in solution in 1857 [12]. These colloids were obtained by reduction of gold salts by white phosphorous [13]. Later, the specific colours of metal colloids were explained by Mie [14].

This phenomenon is due to the well-known plasmon resonance also called local surface plasmon resonance (LSPR), which can be explained by a confinement of the electromagnetic wave associated with the light inside the NPs. Indeed, when its wavelength is greater than the size of the NPs, the

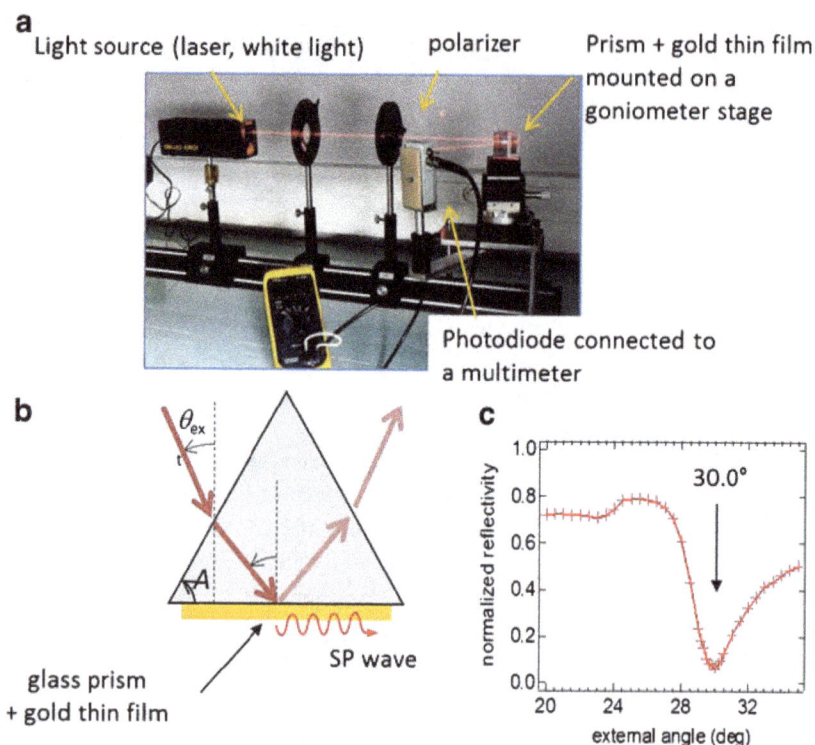

Fig. 1 **a** Image of a setup used for generating surface plasmon waves. **b** The central part of the setup is a glass prism where a 50-nm-thick gold film was deposited. When a *p*-polarized laser beam is shone on this gold film, the reflected intensity exhibits a sharp minimum for one given angle of incidence. **c** The corresponding plot of the reflected intensity as a function of the incidence angle is shown. This minimum indicates that the energy is transferred to the surface plasmon wave

whole NP feels a uniform and oscillating electric field, and consequently electrons oscillate in phase. Nevertheless, this collective oscillation of the electrons is constrained by the reduced dimensions of the NP in which they are confined, leading to a significant absorption of the wavelengths around green. Then, NPs appear with the complementary colour, which is red.

This experiment introduces the audience into the idea that at the nanoscale, basic properties, such as colour, turn out to be very different from what happens in the "macro-world". Therefore, nanoscience focuses on properties linked not only to a material but to its specific size. And sometimes, the changes of properties at the nanoscale have consequences visible at our human scale. In the present case, the assembly of gold atoms into NPs of ca. 1,000 atoms changes the colour of gold. This message is always puzzling for non-scientific audience.

Nowadays, among the different ways of synthesis of spherical Au NPs, Turkevich's method is the most known and the easiest to perform [15]. This method is based on the reduction of an Au(III) salt by sodium citrate, as described in the experimental part (experiment 1). The reaction starts by boiling a gold salt aqueous solution (pale yellow). A few minutes after the addition of sodium citrate, the mixture becomes firstly uncoloured, then grey, violet, and finally burgundy red (Fig. 2a). In this reaction, the sodium citrate is used not only as reducing agent, but also as stabilizing agent. Indeed, its stabilizing role restrains the growth of the NPs and controls the NP diameter. Thus, the size of the NPs can be modulated by changing the concentration of the gold salt solution and the ratio of the quantity of gold salt to the quantity of sodium

citrate [16, 17]. The sizes of the NPs described in this experiment are 15 and 30 nm and has been verified by transmission electronic microscopy (TEM) (Fig. 2b)

Diffusion (or scattering) of light by the Au NPs can be evidenced with a laser beam pointed to the solution (Fig. 2c) and can be compared with a similar experiment using an aqueous solution containing an organic red dye (rhodamine or Congo red for instance) for which diffusion is not detectable (Fig. 2d). When light is scattered by single molecules such as dye molecule, the efficiency of the phenomenon is very low (usually quantified with the scattering cross section) and is barely visible by eyes (Fig. 2d). However, this cross section grows as the volume of the object, and with a particle of 30 nm whose diameter is 100 times larger than a molecule (volume 10^6 larger), the scattering process becomes easily visible as demonstrated in Fig. 2c. Moreover, Fig. 2c shows that if the laser wavelength is tuned to the plasmon resonance, the scattering is even stronger. The green laser beam is quickly damped when going through the NP suspension, whereas the red laser light is scattered less efficiently, but over a longer path.

Many other chemical methods of synthesis exist [18] and can lead to different shapes of NPs: nanorods [19, 20], nanopyramids [21], nanocubes, etc. Physical chemical syntheses such as photolytic and radiolytic methods can also be used, and generally lead to a better control of the size of the NPs [18, 22, 23]. Chemical and optical properties of Au NPs depend on their size, their shape, their aggregation state and their local environment. Such Au NPs have applications for example in catalysis (oxidation of CO), medicine

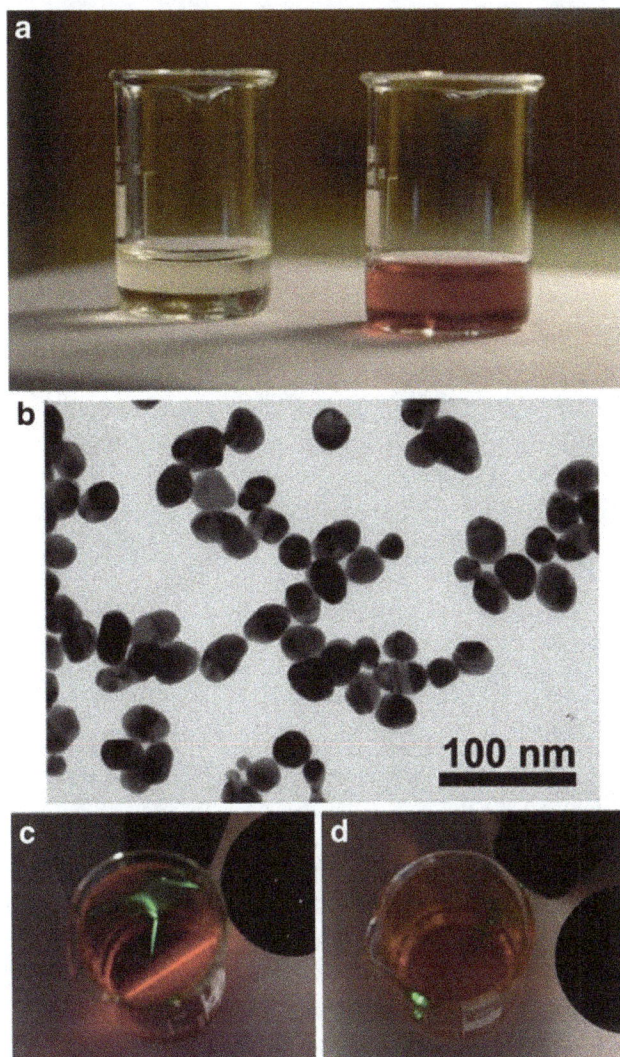

Fig. 2 **a** Pale yellow Au(III) salt (1 mM) (*left*) and red Au NP (*right*) solutions. **b** TEM image of Au NPs with 30 nm of diameter used in solution B. **c** Green and red laser beams diffused in an Au NP solution. **d** Comparison with a similar experiment with a red dye solution

(photothermal therapy and radiotherapy), plasmonics and electronics [9, 24, 25].

Experiment 2. Colour change of nanoparticles by aggregation with a salt

Upon addition of NaCl, the initially red colour of the Au NP solution turn to blue grey (experiment 2).

This experiment illustrates the extreme sensitivity of the coloration property with phenomena occurring at the nanoscale: if NPs are driven close to each other, the plasmon resonance is altered, and the colour changes (Fig. 3a, b).

The surface of Au NPs is charged. This induces repulsive forces between the NPs preventing them from aggregation: the energy barrier is too strong for interaction to occur between particles. However, upon addition of NaCl, this energy

barrier is reduced allowing the Au NPs to interact and aggregate. Dissolved salts, or electrolytes in general, are able to screen the repulsive electrostatic forces caused by the citrate layer: indeed, the positive charges of the electrolyte associate with the negative charges on the surfaces of the NPs. Cryo-TEM experiments, which consist in freezing the NPs after deposition onto an analysis grid by dipping it in a cooling media (liquid nitrogen), allowing to clearly show initially well-dispersed Au NPs in solution (before addition of NaCl) and their aggregation after addition of NaCl, but the nanoparticles keep their shape (Fig. 3c, d). This aggregation causes surface plasmon coupling which induces a shift in the surface plasmon resonance to a higher wavelength, resulting in a change in colour of the solution from red to blue in our case (Fig. 3a). These experiments can be done in an optical cell, and with a spectrophotometer, and students can monitor the NP aggregation by the UV–visible absorption of the solution (Fig. 3b). The students can then associate the absorbed wavelength with the observed colour of the solution.

These experiments show that individual small Au NPs appear red; however, when the particles aggregate together, the plasmon resonances can couple, and the colour changes to blue.

This property is used to develop biosensors. Au NPs can be easily functionalized. Thus, binding of Au NPs to biomolecules offers a promising approach for facile tracking of desired targets in aqueous samples. Antibody/antigen pairs can selectively link to the NPs. By modifying the surfaces of the NPs to incorporate these biomolecules, binding events can be detected by a change in solution colour. The most familiar example of NPs in sensing is the home pregnancy test based on detection of the hormone pregnancy (β-HCG hormone) [26]. Tests for DNA detection take also advantage of surface plasmon resonance changes [27]. For example, one kind of DNA test looks for certain bases. In this test, NPs are present as large aggregates displaying a blue colour. If the complementary DNA base is present, the NPs will preferentially bind to that base instead of each other, and the aggregates will dissolve inducing a deep red colour. We can detect the colour changes associated with DNA binding in about 2 min, making for a rapid assay. Au NPs functionalized with antibodies can be also used for rapid detection of virus (such as A/H5N1) [28]. Another example is the use of Au NPs for sensitive diagnostic tests and novel treatments in the detection of Alzheimer's disease by finding a protein in spinal fluid [29].

Experiment 3. Polymeric film containing gold nanoparticles

A thin film of Au NPs dispersed in PVA can be obtained by simply drying the gold colloid in the presence of dissolved PVA (at high concentration) (experiment 3). Au NPs will be randomly distributed within the PVA film (TEM images not

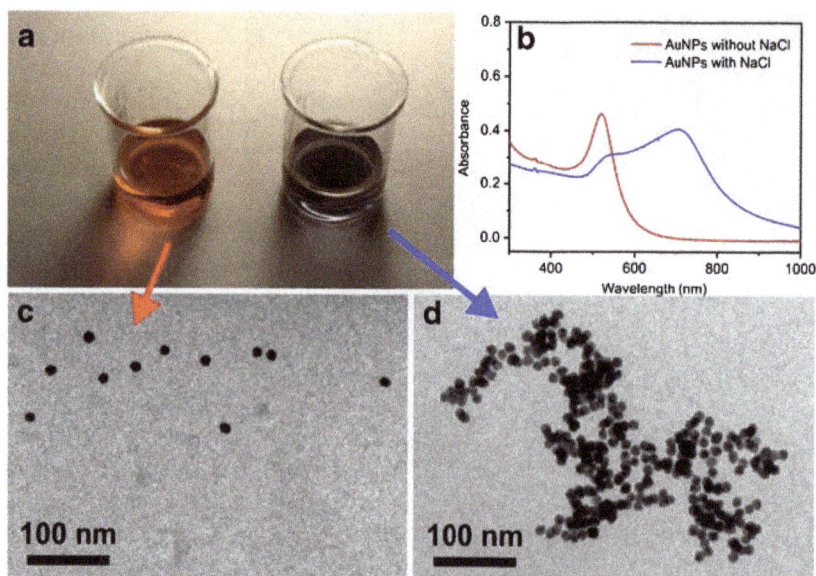

Fig. 3 **a** Au NP solution without NaCl (*left*) and with NaCl (*right*). **b** UV–visible absorption spectra of aqueous solutions of Au NPs with 15 nm of diameter, without and with NaCl (optical path= 2 mm). Cryo-TEM images of **c** non-aggregated Au NPs (without NaCl) and of **d** aggregated Au NPs (with NaCl)

shown here). The film appears purple, because the index of refraction of the medium surrounding the Au NPs is changed: from $n = 1.33$ (in aqueous solution) to $n \sim 1.7$ (in the PVA film) (Fig. 4).

This experiment demonstrates that Au NPs keep their unusual optical properties, even outside of the preparation solution. It makes possible to use their nanoscale properties for various applications in the daily life. Indeed, composite materials made of NPs and polymeric matrices are suitable candidates for biomedical applications, magnetic storage, optoelectronic and electronic devices.

Experiment 4. Formation of a metallic gold film from colloidal gold nanoparticles

While Au NPs are generally red, can we obtain again a golden appearance if we build bulk gold by aggregating and coalescing a lot of NPs? The experiment proposed in this part consists in the formation of a golden metallic film at the interface

between two immiscible liquids (Fig. 5), using a colloidal Au NP solution (experiment 4).

This experiment consists in using a water/heptane biphasic system and adding an ethanol solution of colloidal Au NPs at the interface between the two immiscible liquids [30]. Indeed, when Au NPs are redissolved in ethanol, the surface charge density of the NPs decreases, most likely due to the competitive adsorption of alcohol with citrate on the NPs. The solubility of the Au NPs in the aqueous phase is decreased when they diffuse towards the water/heptane interface. Moreover, when a water-miscible solvent with a lower dielectric constant, such as ethanol, is added at the interface, the interfacial energy between the two phases decreases even more, leading to the formation of the mirror-like gold film at the interface [31].

Fig. 4 PVA film containing Au NPs of 30 nm of diameter

Fig. 5 Images of metallic gold films formed at the interface between the two non-miscible liquids

Experiment 5. Surface plasmon resonance and the colour of gold

The usual colour of bulk gold is yellow, which itself is a non-usual property for metals since they are all grey except gold and copper. This yellow colour is due to the so-called interband transitions of gold that occur in the blue region of the visible spectrum [32]. As a consequence, the reflected light is deprived of its blue component and appears yellow. However, when the thickness of a gold layer decreases down to the nanometer scale, light behaves differently. This phenomenon can be shown in a demonstrative way in the case of the surface plasmon resonance using the setup described in Figs. 1a and 5a. The gold thin film which is attached on the prism should be illuminated with a beam light from a white light lamp. The excitation beam enters the prism through one face, and the reflected beams exits through another face and is observed on a screen (Fig. 6a, b). The experiment consists in slowly rotating the prism: the beam reflected from the thin film goes through a range of different colours that can be easily viewed on a white screen and that are absolutely unusual for gold. This change of colour of a reflected beam is the visible manifestation of the presence of the surface plasmon wave [6]. The energy at one wavelength of the impinging wave is transferred to the surface plasmon wave, and the reflected beam is strongly attenuated. As a consequence, this wavelength is missing in the reflected beam, and the observed colour changes (Fig. 6c). For example, if the angle of the rotation stage is set at 30° in our setup, the surface plasmon wave is generated for the red wavelength, and the reflected beam will be green (complementary colour to red). It helps demonstrate that light behaves very differently when interacting with nano-objects. This experiment can also lead the audience to plasmonics and to the fabrication of some kinds of ultra-sensitive biosensors. An impressive way to demonstrate that this plasmon wave is a very sensitive sensor is to blow gently towards the gold film. The slight optical index change induced by the air flow makes the colours fluctuate. It can help in understanding that if a layer of molecules adsorbs on the gold film, the index change will also be detected by the surface plasmon wave.

Conclusion

A set of five easy experiments has been proposed as an introduction to key concepts of nanoscience targeting a public audience of non-specialists. They are all based on the use of gold either as nanoparticles or in nanometric thin films. The main message is that usual material exhibits unexpected properties when one of their dimensions is nanometric. This article explores more specifically the colour modulations of gold at the nanometric size. The first four experiments are based on gold nanospheres of 15 to 30 nm in diameter that are prepared according to the Turkevich method. In this case, gold adopts a magnificent ruby-red colour. Experiments 1, 2, and 3 describe the chemical preparation, the way for modulating their colour and how to include them into a polymer film. Experiment 4,

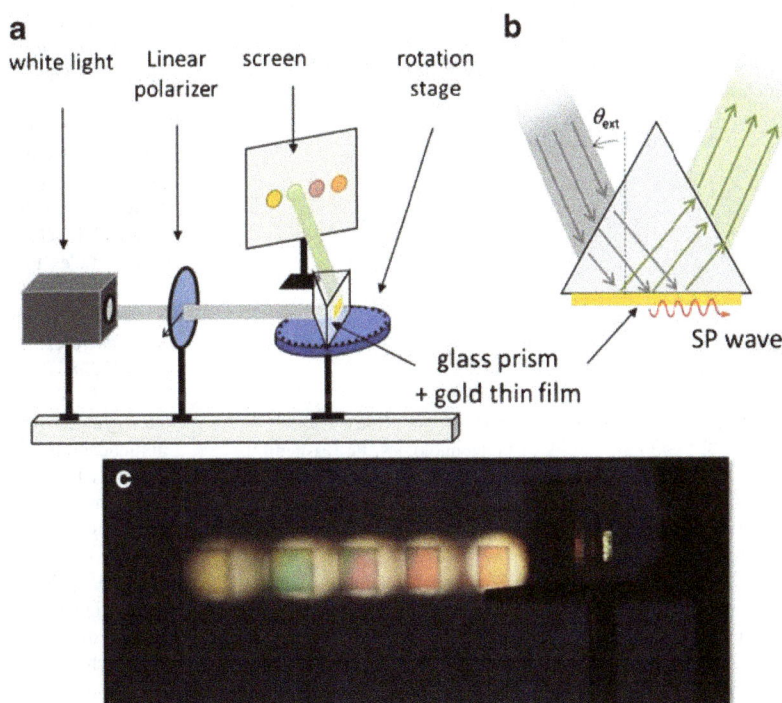

Fig. 6 Experimental setup for exciting surface plasmon wave on a thin gold film (50 nm in thickness) placed on one face of a glass prism. The reflected beam changes colour when the prism is rotated. **a** The excitation is achieved with white light that needs to be *p*-polarized and is reflected by the prism. The resulting colours are observed on the screen. **b** At a given incidence angle, the incoming light is able to excite the surface plasmon wave for one wavelength, the remaining wavelengths being reflected. **c** The photo shows the range of colours reflected by the glass prism (which is visible in the right part). This picture was taken with a long exposure time so that five different angles of the rotation stage were simultaneously captured

based on the use of immiscible liquids, is a striking experiment that shows that starting from these nanoparticles, it is possible to turn them back into a gold film that recovers its conventional yellow colour. Finally, the last experiment plays with the optical properties of the surface plasmon wave. These experiments can be used for teaching in lab courses with undergraduate students after some adaptation. Of course when used in conferences, they are always capturing the public attention if they are conducted as live experiments, although it demands some training so that the effect comes up at the right moment. This is always a very rewarding feeling for us scientists to be able to raise a fascination for scientific work and contribute to attract young people into the scientific community.

It has to be noted that some experiments proposed in this paper have been tested many times with success in school, in partnership with the Nano-School (www.nano-ecole.fr) project. This action aims to educate young people about nanoscience and give them some basics. These experiences around gold nanoparticles are particularly suitable in this context. This partnership has also helped to develop a kit for synthesizing nano particles, for teachers of middle and high schools (www.jeulin.fr).

Acknowledgments The authors thank Patricia Beaunier (Laboratoire de Réactivité de Surface, Université Pierre et Marie Curie) for the TEM observations and Jean-Michel Guigner (Institut de Minéralogie et de Physique des Milieux Condensés, Université Pierre et Marie Curie) for the Cryo-TEM studies. Finally, this paper is in debt to Don Eigler, a prominent physicist, known for his pioneering work in STM. One of the authors attended a keynote given by Don Eigler in 2006 in Basel (Switzerland) entitled "Computation in Small Structures: a Long Road to an Uncertain Destination". The content of the conference was totally unexpected since Don Eigler did not talk about computation at all, but about the urgent need for scientists to address the issue of informing the public, the journalists, the companies and the politics. This paper shows that his call has been heard.

References

1. Carenco S (2012) Développons les nanomatériaux! Rue d'Ulm, Paris
2. IPSOS (2010) Les Français et les nanotechnologies. Poll issued by IPSOS from a sample of 1013 people aged greater than 15
3. Pluchery O (2007) Filmed conference: "Propriétés optiques de l'or nanométrique: la plasmonique" http://www.dailymotion.com/webinsp#video=x9mkdt
4. Sun C, Qu R, Ji C, Meng Y, Wang C, Sun Y, Qi L (2009) Preparation and property of polyvinyl alcohol-based film embedded with gold nanoparticles. J Nanoparticle Res 11(4):1005–1010
5. Gachard E, Remita H, Khatouri J, Keita B, Nadjo L, Belloni J (1998) Radiation-induced and chemical formation of gold clusters. New J Chem 22:1257–1265
6. Novotny L, Hecht B (2006) Principles of nano-optics. Cambridge University Press, Cambridge
7. Pluchery O, Vayron R, Van K-M (2011) Laboratory experiments for exploring the surface plasmon resonance. Eur J Phys 32:585
8. Pluchery O, Ney P (2012) Experimental setup for studying surface plasmon resonance in lab courses. This Lab work equipment is sold by the company DIDA Concept, France. http://www.didaconcept.com. Accessed 6 Nov 2013
9. Schaming D, Remita H (2013) Nanotechnology: from the ancient time to nowadays. Found Chem (in press)
10. Louis C (2012) Gold nanoparticles in the past: before the nanotechnology era. In: Louis C, Pluchery O (eds) Gold nanoparticles for physics, chemistry and biology. World Scientific Publishing, London
11. Freestone I, Meeks N, Sax M, Higgitt C (2007) The Lycurgus Cup—a Roman nanotechnology. Gold Bull 40(4):270–277
12. Faraday M (1857) Experimental relations of gold (and other metals) to gold. Phil Trans R Soc London 147:145–181
13. Thompson D (2007) Michael Faraday's recognition of ruby gold: the birth of modern nanotechnology. Gold Bull 40(4):267–269
14. Mie G (1908) Beiträge zur Optik trüber Medien, speziell kolloidaler Metallösungen. Ann Phys 25:377–445
15. Turkevich J, Stevenson PC, Hillier J (1951) A study of the nucleation and growth processes in the synthesis of colloidal gold. Discuss Faraday Soc 11:55–75
16. Zabetakis K, Ghann WE, Kumar S, Daniel M-C (2012) Effect of high gold salt concentrations on the size and polydispersity of gold nanoparticles prepared by an extended Turkevich-Frens method. Gold Bull 45:203–211
17. Ji X, Song X, Li J, Bai Y, Yang W, Peng X (2007) Size control of gold nanocrystals in citrate reduction: the third role of citrate. J Am Chem Soc 129(45):13939–13948
18. Zhao P, Li N, Astruc D (2013) State of the art in gold nanoparticle synthesis. Coord Chem Rev 257:638–665
19. Jana N, Gearheart L, Murphy C (2001) Seed-mediated growth approach for shape-controlled synthesis of spheroidal and rod-like gold nanoparticles using a surfactant template. Adv Mater 13:1389–1393
20. Nikoobakht B, El-Sayed M (2003) Preparation and growth mechanism of gold nanorods (NRs) using seed-mediated growth method. Chem Mater 15:1957–1962
21. Liu M, Guyot-Sionnest P (2005) Mechanism of silver(I)-assisted growth of gold nanorods and bipyramids. J Phys Chem B 109:22192–22200
22. Abidi W, Remita H (2010) Gold based nanoparticles generated by radiolytic and photolytic methods. Recent Pat Eng 4(3):170–188
23. Torigoe K, Esumi K (1992) Preparation of colloidal gold by photoreduction of tetracyanoaurate(−)-cationic surfactant complexes. Langmuir 8:59–63
24. Hainfeld JF, Slatkin DN, Smilowitz HM (2004) The use of gold nanoparticles to enhance radiotherapy in mice. Phys Med Biol 49(18):N309–315
25. Sardar R, Fuston AM, Mulvaney P, Murray RW (2009) Gold nanoparticles: past, present, and future. Langmuir 25(24):13840–13851
26. Tanaka R, Yuhi T, Nagatani N, Endo T, Kerman K, Takamura Y, Tamiya E (2006) A novel enhancement assay for immunochromatographic test strips using gold nanoparticles. Anal Bioanal Chem 385(8):1414–1420
27. Liu J, Lu Y (2006) Preparation of aptamer-linked gold nanoparticle purple aggregates for colorimetric sensing of analytes. Nat Protoc 1(1):246–252
28. Pham VD, Hoang H, Phan TH, Conrad U, Chu HH (2012) Production of antibody labeled gold nanoparticles for influenza virus H5N1 diagnosis kit development. Adv Nat Sci: Nanosci Nanotechnol 3(4):045017
29. Georganopoulou DG, Chang L, Nam J-M, Thaxton CS, Mufson EJ, Klein WL, Mirkin CA (2005) Nanoparticle-based detection in

cerebral spinal fluid of a soluble pathogenic biomarker for Alzheimer's disease. Proc Natl Acad Sci 102(7):2273–2276

30. Fang P-P, Chen S, Deng H, Scanlon MD, Gumy F, Lee HJ, Momotenko D, Amstutz V, Cortes-Salazar F, Pereira CM, Yang Z, Girault HH (2013) Conductive gold nanoparticle mirrors at liquid/liquid interfaces. ACS Nano 7(10):9241-9248

31. Schaming D, Hojeij M, Younan N, Nagatani H, Lee HJ, Girault HH (2011) Photocurrents at polarized liquid/liquid interfaces enhanced by a gold nanoparticle film. Phys Chem Chem Phys 13:17704–17771

32. Pyykkö P (2004) Theoretical chemistry of gold. Angew Chem Int Ed 43(34):4412–4456

15

Preparation of monodispersed carboxylate-functionalized gold nanoparticles using pamoic acid as a reducing and capping reagent

Md. Abdul Aziz · Jong-Pil Kim · Munetaka Oyama

Abstract A simple preparation method of gold nanoparticles (AuNPs) using pamoic acid (PA; 4,4′-methylene-bis(3-hydroxy-2-naphthalene carboxylic acid)) with NaOH is described. Although PA is insoluble in water, it can be dissolved in the presence of NaOH and function as a capping and reducing reagent to form the AuNPs. The thus-formed AuNPs have a good monodispersity with diameters of 10.8 ±1.2 nm and carboxylate functions that come from the PA. Because PA is a methylene-bridged dimer of 3-hydroxy-2-naphthalene carboxylic acid, the formation of the AuNPs was examined also using the analogous mono-mer molecules, i.e., 3-hydroxy-2-naphthalene carboxylic acid, 2-hydroxy-1-naphtalene carboxylic acid, or 2-naphthol. Consequently, it was found that the case of PA was specific to forming the spherical monodispersed AuNPs while that differently shaped Au nanostructures were formed in the other cases. The present preparation using PA would be an interesting example that stable, monodispersed, carboxylate-functionalized AuNPs could be prepared without using thiols. Also, the present results may provide some insight into molecular designs of capping reagents for prepar-ing functionalized AuNPs without using thiol derivatives.

Keywords Hydrogen tetrachloroaurate · Pamoic acid · Carboxylate-functionalized gold nanoparticles · Capping reagent

Introduction

The chemical preparations of gold nanoparticles (AuNPs) have been attracting considerable attention due to their interesting properties and applicability in nanoscience and nanotechnology. As typical preparation methods of AuNPs in aqueous solutions, two approaches are well known, i.e., citrate ions have been used for the size-controlled synthesis of AuNPs [1] and tannic acid has been used for a rigid control of the size [2]. While a two-phase synthesis with thiol capping reagents [3] has been utilized for preparing AuNPs with smaller sizes and functional capping, the development of preparation methods of AuNPs in an aqueous solution is still active. The green synthesis of AuNPs [4, 5] would be one of the topics including the use of plants [6] and microorganisms [7]. In addition, preparation methods in which one reagent acts as a reducing and capping reagent have been proposed because of their simplicity [8–10].

When we consider the capping action of reagents toward AuNPs, in the case of the thiol derivatives, the Au–S bonding is a definite force for stabilizing the AuNPs. Various function-al cappings have been performed with a variety of thiol

M. A. Aziz (✉)
Center of Research Excellence in Nanotechnology, King Fahd University of Petroleum and Minerals, Dhahran 31261, Saudi Arabia
e-mail: maziz@kfupm.edu.sa

M. A. Aziz · M. Oyama (✉)
Department of Material Chemistry, Graduate School of Engineering, Kyoto University, Nishikyo-ku, Kyoto 615-8520, Japan
e-mail: oyama.munetaka.4m@kyoto-u.ac.jp

J.-P. Kim
Surface Properties Research Team, Korea Basic Science Institute Busan Center, Busan 609-735, South Korea

derivatives. For example, carboxylate-functionalized AuNPs were prepared using some thiol derivatives [11–14]. On the other hand, in the cases of green synthesis as referred above, which were carried out in aqueous solutions, the formation of AuNPs is possible without forming any definite bonding between Au and the capping reagents. This means that, even though there are no peculiar groups to bond with the Au, AuNPs can be prepared in aqueous solutions. Thus, to explore successful examples to form AuNPs would still be meaningful in order to know the molecular interactions to prepare AuNPs and apply the interactions to advanced syntheses of the AuNPs.

In the present paper, we show that pamoic acid (PA; 4,4′-methylene-bis(3-hydroxy-2-naphthalene carboxylic acid); **1**) can work as an effective capping and reducing reagent in the presence of NaOH to form relatively monodispersed AuNPs. PA has two 2-naphthol units, which are bridged at position 1 of 2-naphthol by a methylene group, and each naphthol unit contains one carboxylic acid group at position 3. While PA is usually utilized for the salt formation of some drugs [15], it was also used as a ligand for making dinuclear Ti(IV) complexes [16]. However, to the best of our knowledge, PA has never been used to prepare AuNPs.

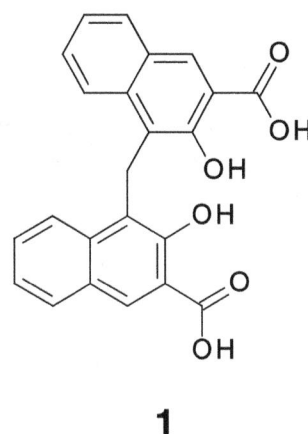

1

While we initially adopted PA as a trial, the use of PA has permitted us to prepare relatively monodispersed ca. 11-nm AuNPs at room temperature by working as a reducing and capping reagent in the presence of NaOH. In addition, the carboxylate functions of the thus-formed AuNPs have been experimentally confirmed. Therefore, to explore the capping actions of PA, the formations of AuNPs were examined also using the analogous monomer molecules, i.e., 3-hydroxy-2-naphthalene carboxylic acid (3H2NCA; **2**), 2-hydroxy-1-naphtalene carboxylic acid (2H1NCA; **3**), and 2-naphtol (**4**).

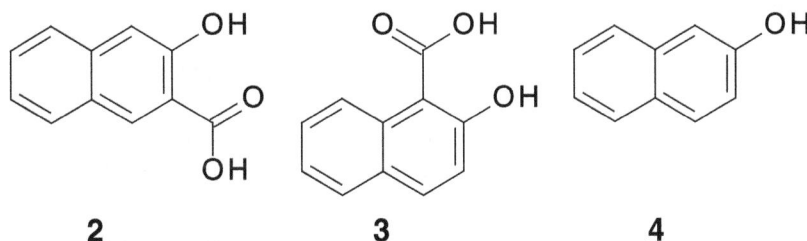

2   **3**   **4**

Kundu and coworkers reported the use of 2-naphtol and 2,7-dihydroxy-naphthalene combined with UV or microwave irradiation for the interesting structural forming of Au or Ag nanocrystals [17–19]. While the naphthols have functions as reducing reagents in their studies, our results present that some naphthol derivatives have apparently less power for nanostructural production in comparison to PA.

The synthesis of AuNPs with a monodispersity under rigid size control have a significant scientific impact: successful size-controlled preparations have been reported for alkanethiol-protected AuNPs [20], and AuNPs were first prepared in an aqueous phase and then transferred to non-polar solvents [21]. In our present trial, it would be characteristic that well-monodispersed AuNPs as in previous studies [20, 21] could be prepared with PA in an aqueous solution.

Experimental

Hydrogen tetrachloroaurate trihydrate (HAuCl$_4$·3H$_2$O), PA, 3H2NCA, 2H1NCA, 2-naphtol, 3-aminopropyl-trimethoxysilane (APTMS), and cadmium nitrate tetrahydrate were purchased from Sigma-Aldrich. All other reagents were purchased from Wako Pure Chemicals, Ltd. Indium tin oxide-coated glasses (ITO) were purchased from Geomatec Co., Ltd. (Yokohama, Japan). All solutions were prepared with ultra-pure water obtained from a water purification system (Millipore WR600A, Yamato Co., Japan).

The UV-visible spectra were recorded by an optical spectrophotometer, USB 2000, Ocean Optics, Inc. The transmission electron microscopic and X-ray photoelectron spectroscopic analyses were performed at the Korea Basic Science Institute, Busan Center, Korea. The scanning electron microscopic (SEM) images were observed using a field-emission scanning electron microscope (FE-SEM; JSM-7400 F, JEOL, Japan).

As a typical preparation method of AuNPs with PA, 7.9 mg PA was placed in a test tube and 9.0 ml of pure water was added followed by sonication for 15 min. Forty microliters of 1.0 M NaOH (aq.) was then added to the solution, and pure water subsequently added to make the volume 10.0 ml. The mixture was then sonicated for 15 min to make the 2.0-mM PA solution clear. Next, 100 μl of 1.0 M NaOH (aq.) was added and sonicated for 1 min; then, 10 ml of a 1.34-mM solution of $HAuCl_4$ (aq.) was added under sonication and stored for 15 min. Finally, the solution was stored for 60 min undisturbed to allow the complete formation of the AuNPs. We checked the necessity of this process by observing the changes in the absorption spectra and found that further stirring after the mixing tended to be unfavorable for preparing the monodispersed AuNPs. Similar preparations were carried out with other reagents, i.e., 3H2NCA, 2H1NCA, or 2-naphtol, but, in these cases, the concentrations were increased to 4.0 mM, i.e., twice that of PA, to make the amount of the naphthol units the same.

In preparing some samples for the TEM and X-ray photo-electron spectroscopy (XPS) measurements, the prepared solution of the AuNPs with PA was centrifuged at 12,000 rpm and the obtained sediment was redispersed in 1 mM NaOH (aq.). The centrifugation and redispersion processes were repeated three times to remove any free or loosely bound molecules. For the TEM analysis, the AuNPs were transferred to a copper grid by dipping it into the purified alkaline solution of the AuNPs. For the XPS analysis, the purified alkaline solution of the AuNPs was dropped onto a cleaned ITO substrate and dried at 40 °C.

In preparing the AuNP-modified ITO, a piece of ITO was immersed overnight in ethanol containing 2 % APTMS (v/v) at room temperature, and the amine-terminated ITO was prepared. After washing with ethanol, the electrode was dried by flowing nitrogen. On the other hand, the solution of the AuNPs was centrifuged at 12,000 rpm and the supernatant was decanted. The obtained sediment was redispersed in water. The centrifugation and redispersion processes were repeated three times. The APTMS-modified ITO was dipped in the purified aqueous AuNPs solution for 2 h. After washing with water, the modified ITO electrode was dried at 40 °C.

Results and discussion

PA is insoluble in water, but its sodium salt can be dissolved. Therefore, in the present study, we first prepared a clear alkaline solution of 2.0 mM PA by sonication followed by the further addition of NaOH to promote the PA function as a reducing reagent. The solution was then mixed with an aqueous solution of $HAuCl_4$ under sonication, and the formation of AuNPs was initiated. After the mixing, the color of the solution changed from yellow to blackish and then deep red, which indicated the formation of the AuNPs. Figure 1a shows the absorption spectra and Fig. 1b (left) shows a photo of the thus-prepared solution. Since the alkaline solution of PA or $HAuCl_4$ does not have any absorption in this wavelength region, the characteristic absorption peak observed at 507 nm in Fig. 1a and the color in Fig. 1b (left) can be attributed to the formation of the AuNPs. The absorption did not change with time, which could be recognized by the absorption spectrum recorded 7 days after the preparation of the solution (Fig. 1a). The color of the solution did not change over several months and no precipitates were formed. These results indicate the high stability of the synthesized AuNPs.

To confirm the capping states of the AuNPs by PA, we added a 0.15-ml aqueous solution of 6.7 mM $Cd(NO_3)_2$, i.e., Cd^{2+}, to a 1.35-ml aqueous solution of the AuNPs, which was prepared by centrifuging and redispersing the AuNPs in water to remove any unbound PA. As a result, the color of the solution has changed to blue as shown in Fig. 1b (right). It is known that such a color change is due to the aggregation of the AuNPs and that, when the carboxylate groups are present on the surfaces of the AuNPs, the aggregation proceeds in the presence of divalent cations, such as Cd^{2+}, Hg^{2+}, and Pd^{2+} [13]. Thus, it was proved that carboxylate groups are present on the surfaces of the present AuNPs prepared with PA. This would be reasonable judging from the molecular structure of PA having two carboxylate groups in one molecule. Because the carboxylate groups can be an anchor group to bind Au

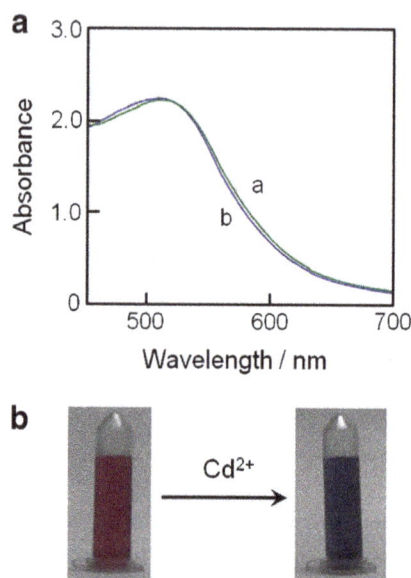

Fig. 1 **a** Visible absorption spectra of the solutions of AuNPs prepared with PA measured (*a*) just after the preparation and (*b*) after 7 days. **b** A photo of a solution of AuNPs prepared with PA (*left*) and the color change after mixing with Cd^{2+} (*right*)

Fig. 2 a A typical TEM image of AuNPs prepared with PA. b A high-magnification image of a. c A histogram depicting the size distribution of the AuNPs. d A typical FE-SEM image of AuNP-modified amine-terminated ITO surface

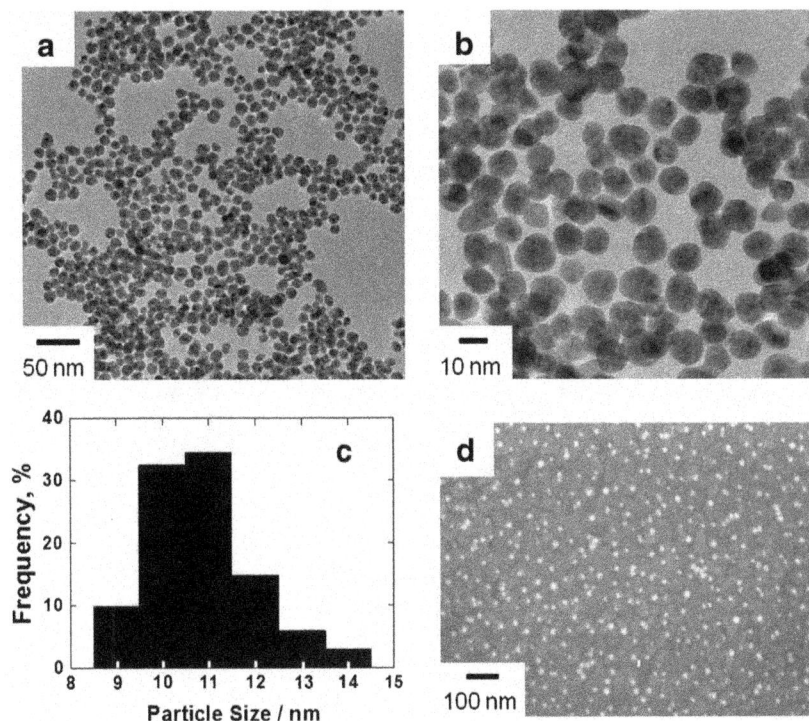

[22], it is considered that one of the carboxylate groups worked by contacting with Au and that the others located on the outer surface.

Next, we obtained TEM images of the AuNPs prepared with PA. The results are shown in Fig. 2a, b. The TEM images clearly indicate that the synthesized AuNPs are spherical and that the mean diameter was 10.8±1.2 nm. The low standard deviation represents the good monodispersity of the prepared AuNPs. The histogram in Fig. 2c shows the size distribution. The diameters of the AuNPs are limited to between 8.5 and 14.5 nm, and the diameters of 81.4 % of the AuNP are in the range between 9.5 and 12.5 nm. This dispersion was almost reproducible in several preparations under the same conditions. The good size regularity was also confirmed in the FE-SEM observation after immobilization of the AuNPs on amine-terminated ITO surfaces (Fig. 2d). We can see that the

AuNPs of the same size were homogeneously dispersed on the surface of the ITO. The attachment should be based on the electrostatic interaction between the amine groups terminated on the ITO surface and the carboxylate groups on each AuNP.

Furthermore, the present AuNPs prepared with PA were analyzed by XPS. The purified AuNPs showed only two peaks at 83.3 and 87.0 in Fig. 3a corresponding to the Au $4f_{7/2}$ and Au $4f_{5/2}$ core-level binding energies which confirmed the presence of only metallic gold (Au^0) in the AuNPs [23]. In Fig. 3b, the C 1s peaks were clearly identified at 284.6, 288.6, and 289.1 eV which correspond to the –C–C–, adsorbed –COO^- group on the Au surfaces, and free –COO^- groups, respectively [24, 25].

As for the formation of the AuNPs with PA, it would be characteristic that the carboxylate-functionalized AuNPs can be easily prepared with a good monodispersity using PA as the

Fig. 3 a, b XPS spectra of AuNPs prepared with PA

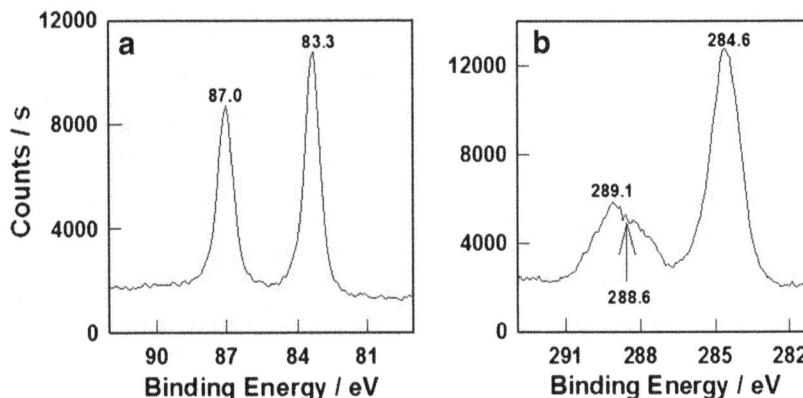

Fig. 4 **a** A typical TEM image of Au nanomaterials prepared with 2H1NCA. **b** A high-magnification image of **a**

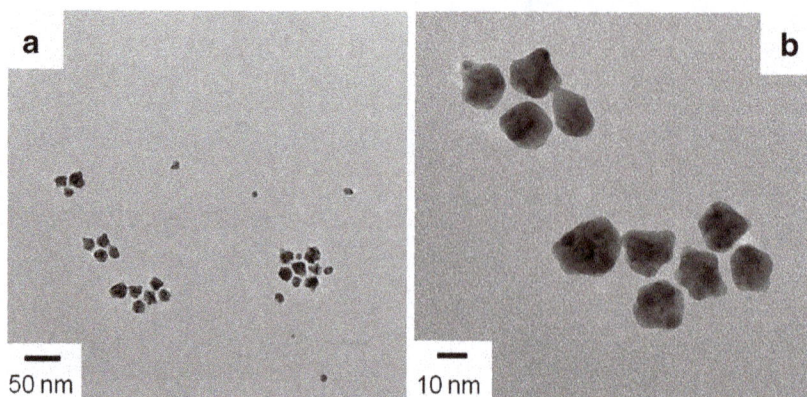

reducing and capping reagent. Because hydroquinone has been reported to have a function to reduce $AuCl_4^-$ with NaOH [8], and because 2-naphtol derivatives work as a reducing reagent [17–19], the two hydroxyl groups of PA should be involved in the reduction of $AuCl_4^-$ to form AuNPs. On the other hand, two carboxylate groups should have the key function for capping AuNPs and for characterizing the nature of the AuNPs by locating at the outer surface.

To compare the results obtained with PA, we next investigated the formation processes with 3-hydroxy-2-naphthalene carboxylic acid (3H2NCA; **2**), 2-hydroxy-1-naphtalene carboxylic acid (2H1NCA; **3**), and 2-naphtol (**4**). Consequently, we could never prepare monodispersed AuNPs as prepared with PA. For example, Fig. 4 shows the TEM results for AuNPs prepared with 2H1NCA. As apparently shown in the images, the shape of the AuNPs was not spherical but irregular, and the size dispersity was apparently not good. In this case, the absorption spectrum exhibited an absorption maximum at 558 nm (data not shown). Because the absorption maximum of the smaller AuNPs (ca. 10 nm) should be around 510 nm as in Fig. 1a, the longer wavelength shift implied the formation of AuNPs with a larger size. In addition, the absorption spectrum changed to an absorption maximum at 541 nm after 7 days, which suggested that the stability of the AuNPs was not good in comparison to the AuNPs prepared with PA. This should be due to the weaker capping of

2H1NCA. Similar results were obtained with 3H2NCA; therefore, it is inferred that the position of a carboxylate group would not be very significant during the formation of the AuNPs. It should now be emphasized that the dimer structure of PA linked with a methylene group would be a key factor to form the monodispersed AuNPs, while the monomer may have a reducing power and weaker capping ability.

We also tried the same reduction of $AuCl_4^-$ with 2-naphthol. As a result, the reduction processes of $AuCl_4^-$ and the formed nanostructures of gold were totally changed in comparison to the other cases. Actually, after mixing with the solution of $HAuCl_4$, the color of the solution turned dark gray and the color was maintained after a 15-min sonication. Figure 5 shows TEM images of the formed materials. Some strange-shaped nanowires, like worms, were formed together with small nanospheres. The diameter of the nanowires was as small as 3.3 ± 0.5 nm. In a previous report, the gray color of the solution has been noted for gold nanochains [26]. Therefore, the present color change matched well with the formation of the nanowires. The smaller nanostructure as in Fig. 5 might be a reflection of the lower reduction power and the unique capping of 2-naphtol. However, the lack of the carboxylate is inferred to be unsuitable for capping; therefore, other capping molecules might be necessary as in the previous studies [17–19].

Fig. 5 **a** A typical TEM image of Au nanowires prepared with 2-naphthol. **b** A high-magnification image of **a**

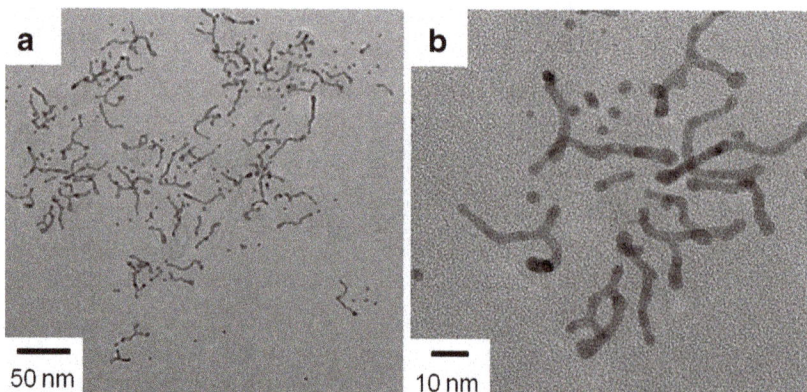

Based on the previous results obtained with 3H2NCA, 2H1NCA, and 2-naphtol instead of PA, we concluded that one naphthalene moiety of the capping reagents was not sufficient to stabilize the formed AuNPs, though they reduced the $AuCl_4^-$. Therefore, it is inferred that the rigid size control (or capping) with PA has been permitted by the methylene-linked two naphthalene moieties. In addition, based on the formation of the wormlike Au nanowires with 2-naphtol and the ill-shaped AuNPs with 3H2NCA and 2H1NCA, the carboxylate group should have the function to stabilize the AuNPs. Thus, the rigid capping of PA is regarded to come from the methylene-linked naphthalene moieties and two carboxylate groups located outside.

Conclusions

During the reduction of $AuCl_4^-$ with PA in the presence of NaOH, monodispersed AuNPs, with diameters of 10.8 ± 1.2 nm, could be easily prepared. Reflecting on the presence of two carboxylate groups in PA, the formed AuNPs have carboxylate functions because only one carboxylate group would be used in the capping interaction. While the anchoring power of a carboxylate group is normally weak [22], due to the help of the naphthalene moieties of PA, it is expected that effective capping could be achieved while maintaining the monodispersity. As for the effects of aromatic rings, some integration with gold have previously been reported [27], and hydroquinone has been reported to form AuNPs [8]. As one remarkable achievement of the present study, the stable, monodispersed, carboxylate-functionalized AuNPs could be prepared without using thiols, while the known carboxylate-functionalized AuNPs had been prepared with thiol anchors [11–14]. Thus, the proposed facile synthesis in an aqueous solution should be worthwhile reporting as a safe and clean synthesis of carboxylate-functionalized AuNPs.

In our trials using the analogous monomer molecules instead of PA, it was found that structural controls were difficult in spite of some progress in the reductions of $AuCl_4^-$. Judging from the formation of the ill-shaped Au nanowires with 2-naphtol, one carboxylate group would be helpful for nanostructural construction. However, the key factors of the rigid size control with PA would be due to the methylene-linked naphthalene moieties and two carboxylate groups. The present results may provide some information about molecular designs for carboxylate-functionalized AuNPs without using thiol derivatives.

Furthermore, in a previous study, heat treatment has been reported to be effective for the precise size control of alkanethiol-protected AuNPs [20]. Also, for the aqueous-phase synthesis of AuNPs to be transferred to nonpolar solvents [21], the ratio of borohydride anions/hydroxyl anions was a key factor for the size control. While our present study has been limited to the preparation of AuNPs of 10.8 ± 1.2 nm in aqueous solutions at room temperature, we are going to explore the possibility of the size control with PA as the next step.

Acknowledgments M. A. A. thanks the Japan Society for the Promotion of Science (JSPS) for the fellowship. This work was supported by JSPS KAKENHI Grant Numbers 20550074 and 21-09245.

References

1. Frens G (1973) Nature Phys Sci 241:20–22
2. Slot JW, Geuze HJ (1985) Eur J Cell Biol 38:87–93
3. Brust M, Walker M, Bethell D, Schiffrin DJ, Whyman R (1994) Chem Commun 801–802
4. Raveendran P, Fu J, Wallen SL (2006) Green Chem 8:34–38
5. Wu CC, Chen DH (2007) Gold Bull 40:206–212
6. Kumar V, Yadav SK (2009) J Chem Technol Biotechnol 84:151–157
7. Das SK, Marsili E (2010) Rev Environ Sci Biotechnol 9:199–204
8. Sirajuddin MA, Torriero AAJ, Nafady A, Lee CY, Bond AM, O'Mullane AP, Bhargava SK (2010) Colloids Surf A 370:35–41
9. Wu CC, Chen DH (2010) Gold Bull 43:234–240
10. Badwaik VD, Bartonojo JJ, Evans JW, Sahi SV, Willis CB, Dakshinamurthy R (2011) Langmuir 27:5549–5554
11. Chen S, Kimura K (1999) Langmuir 15:1075–1082
12. Yonezawa T, Kunitake T (1999) Colloids Surf A 149:193–199
13. Kim Y, Johnson RC, Hupp JT (2001) Nano Lett 1:165–167
14. Lin YC, Yu BY, Lin WC, Lee SH, Kuo CH, Shyue JJ (2009) J Colloid Interface Sci 340:126–130
15. Jørgensen M (1998) J Chromatogr B 716:315–323
16. Baghel GS, Rao CP (2009) Polyhedron 28:3507–3514
17. Kundu S, Panigrahi S, Praharaj S, Basu S, Ghosh SK, Pal A, Pal T (2007) Nanotechnology 18:075712
18. Kundu S, Peng L, Liang H (2008) Inorg Chem 47:6344–6352
19. Kundu S, Wang K, Liang H (2009) J Phys Chem C 113:134–141
20. Shimizu T, Teranishi T, Hasegawa S, Miyake M (2003) J Phys Chem B 107:2719–2724
21. Martin MN, Basham JI, Chando P, Eah SK (2010) Langmuir 26:7410–7417
22. Chen F, Li X, Hihath J, Huang Z, Tao N (2006) J Am Chem Soc 128:15874–15881
23. Aziz MA, Patra S, Yang H (2008) Chem Commun: 4607–4609
24. Han SW, Joo SW, Ha TH, Kim Y, Kim K (2000) J Phys Chem B 104:11987–11995
25. Briggs D, Brewis DM, Dahm RH, Fletcher IW (2003) Surf Interface Anal 35:156–167
26. Polavarapu L, Xu QH (2008) Nanotechnology 19:075601
27. Syomin D, Kim J, Koel BE, Ellison GB (2001) J Phys Chem B 105:8387–8394

16

Human endothelial cell response to polyurethane–gold nanocomposites

Tung-Tso Ho · Yu-Chun Lin · Shan-hui Hsu

Abstract Gold (Au) is long considered as a biocompatible metal, and gold nanoparticles (AuNPs ~5 nm) were recently reported to scavenge free radicals. The effect of Au embedded in a polymeric material is less investigated compared with that of silver. In this study, nanocomposites from polyurethane (PU) and 43.5 or 174 ppm of AuNPs were prepared from a waterborne process. The response of endothelial cells (ECs) to the PU-Au nanocomposites was investigated in vitro and in vivo. ECs on PU-Au nanocomposites showed lamellipodia formation and better cell proliferation. The activation of proteins in ECs grown on PU-Au nanocomposites was analyzed by two-dimensional gel electrophoresis and confirmed by Western blot. The new protein identified through this procedure was valosin-containing protein (VCP), which is known to have immunomodulating effect. VCP was upregulated by PU-Au 43.5 ppm and PU-Au 174 ppm, but more in PU-Au 43.5 ppm. This suggested that the dispersion of AuNPs in the polymer matrix may be more important than the loading amount. PU-Au catheters implanted in rat blood vessels showed less foreign body reaction and more extensive EC coverage than the control PU catheters. The good in vivo biocompatibility of PU-Au may be associated with the anti-inflammatory effect of PU-Au. Based on this study, AuNPs may serve as an antioxidant additive for biomedical polymers.

Keywords Gold nanoparticles (AuNPs) · Polymer nanocomposites · Two-dimensional gel electrophoresis · Anti-inflammatory

T.-T. Ho · S.-h. Hsu (✉)
Institute of Polymer Science and Engineering,
National Taiwan University,
No. 1, Sec. 4 Roosevelt Road,
Taipei 10617 Taiwan, Republic of China
e-mail: shhsu@ntu.edu.tw

T.-T. Ho
Institute of Biomedical Engineering,
National Chung Hsing University,
Taichung, Taiwan, Republic of China

Y.-C. Lin
Department of Surgery,
Fong-Yuan Hospital Department of Health Executive Yuan,
Taichung, Taiwan, Republic of China

Introduction

Recent progress in nanotechnology has brought some new insights regarding the surface design of a biomaterial. When a biomaterial is applied, it is the surface of the material that is in contact with the human body. The physicochemical features of a biomaterial in nanometric scale often have an impact on the physiological behavior of the human cells attached to its surface [1, 2]. This concept is now widely recognized and used in designing a biomaterial surface for controlling the cell function [3–5].

Bulk gold (Au) is an inert and biocompatible material. Gold nanoparticles (AuNPs) when added to a few biomaterials in small amounts (from several parts per million to 1 %) can modify the properties and biological performance of these materials through the introduction of nanometric surface features [6–8]. For example, polyurethane (PU), as an industrial elastic polymer with pervasive influence on modern human life, displayed a rise of thermal stability upon the addition of AuNPs (~5 nm) and demonstrated different surface morphologies comprised of nanosized harder and softer domains [7]. A previous study showed that PU containing 43.5 ppm AuNPs may activate focal adhesion kinase (FAK) and the phosphatidylinositol 3-kinase/serine–threonine kinase (PI3K/Akt) signaling pathway in bovine endothelial cells (ECs), leading to cell proliferation and migration [9]. An overloading of AuNPs (e.g., 174 ppm) can cause the loss of the positive effect because of NP aggregation.

Two-dimensional gel electrophoresis (2-DGE) is a core platform technology in the study of proteomics. It involves the separation of protein complex mixtures from cells or tissue by the isoelectric point (pI) and molecular size of proteins. It has been used in many research areas such as protein analysis of mesenchymal stem cells [10], search of biomarkers [11], and detection of cell surface proteins [12].

In this study, the polymer PU and PU nanocomposites with two different doping concentrations of AuNPs were used and served as a model to investigate how the nanostructure brought about by AuNPs may have affected the protein expression levels of human umbilical vein endothelial cells (HUVECs). 2-DGE was used as a novel tool to identify new possible proteins and signaling pathways responsible for the changes of HUVEC behavior on AuNPs-modified PU. We also confirmed the better biocompatibility of this PU-Au nanocomposite in vivo.

Materials and methods

Preparation of PU-Au nanocomposites

AuNPs were manufactured by Gold Nano Tech, Inc., Taiwan. AuNPs (~5 nm in size) were sonicated for 3 h at a temperature below 20 °C. The concentration of AuNPs was confirmed by inductively coupled plasma.

The PU emulsion and cross-linker were obtained from Great Eastern Resins Industrial Co., Taiwan. The PU emulsion (50 % solid content in distilled water) was synthesized using hexamethylene diisocyanate (HDI) and the macrodiol poly (butadiene adipate) (average molecular weight, 2,000) at a molecular ratio of 3:1 and chain-extended by ethylenediamine sulfonate sodium salt and ethylenediamine. The cross-linker was a mixture of the isocyanurate trimer of hexamethylene diisocyanate (HDI trimer) and 6 % Bayer hardener (made from HDI trimer and polyethylene glycol). PU polymer was prepared by mixing 1 % cross-linker in the PU emulsion and further dried. The chemical composition of the final PU polymer is illustrated in Scheme 1 [7]. For PU-Au nanocomposites, AuNPs were added to 10 % PU emulsion, so that the final concentration of AuNPs was 43.5 or 174 ppm. To fabricate materials into films, the mixture containing 0, 43.5, or 174 ppm AuNPs was cast on 15 or 32 mm coverslip glass by a spin coater (PM-490, Synrex, Taiwan). The cast films were put in a 60 °C oven for 48 h and in a 60 °C vacuum oven for

another 48 h to remove water. The final film samples were abbreviated as PU, PU-Au 43.5, and PU-Au 174.

Cell culture

HUVECs were purchased from Bioresource Collection and Research Center (No. H-UV001). Cells were cultured in medium M199 (Gibco) with 10 % fetal bovine serum (Gibco), 25 U/ml heparin (Sigma), 30 μg/ml endothelial cell growth supplement (Upstate), 2 mM L-glutamine (Gibco), and 1 % penicillin/streptomycin (Gibco) in a T-flask coated with 1 % gelatin (Sigma Catalog No. G-9382). The cultures were maintained in a 5 % CO_2/37 °C incubator. The seeding density on material samples was 2×10^4 cells for each 15 mm diameter films (in 24-well culture plates) and 8×10^4 cells for each 32 mm diameter films (in 6-well culture plates). A blank cell tissue culture polystyrene (TCPS) served as the control.

Cellular response

HUVECs were incubated on the material for 48 h. The seeding density was 2×10^4 cells for each 15 mm diameter material films fit in each well of 24-well culture plates. Cells were trypsinized and counted for cell proliferation using a hemacytometer combined with an inverted phase contrast microscope (TE-300, Nikon, Japan). For cytoskeleton staining, cells were incubated on the material for 16 h, fixed with 4 % paraformaldehyde (Sigma), and stained by phalloidin (Sigma). Cell morphology and cytoskeleton were examined by an optical microscope (Labophot-2, Nikon, Japan). Particularly, the number of lamellipodia was counted under the microscope. The percent chance of lamellipodia formation was defined as the number of lamellipodia divided by the total number of cell–material contacts over an area of 0.6 mm^3 (more than 60 contacts for each material).

Protein preparation and 2-DGE

Proteins were extracted from cells cultured on each material for 2 days by the liquid nitrogen freezing and 37 °C thawing process. Protein extracts were centrifuged at 13,000 rpm and 4 °C for 10 min to remove particulates. Protein extracts (200 μl) were added with 62.5 μl of 100 mM dithiothreitol (DTT) and 65 μl of 98 % trichloroacetic acid without shaking and placed in −20 °C overnight. Protein extracts were then centrifuged at 13,000 rpm and 4 °C for 15 min to remove the supernatant. The

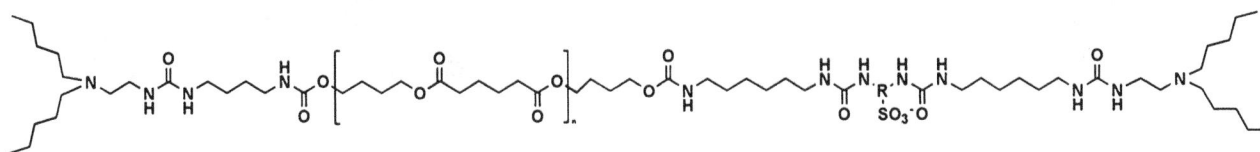

Scheme 1 The chemical composition of the PU polymer prepared in this study

precipitation was washed by acetone at −20 °C. The acetone was removed by centrifugation at 13,000 rpm and 4 °C for 10 min. Proteins were solubilized in 200 µl of sample buffer (containing 8 M urea, 2 % Triton X-100, and 40 mM DTT). The concentration of proteins was detected by the Quant-iT™ Protein Assay Kit (Invitrogen). The clear solution containing 100 µg protein was resuspended in 250 µl of the rehydration buffer [8 M urea, 2 % Triton X-100, 2 % Immobiline™ Dry-Strip gels buffer (IPG buffer) pH 3–10, and 0.002 % bromophenol blue]. Samples were loaded overnight onto 13 cm IPG strips, pH 3–10 (G.E. Healthcare) and rehydrated at 20 °C for

12 h. Sample strips were then resolved by isoelectric focusing using the following program: voltage set at 500 V for 2 h, accumulation to 1,000 V for 2 h, accumulation to 8,000 V for 5 h, and at 8,000 V for 1 h. Sample strips were then equilibrated for 15 min in each equilibration buffer (6 M urea, 75 mM Tris–HCl, 29.3 % glycerol, 2 % sodium dodecyl sulfate, 0.002 % bromophenol blue) containing 10 mg/ml DTT and 25 mg/ml iodoacetamide. Following equilibration, strips were loaded onto 10 % sodium dodecyl sulfate polyacrylamide gel electrophoresis (SDS-PAGE) and resolved in two dimensions at 45 mA per gel. Proteins on gels were visualized using a silver stain kit (Investigator Silver Stain Kit, Genomic Solutions) following the manufacturer's instructions [13, 14].

The protein expression profiles for each sample were scanned using the Imagescanner III (G.E. Healthcare) and the images were analyzed by the ImageMaster 2D Platinum computer software. During the image analysis, the software adjusted the location and expression according to the landmarks on the pictures which serve as the internal control.

In gel digestion and protein identification

The protein spots with the expression levels distinguished from the reference by more than 50 % were excised from the gels and placed in the double-distilled water. The excised protein spots were then subjected to the in gel digestion and further analysis by matrix-assisted laser desorption/ionization time-of-flight (MALDI-TOF) mass spectrometry. Spectra were collected on an ABI Voyager-DE PRO MALDI-TOF mass spectrometer. The University of California, San Francisco database search program MS-Fit (http://prospector2.ucsf.edu/prospector/mshome.htm) was employed to match the peptide m/z values with the databases SwissProt and NCBInr.

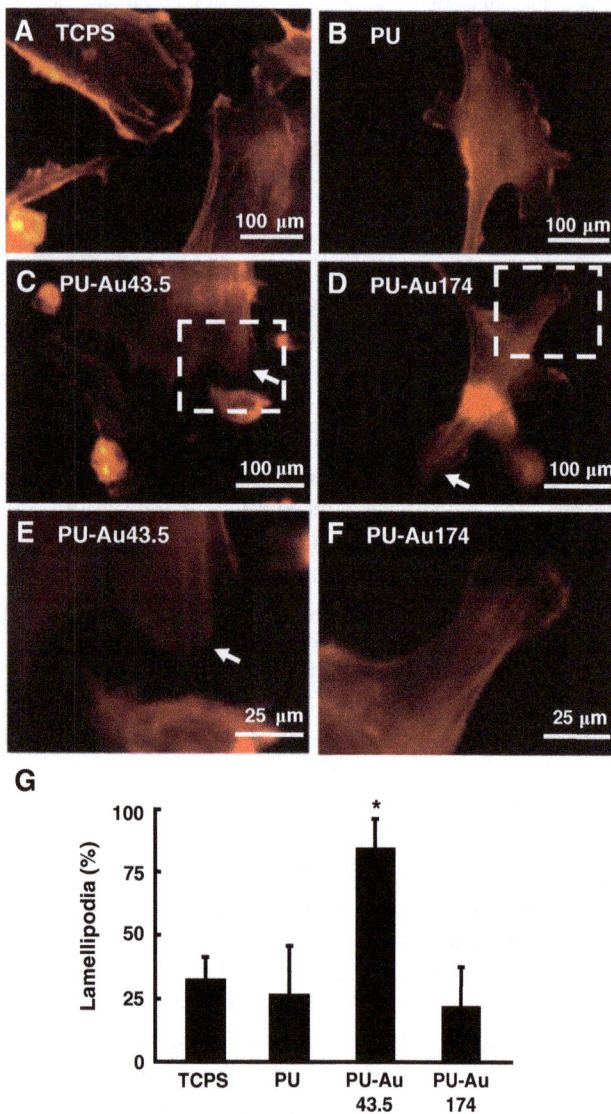

Fig. 1 The cell morphology of HUVECs on different substrates: **a** TCPS; **b** PU; **c**, **e** PU-Au 43.5 ppm; **d**, **f** PU-Au 174 ppm. Actin was stained by phalloidin. *Arrows* in **c**, **d**, **e** indicate the presence of lamellipodia, in contrast to the regular adhesion plaque demonstrated in **f**. The areas marked as *boxes* in **c**, **d** are shown in higher magnification in **e**, **f**. The number of lamellipodia normalized to the total number of contacts (i.e., percent chance of lamellipodia formation) is depicted in **g**. *$p < 0.05$ compared to the other groups

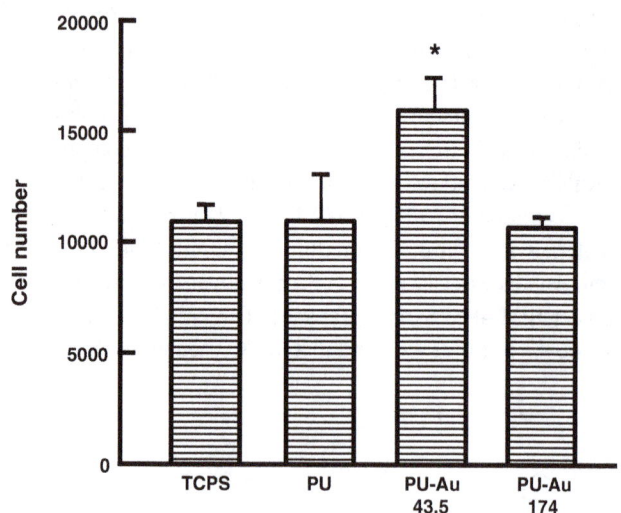

Fig. 2 The cell numbers of HUVECs for 48 h on different substrates. *$p < 0.05$ compared to the other groups

Gene expression and Western blot

Total RNA were extracted from cells cultured on each material for 2 days by the Trizol reagent (Invitrogen). Five hundred nanograms of mRNA was used as a template for cDNA synthesis using a reverse transcriptase (Fermentas). The zinc finger protein 792 (ZNF792), human SAM domain and HD domain-containing protein 1 (SAMHD1), human glucosamine-fructose-6-phosphate aminotransferase 1 (GFPT1), human transitional endoplasmic reticulum ATPase (also called valosin-containing protein [VCP]), and endothelial cell nitric oxide synthase (eNOS) genes were used as templates for mRNA level quantification and β-actin genes were used as control using the PCR Master

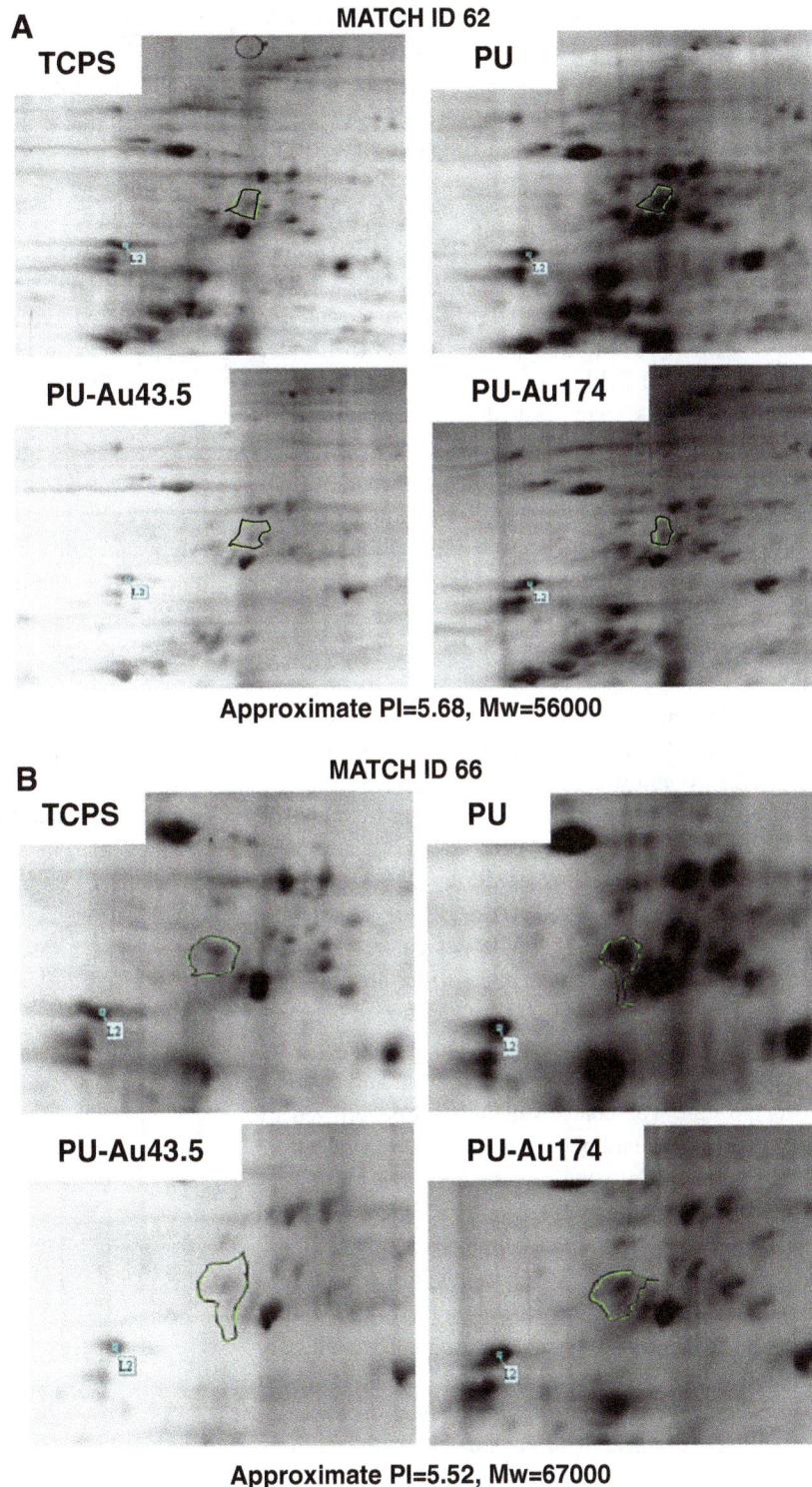

Fig. 3 The spots detected with distinct protein expression levels on each 2-DGE gel. The spots were indicated by the *black circular contour* on each gel. **a** The match ID 62. **b** The match ID 66. *L2* is one of the landmarks which serve as the basis for location adjustment and as an internal control for protein expression by the 2-DGE analysis software

Mix (Gene Mark). The primers for ZNF792 were agactcctg-tactgcgatgtga and tcacatatgcctcacacacaaa, with annealing template of 50 °C. The primers for SAMHD1 were atta-caggtgctggaggaaaaa and ttctctggcagaagttgtgaaa, with annealing template of 50 °C. The primers for GFPT1 were ggaaaagttaaggcactggatg and ttcactccgtacaccaatcaac with annealing template of 50 °C. The primers for VCP were tcctagcccttattgcattgtt and gatcaagaagaagaaggctcca, with annealing template of 50 °C. The primers for eNOS were gagatggtcaactatttcctgt and gaagcggatcttataactcttg, with annealing template of 53 °C. The primers for β-actin were tccctcaagattgtcagcaa and agatccacaacggatacatt, with anneal-ing template of 55 °C. The protein expression of VCP was analyzed by Western blot at 48 h post seeding. Proteins were separated by SDS-PAGE (10 % gels) and transferred to polyvinylidene difluoride membranes (Pall, USA). Mem-branes were blocked for 1 h with 5 % milk in phosphate-buffered saline (PBS) containing 0.05 % Tween-20 and then hybridized primary antibody with anti-VCP (Millipore, USA) or anti-β-actin (Millipore, USA) (1:1,000 dilution). After the membranes were washed with PBS containing 0.05 % Tween-20, primary antibodies were detected using horseradish peroxidase (HRP)-conjugated secondary anti-bodies (Jackson ImmunoResearch, USA) (1:5,000 dilution) and signals were developed by the Immobilon Western Chemiluminescent HRP Substrate (Millipore).

In vivo biocompatibility

All use and care of animals in this study were approved by the university Animal Care and Use Committee with guide-lines for the care and use of laboratory animals observed. Commercial indwelling venous PU catheters (external di-ameter 1.1 mm and length 32 mm; Terumo, Germany) were employed. A few catheters were additionally coated with PU-Au 43.5 ppm. A few coated catheters were additionally cultured with rat ECs (1×10^5 cells per catheter) on the surface for 3 days. To coat the catheters, the outer surface of the catheters was first activated by air plasma (Plasma-treat, Germany). The catheters were then immersed in the mentioned polymer solution for 5 s, removed, and baked at 60 °C for 30 min. The coated catheters were sterilized by 75 % ethanol, rinsed, and placed in PBS. The rat ECs were isolated from rat caudal vena cava by cutting it into pieces and digested with collagenase II. These cells were identified by von Willebrand factor and eNOS. The model of rat jugular vein implantation followed a previous method [15] with modification. Adult Sprague–Dawley rats (400–450 g) were anesthetized before implantation. Animals were deeply anesthetized with isoflurane (Halocarbon, USA) throughout the surgical procedures. Each rat was shaved on the neck under aseptic conditions (70 % alcohol). A small venotomy was made through which the catheter was introduced into

the right jugular vein (30 mm depth from the venotomy). The external end of the catheter was closed by 4-0 nylon suture. The skin incision was closed layer by layer and the implanted catheter was not exposed. After 3 months, the rats were euthanatized by CO_2 overdose treatment. The implants with surrounding tissue were excised and fixed in 10 % natural formalin solution, embedded in paraffin, thin-sectioned, and stained by hematoxylin and eosin (H&E) for histological analysis using an optical microscope (Lab-ophot-2, Nikon, Japan). The images were analyzed quanti-tatively by the Image Pro Plus 4.5 software.

Statistics

2-DGE data for PU-Au 43.5 ppm were repeated three times and there were no significant differences among the three repeats. 2-DGE for TCPS, PU, and PU-Au 174 ppm was repeated twice and there were no significant differences among these groups. There was a significant difference between PU-Au 43.5 ppm and other groups.

Results

After the cells attached to the materials for 16 h, their morphology had changed. As shown in Fig. 1c, most cells on PU-Au 43.5 ppm showed lamellipodia, as indicated by the arrow. Lamellipodial formation was not as frequently observed for cells on TCPS, PU, or PU-Au 174 ppm (Fig. 1a, b, d). Lamellipodia were cytoskeletal projections with extensively oriented actin bundles ("microspikes," arrows in Fig. 1c, d, e) in contrast to the regular cell–material contact (adhesion plaque) which has partially rounded margin and less oriented cytoskeleton (Fig. 1f). The percent chance of lamellipodial formation is shown in Fig. 1g. It was evident that cells on PU-Au 43.5 ppm had

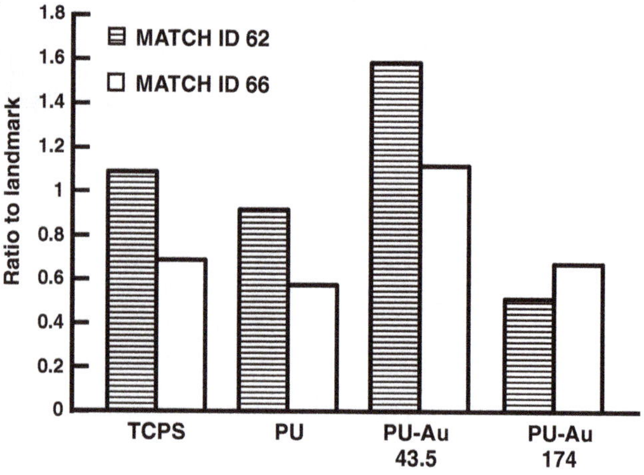

Fig. 4 The expression level of distinct protein spots relative to the landmark on each gel

significantly more lamellipodial formation than cells on the other materials. The number of HUVECs on different materials after 48 h is shown in Fig. 2. PU-Au 43.5 ppm also demonstrated the greatest number of cells.

Figure 3a, b shows the spots detected with distinct protein expression levels, which were indicated by the black circular contour on each gel. The ID numbers were provided by the software. The match ID 62 corresponded to PI=5.68 and Mw= 56,000. The match ID 66 corresponded to PI=5.52 and Mw=

67,000. Figure 4 shows the relative intensity of the spots with distinct protein expression levels on each gel for the match IDs 62 and 66. Figure 5 shows the fingerprint of protein spots by the mass spectrometer. A search by the databases SwissProt and NCBInr suggested that the match ID 62 may be ZNF792 and the match ID 66 may be VCP, SAMHD1, or GFPT1.

To identify the protein, we conducted the gene expression analysis by reverse transcription polymerase chain reaction to screen the possible proteins first. It was found that the

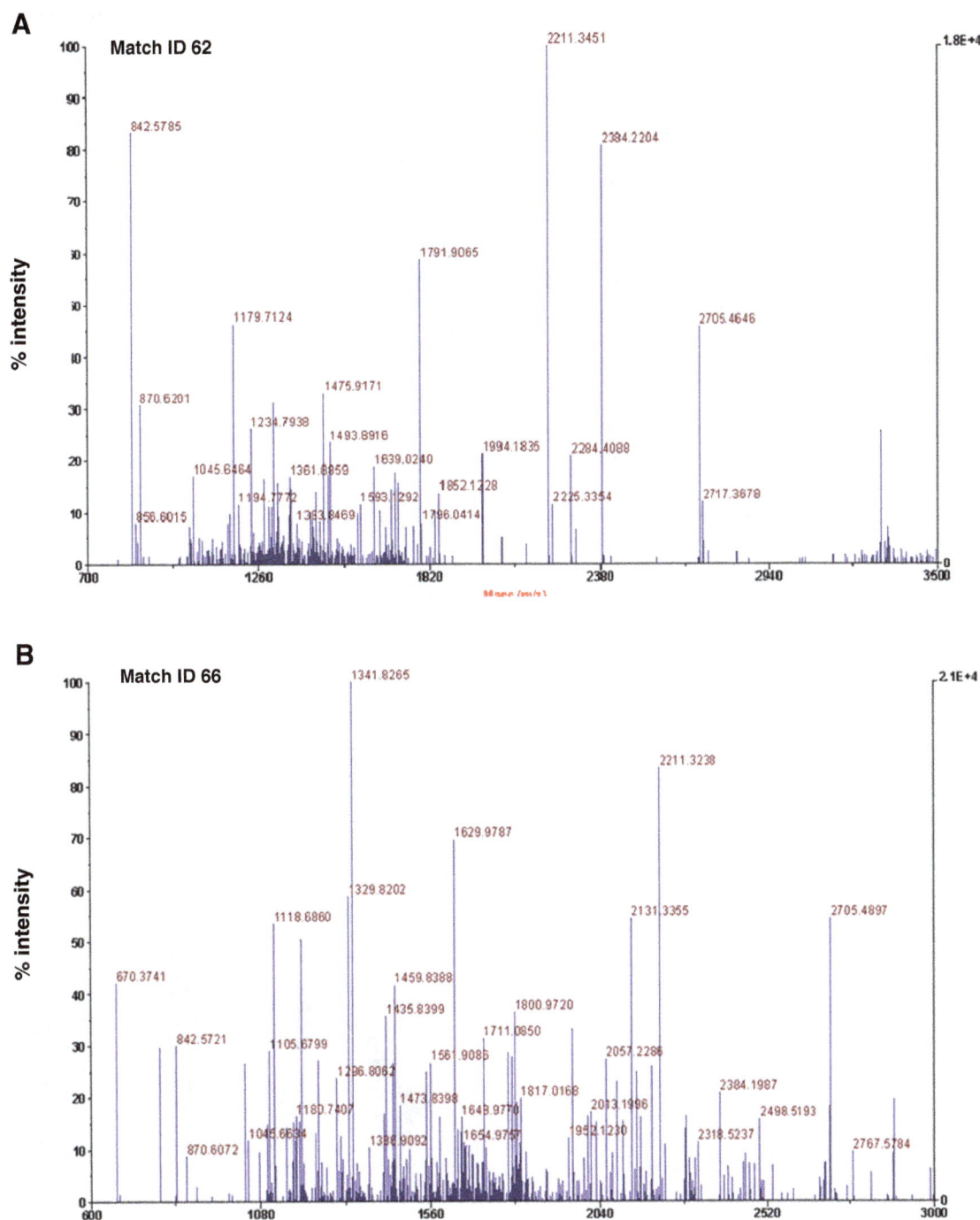

Fig. 5 The fingerprint of protein spots analysis by the mass spectrometer. **a** The match ID 62. **b** The match ID 66

gene expression level for human ZNF792 was very low, suggesting that match ID 62 may not be this protein. The expression levels of SAMHD1 and GFPT1 were also very low. Only the gene expression of VCP was visible (data not shown). Therefore, we went further to identify VCP by Western blot. As shown in Fig. 6, the VCP protein expression profile followed a similar trend to that by 2-DGE, suggesting that match ID 66 may be the protein VCP.

The in vivo biocompatibility results of PU-Au are shown in Fig. 7. Since catheter insertion may elicit an inflammatory reaction of the vein, the venous wall thickening and fibrous capsule formation may serve as indexes for the evaluation of the biocompatibility of the catheters. The normal jugular vein (before catheter insertion) has an average wall thickness of 43 μm. After insertion, the control group (PU catheters; Fig. 7a, b) showed vein thickening and endothelium injury. The fibrous capsule was very thick and connected with the damaged endothelium. The average thickness of the venous wall was about 451 ± 23 μm and that of the fibrous capsule was about 123 ± 4 μm in the control group. On the other hand, PU-Au catheters caused less tissue injury than the control PU catheters. A lumen space existed between PU-Au-coated catheters and the intact endothelium (Fig. 7c, d), though some thrombus could be found. The average thickness of the venous wall was about 49 ± 8 μm and that of the fibrous capsule was about 19 ± 1 μm in the PU-Au catheter group. PU-Au catheters pre-seeded with ECs also showed less tissue injury than the control PU catheters (Fig. 7e, f). However, they did not perform better than the non-seeded PU-Au catheters. The average thickness of the venous wall was about 81 ± 8 μm and that of the fibrous capsule was about 44 ± 7 μm. Considering the extents of vein thickening, fibrous capsule, and endothelium injury, the overall biocompatibility was ranked as PU-Au ≥ PU-Au + ECs > control PU catheters.

Discussion

Traditional PU is solvent borne, which involves the use of a significant amount of toxic organic solvents. The current study employed a waterborne, more environmentally friendly process to produce PU and its nanocomposites. Based on the in vivo studies, vascular catheters coated with PU-Au were found to cause very minor wall thickening of the blood vessels, compared to the blank catheters. This result indicated that the PU-Au nanocomposite was anti-inflammatory, which increased the biocompatibility of the implants in vivo. Therefore, we suggested that PU-Au nanocomposites may be used as an elastic and anti-inflammatory coating material for cardiovascular devices such as stents.

In our previous in vitro study, PU-Au nanocomposites increased the migration of ECs through the PI3K/Akt and FAK signal transduction of bovine ECs [9]. In this study, the

cytoskeleton of HUVEC was found to be more spread on PU-Au, which was consistent with the earlier observation. By 2-DGE, we tried to identify other downstream regulatory molecules for human vascular ECs in the present study. One of them was VCP. VCP is a multifunctional protein with ATPase activity. It is associated with many cellular functions such as membrane transport, protein folding, ubiquitin–proteasome degradation, apoptosis, as well as antigen processing and presentation to lymphocytes [16, 17]. It is also one of the direct regulatory targets of the Akt protein in the Akt pathway [18]. The Akt pathway influences the activity of eNOS which controls the synthesis of nitric oxide (NO) [19]. NO directly influences the proliferation, migration, and vascularization of ECs. The finding regarding the involvement of VCP suggested that PU-Au nanocomposites may upregulate the VCP expression of HUVECs to decrease their apoptosis [20]. VCP was a new protein identified by 2-DGE in this study. Therefore, the superior biocompatibility of PU-Au in vivo may be associated with the immunomodulation/anti-inflammation effects of this material through the stimulation of VCP expression. However, the actual role of VCP in PU-Au biocompatibility remains undefined. The contribution of VCP to biocompatibility will be investigated more carefully using technology such as VCP knockdown in the future studies.

2-DGE has been used to analyze the different patterns of protein expression between the differentiated and undifferentiated stem cells [21]. It has also been used to analyze the

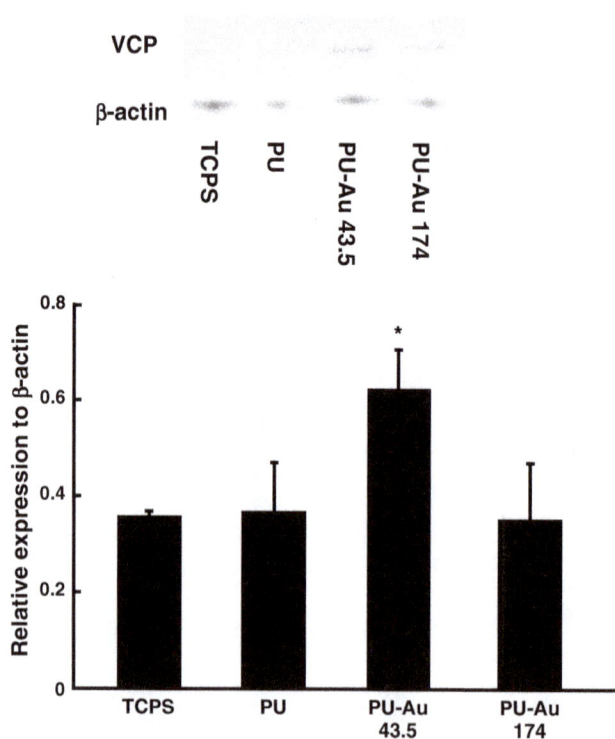

Fig. 6 The protein expression of the suspected protein, VCP. *$p<0.05$ compared to the other groups

differences in protein expression for stem cells from various sources, as well as those with different differentiation status and donors [22, 23]. In addition, 2-DGE can be used to identify the biomarkers for disease diagnosis in human patients [24]. In this study, we tried to use this technology on analyzing the cellular physiology on different biomaterials, which was rarely reported. We expected to establish new proteins that were responsible for changed cell behaviors on biomaterials. The advantage of using 2-DGE as an analytic tool for this purpose is that abundant information can be obtained. Unlike the traditional cDNA array that directly determines the genes of interest, 2-DGE requires the use of MALDI-TOF to correctly identify the proteins. After protein identification by MALDI-TOF, confirmation action by Western blots is also needed [25]. Nevertheless, the gene expression profile obtained by cDNA array not necessarily reflects the actual protein expression. Therefore, 2-DGE is still a useful tool to discover new proteins involved in the cellular response to a material.

Literature has indicated that nanotopography can affect cell attachment, proliferation, migration, and differentiation [26]. Fibroblasts on 13-nm-high nanoislands showed upregulated cell attachment, proliferation, rearrangement of cytoskeleton, and extracellular matrix. Fibroblasts on 95-nm-high nanoislands, on the other hand, decreased cell attachment and cytoskeleton forming [27]. Studies using gene chips have pointed out that nanotopography can upregulate 584 genes which involve cytoskeleton, proliferation, gene transcription and translation, extracellular matrix secretion, and adjust cell signaling [28]. Biomaterials can regulate signaling pathways to promote wound healing. Using 2-DGE to determine how materials affect wound repair, e.g., vascular repair, is a subject worth exploring.

Here in this study, we used 2-DGE as a tool to identify a protein molecule, VCP, in vascular ECs that was activated to various extents by different materials. The research tool 2-DGE is normally used for feasibility studies. Further efforts may be focused on the anti-inflammatory mechanism of PU-Au. Studies may be extended to use a PI3K inhibitor to determine the location of this protein in the signaling pathway, so the relation between the material and ECs can be further configured. This will aid in the design of a biomaterial that effectively enhances angiogenesis with less foreign body reaction.

The increased vascular wall thickness (overgrowth of vascular smooth muscle cells) is normally caused by EC injury, which was an undesired result in our in vivo study.

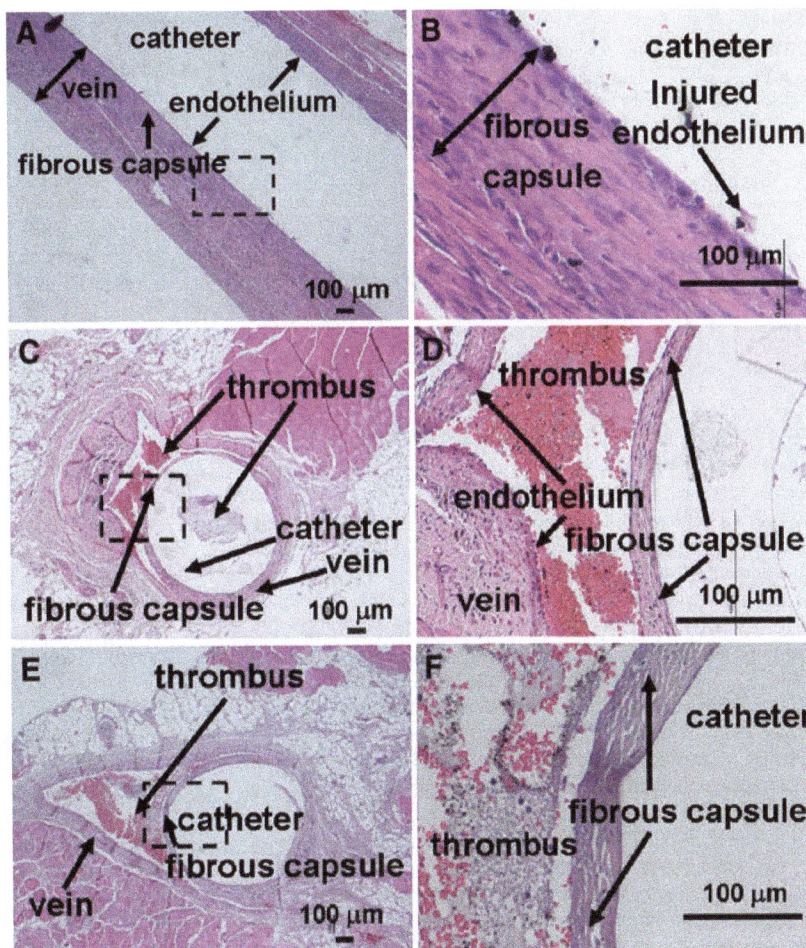

Fig. 7 The H&E staining for the explanted catheters in a, b the control PU catheter group; c, d the group of PU-Au-coated catheters; e, f the group of PU-Au-coated catheters pre-seeded with ECs

The vascular endothelium was not injured by the PU-Au catheters so no increase in wall thickness was observed. The rationale for seeding ECs was to further enhance the biocompatibility of PU-Au. Although PU-Au-coated catheters pre-seeded with ECs did perform better than the control PU group, they showed more foreign body reaction than those without pre-seeding. We thus hypothesized that the ECs pre-seeded on PU-Au may have prevented the direct contact of PU-Au with the vascular lumen and decreased the positive effect of PU-Au on the host ECs in vivo.

AuNPs are widely considered safe. When mixed in materials, AuNPs do not induce oxidative DNA damage [29]. AuNPs possess certain antibacterial activities but do not trigger the reactive oxygen species (ROS) when killing bacteria [30], an antibacterial mechanism used by silver NPs. AuNPs can even scavenge ROS [31]. Adding a small amount of AuNPs in PU may enhance EC growth and provoke a smaller foreign body reaction without ROS damage. In addition to the free radical scavenging effect arising from the presence of AuNPs, PU-Au increases the expression of VCP that can prevent cell apoptosis. Both of these may contribute to the anti-inflammatory mechanism of PU-Au in this study. The anti-inflammatory effect of Au–polymer nanocomposites makes them good coating materials for medical implants.

Acknowledgments This research was supported by the National Research Program for Nanoscience and Technology sponsored by the National Science Council (100-2120-M-002-006). Gold nanoparticles were provided by Gold Nanotech.

References

1. Clark P, Connolly P, Curtis ASG, Dow JAT, Wilkisson CDW (1990) Topographical control of cell behavior: II. Multiple grooved substrata. Development 108:635–644
2. Clark P, Connolly P, Curtis ASG, Dow JAT, Wilkisson CDW (1987) Topographical control of cell behavior: I. Simole step cues. Development 99:439–448
3. Affrosssman S, Hemn G, O'Neill SA, Pethrick RA, Stamm M (1996) Surface topography and composition of deuterated polystyrene–poly(bromostyrene) blends. Macromolecules 29:5010–5016
4. Thapd AT, Webster TJ, Haberstroh KM (2003) Nano-structured polymers enhance bladder smooth muscle cell function. Biomaterials 24:2915–2926
5. Miller DC, Thapa A, Haberstroh KM, Webster TJ (2004) Endothelial and vascular smooth muscle cell function on poly(lactic-co-glycolic acid) with nano-structured surface features. Biomaterials 25:53–61
6. Hsu S, Yen H, Tsai C (2007) The response of articular chondrocytes to type II collagen-Au nanocomposites. Artif Organs 31:854–868
7. Hsu S, Tang C, Tseng H (2008) Gold nanoparticles induce surface morphological transformation in polyurethane and affect the cellular response. Biomacromolecules 9:241–248
8. Hsu S, Chang Y, Tsaie C, Fua K, Wang S, Tseng H (2011) Characterization and biocompatibility of chitosan nanocomposites. Colloids Surf B Biointerfaces 85:198–206
9. Hung H, Wu C, Chien S, Hsu S (2009) The behavior of endothelial cells on polyurethane nanocomposites and the associated signaling pathways. Biomaterials 30:1502–1511
10. Provansal M, Jorgensen C, Lehmann S, Roche S (2011) Two dimensional gel electrophoresis analysis of mesenchymal stem cells. Methods Mol Biol 698:431–442
11. Cruz-Topete D, Jorgensen JO, Christensen B, Sackmann-Sala L, Krusenstjerna-Hafstrøm T, Jara A, Okada S, Kopchick JJ (2011) Identification of new biomarkers of low-dose GH replacement therapy in GH-deficient patients. J Clin Endocrinol Metab 96:2089–2097
12. Zhang W, Liu G, Tang F, Shao J, Lu Y, Bao Y, Yao H, Lu C (2011) Pre-absorbed immunoproteomics: a novel method for the detection of Streptococcus suis surface proteins. PLoS One 6:e21234
13. Guo X, Zhao C, Wang F, Zhu Y, Cui Y, Zhou Z, Huo R, Sha J (2010) Investigation of human testis protein heterogeneity using two-dimensional electrophoresis. J Androl 31:419–429
14. Thompson LJ, Wang F, Proia AD, Peters KG, Jarrold B, Greis KD (2003) Proteome analysis of the rat cornea during angiogenesis. Proteomics 3:2258–2266
15. Klement P, Du YJ, Berry LR, Tressel P, Chan AKC (2006) Chronic performance of polyurethane catheters covalently coated with ATH complex: a rabbit jugular vein model. Biomaterials 27:5107–5117
16. Dai RM, Li CC (2001) Valosin-containing protein is a multi-ubiquitin chain-targeting factor required in ubiquitin–proteasome degradation. Nat Cell Biol 3:740–744
17. Larsen CN, Price JS, Wilkinson KD (1996) Substrate binding and catalysis by ubiquitin C-terminal hydrolases: identification of two active site residues. Biochemistry 35:6735–6744
18. Vandermoere F, El Yazidi-Belkoura I, Slomianny C, Demont Y, Bidaux G, Adriaenssens E, Lemoine J, Hondermarck H (2006) The valosin-containing protein (VCP) is a target of Akt signaling required for cell survival. J Biol Chem 281:14307–14313
19. Karar J, Maity A (2011) PI3K/AKT/mTOR pathway in angiogenesis. Front Mol Neurosci 4:51
20. Wójcik C, Yano M, DeMartino GN (2004) RNA interference of valosin-containing protein (VCP/p97) reveals multiple cellular roles linked to ubiquitin/proteasome-dependent proteolysis. J Cell Sci 117:281–292
21. Zhang AX, Yu WH, Ma BF, Yu XB, Mao FF, Liu W, Zhang JQ, Zhang XM, Li SN, Li MT, Lahn BT, Xiang AP (2007) Proteomic identification of differently expressed proteins responsible for osteoblast differentiation from human mesenchymal stem cells. Mol Cell Biochem 304:167–179
22. Kheterpal I, Ku G, Coleman L, Yu G, Ptitsyn AA, Elizabeth Floyd Z, Gimblez JM (2011) Proteome of human subcutaneous adipose tissue stromal vascular fraction cells versus mature adipocytes based on DIGE. J Proteome Res 10:1519–1527
23. Lazzarotto-Silva C, Binato R, Rocher BD, Costa JA, Pizzatti L, Bouzas LF, Abdelhay E (2009) Similar proteomic profiles of human mesenchymal stromal cells from different donors. Cytotherapy 11:268–277
24. Xu G, Hou CR, Jiang HW, Xiang CQ, Shi N, Yuan HC, Ding Q, Zhang YF (2010) Serum protein profiling to identify biomarkers for small renal cell carcinoma. Indian J Biochem Biophys 47:211–218
25. McNamara LE, Dalby MJ, Riehle MO, Burchmore R (2010) Fluorescence two-dimensional difference gel electrophoresis for biomaterial applications. J R Soc Interface 7:S107–S118
26. Dalby MJ, Riehle MO, Johnstone H, Affrossman S, Curtis AS (2004) Investigating the limits of filopodial sensing: a brief report using SEM to image the interaction between 10 nm high nano-topography and fibroblast filopodia. Cell Biol Int 28:229–236

27. Dalby MJ, Pasqui D, Affrossman S (2004) Cell response to nano-islands produced by polymer demixing: a brief review. IEE Proc Nanobiotechnol 151:53–61

28. Dalby MJ, Riehle MO, Sutherland DS, Agheli H, Curtis ASG (2004) Use of nanotopography to study mechanotransduction in fibroblasts—methods and perspectives. Eur J Cell Biol 83:159–169

29. Nelson BC, Petersen EJ, Marquis BJ, Atha DH, Elliott JT, Cleveland D, Watson SS, Tseng IH, Dillon A, Theodore M, Jackman J (2011) NIST gold nanoparticle reference materials do not induce oxidative DNA damage. Nanotoxicology.

30. Cui Y, Zhao Y, Tian Y, Zhang W, Lü X, Jiang X (2012) The molecular mechanism of action of bactericidal gold nanoparticles on *Escherichia coli*. Biomaterials 33:2327–2333

31. Srinivas P, Patra CR, Bhattacharya S, Mukhopadhyay D (2011) Cytotoxicity of naphthoquinones and their capacity to generate reactive oxygen species is quenched when conjugated with gold nanoparticles. Int J Nanomedicine 6:2113–2122

17

Highly sensitive voltammetric detection of DNA hybridization in sandwich format using thionine-capped gold nanoparticle/reporter DNA conjugates as signal tags

Xuhui Liu · Rui Zhang · Xiaqing Yuan · Lu Liu ·
Yiying Zhou · Qiang Gao

Abstract A highly sensitive sandwich DNA detection method based on voltammetric detection of thionine-capped gold nanoparticle (AuNP)/reporter DNA conjugate tags on gold particle-modified screen-printed carbon electrode (SPCE) was developed. The SPCE was modified with gold particle by electrodeposition of gold on SPCE surface. The DNA sensor was prepared by self-assembly of a thiolated DNA probe on gold particle-modified SPCEs. The sandwich-type system was formed by specific recognition of biosensor surface-confined probe DNA to target DNA, followed by attachment of thionine-capped AuNPs/reporter DNA conjugates. The biosensor is very sensitive because of the large number of electroactive thionine molecules in the thionine-capped AuNPs/reporter DNA conjugates. Under optimal conditions, the dynamic detection range of target DNA was from 1.0×10^{-16} to 1.0×10^{-14} mol L^{-1}, and the detection limit was 0.5×10^{-16} mol L^{-1}. The DNA sensor exhibited selectivity against single-base mismatched DNA.

Keywords DNA · Sandwich · Thionine · Gold nanoparticles/ reporter DNA conjugates · Screen-printed carbon electrode

Introduction

High-sensitivity detection of nucleic acids is essential in clinical diagnosis, pathology, and genetics [1, 2]. Sensitivity of DNA biosensors is usually determined by signal variation amplitude of hybridization event. Increasing efforts have been focused on the improving analysis of DNA by signal amplification to enhance the sensitivity of DNA biosensors [3–5].

Electrochemical method has received considerable attention in the development of DNA biosensors because it is a simple, inexpensive, and sensitive platform [6]. The feasible sensitivity of an electrochemical detection scheme is generally dependent on the amount of electrical charge provided by the labels. Therefore, the sensitivity of detection can be enhanced by increasing the signal elements attached to each oligonucleotide target [7–9]. Gold nanoparticles (AuNPs) have recently drawn great interest in biosensor development because they are biocompatible and inert [10]. AuNPs have been employed to amplify signals by forming a nanoparticle ensemble substrate on electrode, thereby increasing the amount of immobilized probe DNA on the electrode surface [11]. AuNPs are also extensively used as labels because each AuNP can provide thousands of electrochemically detectable elements either by itself [12] or as a carrier for enzymes to produce electroactive materials [13]. Wang et al. used 6-ferrocenylhexanethiol to cap the AuNPs to amplify voltammetric detection of DNA hybridization [14]. In this method, AuNPs were used as carrier for electroactive materials. However, streptavidin was needed to functionalize AuNPs, so that 6-ferrocenylhexanethiol-capped AuNPs could bind to biotin-labeled reporter DNA. In these methods, complex detection procedures necessitate a robust and simple signal-amplification approach.

Thionine is a small electroactive molecule [15] that is usually used as an electrochemical indicator for DNA hybridization because it possesses strong interaction with DNA [16]. Nitrogen atoms of the NH$_2$ moieties of thionine bind strongly to AuNP surfaces. Therefore, AuNPs can be used as labels for the accumulation of thionine on the electrode surface [17, 18]. Through thionine modified AuNPs was prepared and used as tags, the aggregation of AuNPs occurred due to surface charge

X. Liu · R. Zhang · X. Yuan · L. Liu · Y. Zhou · Q. Gao (✉)
Key Laboratory of Applied Surface and Colloid Chemistry, Ministry of Education, School of Chemistry and Chemical Engineering, Shaanxi Normal University, Xi'an 710062, China
e-mail: gaoqiang@snnu.edu.cn

neutralization of cationic thionine molecules and negative citrate-protected AuNPs [19]. Since their introduction in 1996, the use of oligonucleotide-AuNP conjugates (DNA–AuNPs) in DNA sensing has been extensively studied [20]. After the modification of AuNPs with oligonucleotides, the AuNP solution is stable even in the presence of NaCl [21] (Scheme 1).

In this study, we report the preparation of thionine-capped AuNP/reporter DNA conjugates and their application in sandwich-type electrochemical detection of target DNA at low levels. The sandwich-type system is formed by specific recognition of biosensor surface-confined probe DNA to target DNA, followed by successive attachment of thionine-capped AuNPs/reporter DNA conjugates. Fabrication and performance of the DNA sensor are presented.

Experimental

Chemicals

All oligodeoxynucleotides were purchased from Shanghai Sangon Biological Engineering Technology and Services Co., Ltd. (China). Oligodeoxynucleotide concentration was determined by its molar extinction coefficient at 260 nm using UV spectrophotometer (UV-2450, Shimadzu Corporation, Japan). The oligonucleotide sequences for both the probes and the targets are shown in Table 1.

Bovine serum albumin (BSA), thionine, 6-mercaptohexanol (MCH), and $HAuCl_4 \cdot 3H_2O$ were purchased

Table 1 DNA probes and their targets

Name	Sequence (5′ to 3′)
Probe DNA	SH-$(CH_2)_6$-GCTGCTCTGGGTCTCAATGG
Reporter DNA	SH-[PEG]-TCCCCGGCGCCACTGGCCAC
Diblock-DNA	AAAAAAAAAAAAAATTT
Target DNA	GTGGCCAGTGGCGCCGGGGAGGCAGCCA TTGAGACCCAGAGCAGC
Single mismatch DNA	GTGGCCAGTCGCGCCGGGGAGGCAGCCA TTGAGACCCAGAGCAGC
Noncomplementary DNA	AACCAGTCACATATAAAATTCCATCAGTCG CTATGGGTCTCTTAG

from Sigma-Aldrich. Carbon ink (Electrodag 423SS) was obtained from Acheson Colloids (Japan). Phosphate buffer solution (PBS) consisted of 0.01 M phosphate-buffered saline, 0.1 M NaCl, and 3 mM KCl (pH 7.4). Other chemicals and reagents were commercially available and were of analytical grade. Water was obtained from Millipore Milli-Q purification system.

Instrumentation

Cyclic voltammetry (CV) and differential pulse voltammetry (DPV) experiments were performed on a CHI 832 electrochemical workstation (Shanghai, China). All experiments were performed with a conventional three-electrode system using screen-printed carbon electrode (SPCE) as working

Scheme 1 Scheme for sandwich DNA hybridization detection using thionine-capped AuNP/reporter DNA conjugates as signal tags

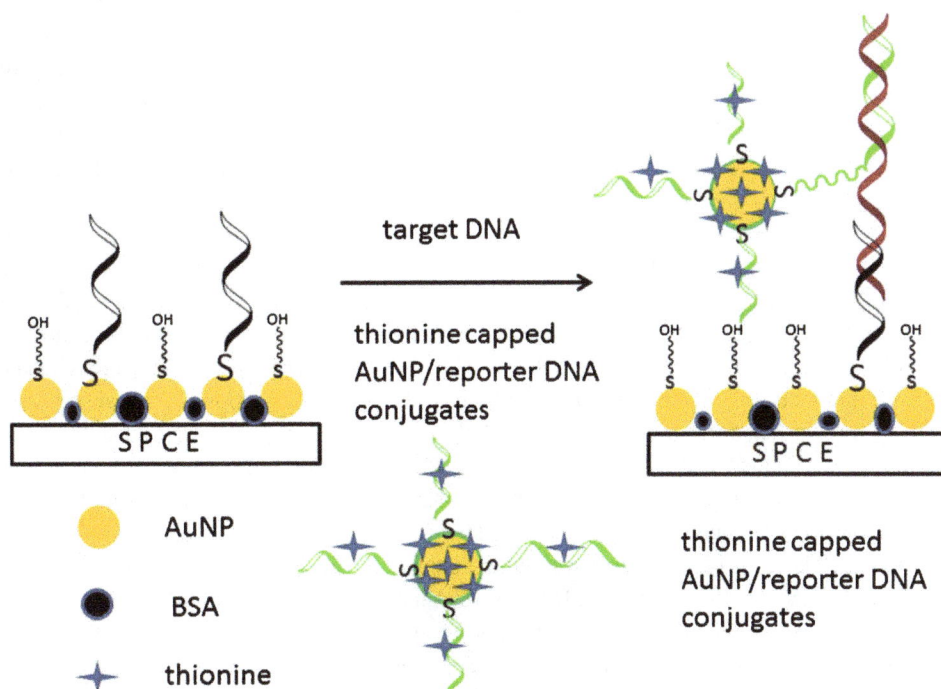

electrode, a platinum foil as counter electrode, and an Ag/AgCl (saturated KCl) as reference electrode. All the potentials were presented in terms of Ag/AgCl electrode potentials.

UV visible spectra from 200 to 800 nm were measured using a Shimadzu UV-2450 spectrophotometer. Transmission electron microscopy (TEM) images were obtained using a JEM-2100 microscope with an accelerating voltage of 200 kV. Scanning electron microscopy (SEM) was performed using a Quanta 200 (FEI, Philips) apparatus. The acceleration voltage was 20 kV.

Synthesis of thionine-capped AuNP/reporter DNA conjugates

AuNPs (13 nm diameter) were prepared by reducing of $HAuCl_4$ with trisodium citrate [22]. Conjugates of reporter DNA-conjugated AuNPs (AuNP/reporter DNA) were synthesized following a published protocol [23, 24]. Briefly, AuNP/reporter DNA conjugates were synthesized by incubating ssDNA, including 1.0 μM reporter DNA and 4.0 μM diblock-DNA, in 1 mL AuNP solution (13 nM). The AuNP/reporter DNA conjugates were left for 16 h and "aged" in salts (0.1 M NaCl, and 10 mM Tris-HCl, pH 7.0) for 24 h. Excess reagents were removed by centrifugation at 15,000 rpm for 30 min. The red precipitate was washed, recentrifuged, and then dispersed in 1 mL water.

AuNP/reporter DNA conjugate solution (1 mL) was mixed with 0.05 mL thionine aqueous solution (20 mM) and stirred effectively for 24 h. The solution was then centrifuged at 15,000 rpm to obtain the precipitate of thionine-capped AuNP/reporter DNA conjugates. Then, the precipitate was washed several times with cold water. The precipitate was redispersed easily in water through sonication. Unbound thionine were removed from the DNA-AuNP solutions by repeated washing and centrifugation.

Immobilization of probe DNA

SPCE was used as base electrode and the electrodeposition of gold on SPCEs was according to the Ref [25]. Probe DNAs were immobilized on gold particle-modified SPCEs by self-assembly. Briefly, 50 μL of 0.5 μM probe DNA in PBS (10 mM phosphate, pH 7.4, and 1 M NaCl) was added to the electrode and incubated overnight at 4 °C. Subsequently, the electrode was washed thrice with PBS, and then 250 μL of 50 mM MCH was added and incubated for 3 h at room temperature. In the present research, MCH was used to block the surface of AuNPs through self-assembly to reduce non-specific binding. Then, the electrode was washed three times with PBS and immersed in PBS containing BSA for 10 min to block electrode surface. Finally, the electrode was washed once to remove the unbound BSA. The resulting electrode was used as DNA sensor.

Target DNA detection and analytical signal recording

Hybridization was performed by incubating the DNA sensor in a target DNA solution for 30 min. After hybridization, the sensor was washed three times with PBS buffer to remove the physically adsorbed target DNA. Subsequently, the thionine-capped AuNP/reporter DNA conjugate solution was coated onto the resultant electrode surface, and the interaction was kept at room temperature for 30 min to obtain a sandwich sensing system. Finally, the DNA sensor was again washed with PBS buffer three times to remove unbound thionine-capped AuNP/reporter DNA conjugates.

Results and discussion

Preparation of thionine-capped AuNP/reporter DNA conjugates

While DNA–AuNPs conjugates typically exploit the well-established strong Au–S chemistry to self-assemble thiolated oligonucleotides at the surface of AuNPs, it remains challenging to precisely control the orientation and conformation of surface-tethered oligonucleotides and finely tune the hybridization ability [23]. Studies with planar Au substrates have shown that poly adenine (polyA) sequences containing multiple consecutive adenines preferentially adsorb Au with high affinity, even comparable to Au–S chemistry. Pei et al. [24] reported a bioconjugation of AuNPs with diblock-DNA containing polyA. The strategy provided a reproducible means to prepare nanoconjugates with well-defined surface density and favorable hybridization ability. Further, unlike thiolated oligonucleotides, diblock oligonucleotides are natural sequences that are essentially free of any modification, hence the synthesis cost is reduced and possible contamination is prevented. Therefore, the diblock-DNA (contains a poly T and a polyA sequence) was used to decrease the density of reporter DNA on AuNP and improve the hybridization of reporter DNA to target DNA. Formation of thionine-capped AuNP/reporter DNA conjugates was confirmed using UV–Vis spectroscopy and TEM. The spectrum of thionine (curve a) in water exhibits characteristic absorption bands at 598 nm, which is a characteristic absorption feature of its monomeric form (Fig. 1) [26]. Curve b in Fig. 1 is the absorption spectrum of AuNP solution, in which the band at 520 nm is a characteristic absorption feature of surface plasmon resonance (SPR) of AuNPs. The characteristic absorbance of reporter ssDNA appeared at 260 nm. Two characteristic absorbances (520 and 260 nm) were observed on the AuNP-ssDNA conjugation (curve c), which proved that the reporter ssDNA was successfully bound to the surface of AuNPs by the formation of stable Au–S bond [27]. Curve d in Fig. 1 is the absorption spectrum of the prepared thionine-capped AuNP/reporter DNA conjugates,

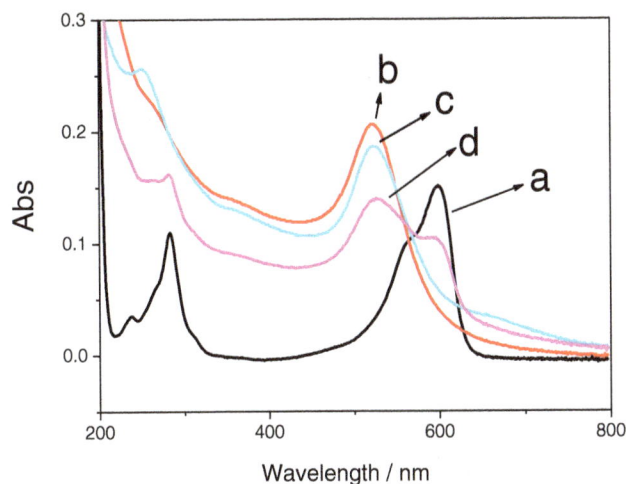

Fig. 1 Absorption spectra. (*a*) thionine, (*b*) AuNP, (*c*) AuNP/reporter DNA conjugates, and (*d*) thionine-capped AuNP/reporter DNA conjugates

in which three characteristic absorbances (thionine at 598 nm, ssDNA at 260 nm and AuNP at 520 nm) are observed. This finding, clearly indicates the adsorption of thionine on AuNP/reporter DNA conjugates, and that the position of SPR exhibited by AuNPs is not influenced by thionine adsorption.

Uniform dispersion of thionine-capped AuNP/reporter DNA conjugates was further confirmed using TEM. Figure 2 shows the TEM images and the corresponding size distribution histograms of AuNP/reporter DNA conjugates before and after thionine addition. The prepared AuNP/reporter DNA conjugates are almost spherically shaped and separated from

one another (Fig. 2a). The AuNP/reporter DNA conjugates still exhibit similar particle sizes (approximately 13 nm) without obvious aggregation after thionine addition, indicating that adsorption of DNA on AuNPs surface can protect AuNPs from aggregation even with the presence of thionine (Fig. 2b).

Detection of target DNA

The morphology of the Au particles modified SPCEs were characterized by SEM. Figure 3 displays the typical SEM images of a bare SPCE (A) and an SPCE after Au electrodeposition (B). A rough and jagged structure with randomly distributed carbon particles was observed for the bare SPCEs (Fig. 3a). After the electrodeposition of Au on the SPCEs, the surfaces of the electrodes were mostly covered with homogeneous particles (Fig. 3b), indicates that Au particles have been electrodeposited on the SPCE surface for the immobilization of probe DNA.

Figure 4 shows the CVs of the DNA sensor before and after hybridization to target DNA as well as the formation of a sandwich complex. No redox current response was observed in the absence of target DNA (curve a). After hybridization to

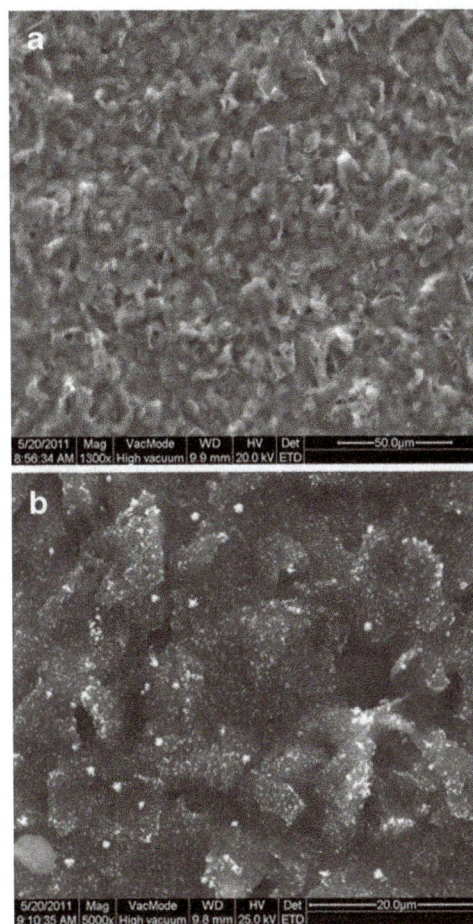

Fig. 2 TEM images of AuNP/reporter DNA conjugates before (**a**) and after (**b**) thionine addition

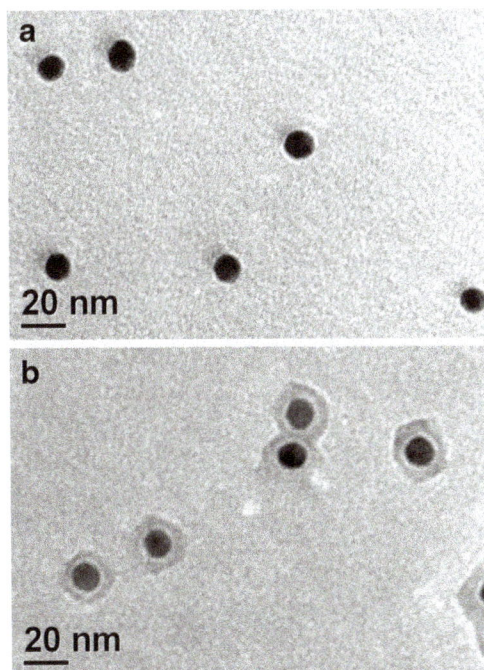

Fig. 3 SEM images of SPCE before (**a**) and after (**b**) electrochemical Au deposition

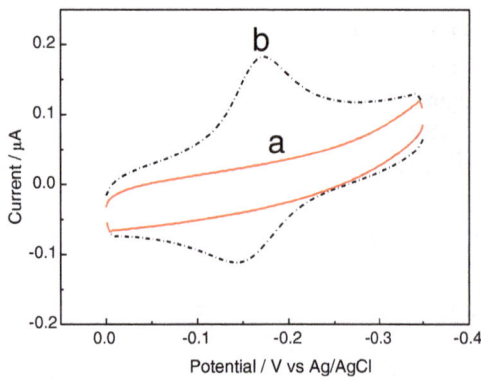

Fig. 4 Cyclic voltammograms of DNA sensors before (*a*) and after (*b*) hybridization to target DNA and formation of sandwich complex

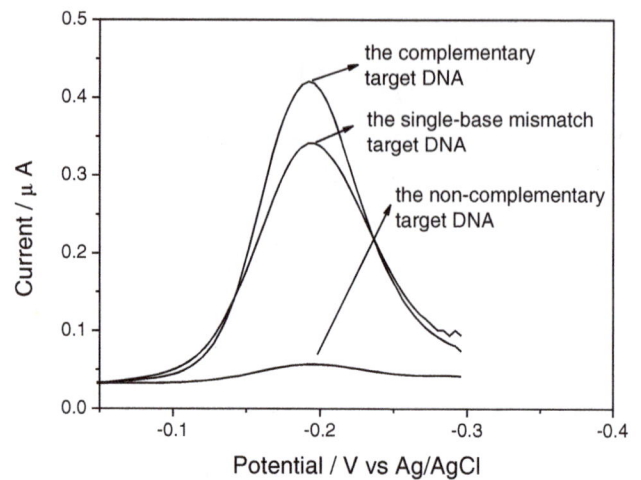

Fig. 6 DPV responses of DNA sensor to the complementary target DNA, single-base mismatch target DNA, and noncomplementary target DNA

the target DNA (100 fM) and formation of a sandwich complex, a well-defined CV corresponding to thionine was observed at formal potential of −0.17 V versus Ag/AgCl. When the sweep rate was 20 mV/s, thionine showed an oxidation peak at −0.15 V and a reduction peak at −0.19 V, the peak-to-peak separation was found to be about 40 mV. These results suggest that the thionine-capped AuNP/reporter DNA conjugates were bound to the DNA sensors after hybridization of target DNA to the immobilized probe DNA and reporter DNA.

Figure 5 shows the DPV response of the DNA sensor for the target DNA with an increase in concentration from 0.1 to 500 fM. Well-defined DPV curves were observed. The currents were proportional to the target DNA concentrations. The DPV response was linearly related to the target concentration across a range of 0.1 to 100 fM. The regression equation was I

(μA)=0.27+0.17 lg [target DNA] (fM), and the correlation coefficient was 0.993. The detection limit was 0.05 fM (S/N=3), lower than 0.5 fM for enzyme-based electrochemical DNA sensors [28] but similar to the lowest values obtained in bioassays based on multifunctional encoded DNA-Au bio bar code amplification [29]. Hence, a method of improving sensitivity and lowering the detection limit is obtained.

Measurement reproducibility, which is estimated as the relative standard deviation of three measurements with different electrodes at different concentrations, was investigated. Results are shown in Fig. 5 and its inset. For a DNA concentration of 100 fM, the relative standard deviation for six

Fig. 5 DPV responses of the DNA sensor to different concentrations of target DNA. Concentrations of target DNA (*bottom to top*): 0.1, 0.3, 1.0, 5.0, 10, 50, 100, and 500 fM. *Inset* linear detection range of 0.1 to 100 fM

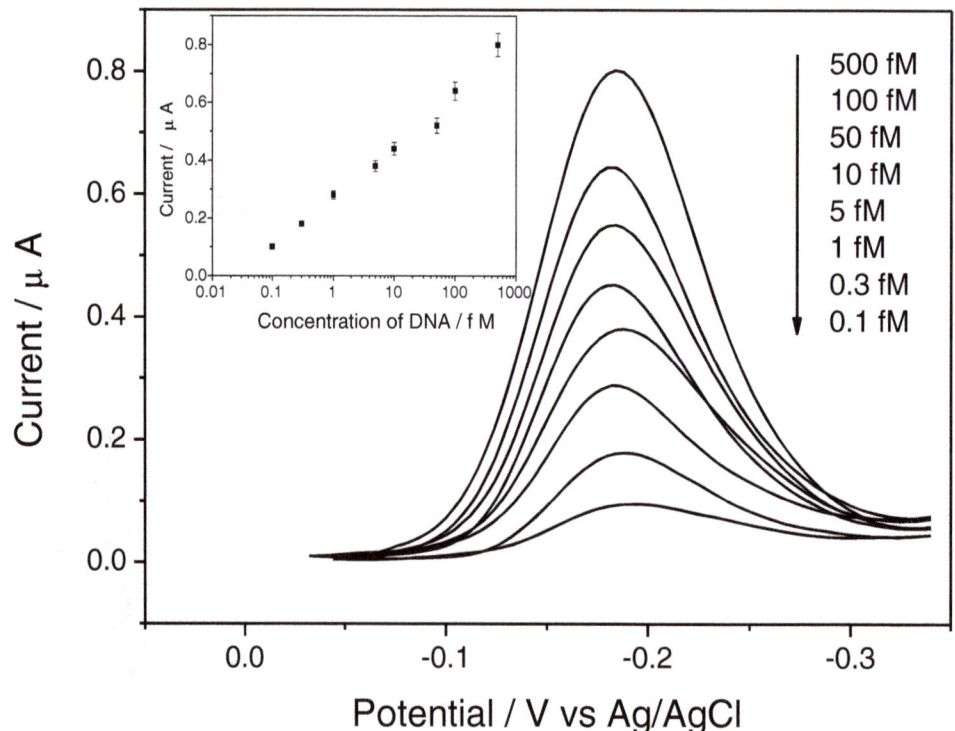

detections with six different sensors was 6.8 %, indicating that measurements using the DNA sensor are highly reproducible.

Detection of sequence-selective hybridization

The DNA sensor was exposed to noncomplementary DNA and single-base mismatch DNA to illustrate its selectivity. Almost no DPV peak was observed in the presence of non-complementary target DNA (Fig. 6). This result demonstrates that the DNA sensor did not respond to the noncomplementary target DNA. Though the DNA sensors responded to the single-base mismatch DNA, the response was weaker than that of the complementary target DNA. The current increments were 0.38 and 0.44 µA for the 10 fM single-base mismatch and complementary target DNA, respectively, indicating that the mismatched target DNA can be distinguished from the complementary targets. These results demonstrate the selectivity of the current assay.

Conclusion

A signal-amplified sandwich-type electrochemical DNA biosensor for the detection of target DNA was developed. Thionine-capped AuNP/reporter DNA conjugates were prepared by attaching electroactive thionine to ssDNA protected AuNP surface. Relatively low detection limits were obtained because of the signal-amplification effect of thionine-capped AuNP/reporter DNA conjugates. Compared with other electrochemical sensors, our system exhibits two main advantages. First, the AuNPs were coated with reporter DNA, which increased the stability of AuNPs, and more thionine can adsorb to AuNPs surface. Second, a simplified detection procedure was used. Thionine-capped AuNP/reporter DNA conjugates were used directly to enhance the signal. This eliminates the need for further procedure such as enzymatic electroactive materials after the formation of a sandwich complex. We believe that our method would be valuable in electrochemical detection of DNA.

Acknowledgments The authors gratefully acknowledge the financial support from the National Nature Science Foundation of China (Nos. 21175089 and 21027007), Program for Changjiang Scholars and Innovative Research Team in University (IRT 1070), and Creative Experimental Project of National Undergraduate Student (CX12022).

References

1. Liu AL, Wang K, Weng SH, Lei Y, Lin LQ, Chen W, Lin XH, Chen YZ (2012) Development of electrochemical DNA biosensors. TrAC Trends Anal Chem 37:101–111
2. Luan QF, Xue Y, Yao X (2010) A simple hairpin DNA sensor for label-free detection of sub-attomole DNA target. Sensors Actuators B Chem 147:561–565
3. Zhang B, Liu BQ, Tang DP, Niessner R, Chen GN, Knopp D (2012) DNA-based hybridization chain reaction for amplified bioelectronic signal and ultrasensitive detection of proteins. Anal Chem 84:5392–5399
4. Su XD, Teh HF, Aung KMM, Zong Y, Gao ZQ (2008) Femtomol SPR detection of DNA–PNA hybridization with the assistance of DNA-guided polyaniline deposition. Biosens Bioelectron 23:1715–1720
5. Huang YQ, Liu XF, Fan QL, Wang LH, Song SP, Wang LH, Fan CH, Huang W (2009) Tuning backbones and side-chains of cationic conjugated polymers for optical signal amplification of fluorescent DNA detection. Biosens Bioelectron 24:2973–2978
6. Sadik OA, Mwilu SK, Aluoch A (2010) Smart electrochemical biosensors: from advanced materials to ultrasensitive devices. Electrochim Acta 55:4287–4295
7. Yu FL, Li G, Qu B, Cao W (2010) Electrochemical detection of DNA hybridization based on signal DNA probe modified with Au and apoferritin nanoparticles. Biosens Bioelectron 26:1114–1117
8. Soreta TR, Henry OYF, OıSullivan CK (2011) Electrode surface nanostructuring via nanoparticle electronucleation for signal enhancement in electrochemical genosensors. Biosens Bioelectron 26:3962–3966
9. Li W, Wu P, Zhang H, Cai CX (2012) Signal amplification of graphene oxide combining with restriction endonuclease for site-specific determination of DNA methylation and assay of methyl-transferase activity. Anal Chem 84:7583–7590
10. Li Y, Schluesener HJ, Xu S (2010) Gold nanoparticle-based biosensors. Gold Bull 43:29–41
11. Song YZ, Song Y, Zhong H (2011) Gold nanoparticle/double-walled carbon nanotube-modified glassy carbon electrode and its application. Gold Bull 44:107–111
12. Kerman K, Saito M, Morita Y, Takamura Y, Ozsoz M, Tamiya E (2004) Electrochemical coding of single-nucleotide polymorphisms by monobase-modified gold nanoparticles. Anal Chem 76:1877–1884
13. Li XM, Fu PY, Liu JM, Zhang SS (2010) Biosensor for multiplex detection of two DNA target sequences using enzyme-functionalized Au nanoparticles as signal amplification. Anal Chim Acta 673:133–138
14. Wang J, Li JH, Baca AJ, Hu JB, Zhou FM, Yan W, Pang DW (2003) Amplified voltammetric detection of DNA hybridization via oxidation of ferrocene caps on gold nanoparticle/streptavidin conjugates. Anal Chem 75:3941–3945
15. Zhu LM, Luo LQ, Wang ZX (2012) DNA electrochemical biosensor based on thionine-graphene nanocomposite. Biosens Bioelectron 35:507–511
16. Dohno C, Stemp EDA, Barton JK (2003) Fast back electron transfer prevents guanine damage by photoexcited thionine bound to DNA. J Am Chem Soc 125:9586–9587
17. Liu SN, Wu P, Li W, Zhang H, Cai CX (2011) Ultrasensitive and selective electrochemical identification of hepatitis C virus genotype 1b based on specific endonuclease combined with gold nanoparticles signal amplification. Anal Chem 83:4752–4758
18. Wang JL, Munir A, Li ZH, Zhou HS (2010) Aptamer-AuNPs conjugates-accumulated methylene blue for the sensitive electrochemical immunoassay of protein. Talanta 81:63–67
19. Zhang XA, Teng YQ, Fu Y, Xu LL, Zhang SP, He B, Wang CG, Zhang W (2010) Lectin-based biosensor strategy for electrochemical assay of glycan expression on living cancer cells. Anal Chem 82:9455–9460

20. Mirkin CA, Letsinger RL, Mucic RC, Storhoff JJ (1996) A DNA-based method for rationally assembling nanoparticles into macroscopic materials. Nature 382:607–609

21. Wang J, Wang LH, Liu XF, Liang ZQ, Song SP, Li WX, Li GX, Fan CH (2007) A gold nanoparticle-based aptamer target binding readout for ATP assay. Adv Mater 19:3943–3946

22. Liu T, Zhao J, Zhang DM, Li GX (2010) Novel method to detect DNA methylation using gold nanoparticles coupled with enzyme-linkage reactions. Anal Chem 82:229–233

23. Hurst SJ, Lytton-Jean AKR, Mirkin CA (2006) Maximizing DNA loading on a range of gold nanoparticle sizes. Anal Chem 78:8313–8318

24. Pei H, Li F, Wan Y, Wei M, Liu H, Su Y, Chen N, Huang Q, Fan C (2012) Designed diblock oligonucleotide for the synthesis of spatially isolated and highly hybridizable functionalization of DNA-gold nanoparticle nanoconjugates. J Am Chem Soc 134:11876–11879

25. Liu J, Yuan XQ, Gao Q, Qi HL, Zhang CX (2012) Ultrasensitive DNA detection based on coulometric measurement of enzymatic silver deposition on gold nanoparticle-modified screen-printed carbon electrode. Sensors Actuators B Chem 162:384–390

26. Das S, Kamat PV (1999) Can H-aggregates serve as light-harvesting antennae? Triplet-triplet energy transfer, between excited aggregates and monomer thionine in aersol-OT solutions. J Phys Chem B 103:209–215

27. Li GJ, Liu LH, Qi XW, Guo YQ, Sun W, Li XL (2012) Development of a sensitive electrochemical DNA sensor by 4-aminothiophenol self-assembled on electrodeposited nanogold electrode coupled with Au nanoparticles labeled reporter ssDNA. Electrochim Acta 63:312–317

28. Liu YH, Li HN, Chen W, Liu AL, Lin XH, Chen YZ (2013) Bovine serum albumin-based probe carrier platform for electrochemical DNA biosensing. Anal Chem 85:273–277

29. Hu KC, Lan DX, Li XM, Zhang SS (2008) Electrochemical DNA biosensor based on nanoporous gold electrode and multifunctional encoded DNA-Au bio bar codes. Anal Chem 80:9124–9130

18

Probing the surface oxidation of chemically synthesised gold nanospheres and nanorods

Blake J. Plowman · Nathan Thompson ·
Anthony P. O'Mullane

Abstract In this study, the electrochemical behaviour of commercially available gold spheres and rods stabilised by carboxylic acid and cetyl trimethyl ammonium bromide (CTAB) moieties, respectively, are investigated. The cyclic voltammetric behaviour in acidic electrolyte is distinctly different with the nanorods exhibiting unusual oxidative behaviour due to an electrodissolution process. The nanospheres exhibited responses typical of a highly defective surface which significantly impacted on electrocatalytic activity. A repetitive potential cycling cleaning procedure was also investigated which did not improve the activity of the nanorods and resulted in deactivating the gold spheres due to decreasing the level of surface defects.

Keywords Gold nanoparticles · Electrocatalysis · Active sites

Introduction

The electrochemical behaviour of metallic nanoparticles has received significant attention due to their applicability as electrocatalysts for a variety of technologically important reactions, in particular, related to fuel cells and electrochemical sensing [1–7]. There are numerous approaches to creating nanoparticles; however, chemical synthesis is particularly attractive given the high degree of control that can be achieved over their size, shape and monodispersity [8]. Central to this is the use of capping agents to ensure that nanoparticles do not agglomerate in solution. The effect of size and shape on electrocatalytic performance of metal nanoparticles has been well documented [9–12]. However, there are other factors that need to be considered when assessing the applicability of an electrocatalyst and that is the presence of defect sites and capping agents. In general, it is agreed that the removal of stabilising species from the surface of nanoparticles is required to ensure access to surface active sites that are responsible for electrocatalytic activity [2]. In many cases, this can be quite involved and may perturb the shape and size of the nanomaterial in question. Electrochemical cleaning routes are often used which employ oxidation of the surface to remove the capping agent [13, 14]. Indeed, repetitive cycling in acidic electrolyte is often used even for bare gold surfaces and nanoparticles to generate a clean sample [15, 16]. Given the extensive use of gold nanomaterials in catalysis, electrocatalysis, photocatalysis and sensing and their widespread commercial availability, we, therefore, in this work, explore the effect of electrochemical pretreatment of gold nanospheres and nanorods immobilised on a glassy carbon support electrode that are capped with carboxylic acid and cetyl trimethyl ammonium bromide (CTAB) via repetitive potential cycling. Significantly, cyclic voltammetric characterisation in supporting electrolyte alone gives important insights in relation to their oxidation potential as well as their behaviour as electrocatalysts for reactions such as H_2O_2 oxidation and reduction and hydrazine oxidation.

B. J. Plowman · N. Thompson · A. P. O'Mullane (✉)
School of Applied Sciences, RMIT University, GPO Box 2476V,
Melbourne, VIC 3001, Australia
e-mail: anthony.omullane@qut.edu.au

A. P. O'Mullane
School of Chemistry, Physics and Mechanical Engineering,
Queensland University of Technology, GPO Box 2434, Brisbane,
QLD 4001, Australia

Experimental

Gold nanoparticles stabilised with carboxylic acid (10 ± 2 and 25 ± 2 nm diameter and monodispersity <2 % coefficient of variation (CV)) and gold nanorods stabilised with CTAB (10×50 nm and 25×256 nm, 95 % rods and <10 % CV) as quoted by the manufacturer were purchased from Nanopartz™ Accurate™. Aqueous 1 M H_2SO_4 (Merck) solutions were made with deionised water (resistivity 18.2 MΩ cm) purified using a Milli-Q reagent deioniser (Millipore). Hydrazine (BDH), hydrogen peroxide 30 % (w/w) (Sigma) and $KAuBr_4$ (Sigma) were used as received.

Electrochemical experiments were conducted with a CH Instruments (CHI760C) potentiostat at 20±2 °C. A glassy carbon (GC) plate electrode (0.158 cm^2 HTW) was used as the supporting working electrode and polished using an aqueous 0.3-μm alumina slurry, sonicated in deionised water, and dried with a flow of nitrogen gas. Gold nanomaterials with an equivalent optical density of 1 were dropcast (5 μL of 10 nm (9 nM), 25 nm (0.5 nM), 10×50 nm (0.9 nM), 25×256 nm (0.04 nM)) on the GC electrode and allowed to air dry. The bulk gold electrode (BAS) was 1.6 mm in diameter and polished using an aqueous 0.3-μm alumina slurry, sonicated in deionised water, and dried with a flow of nitrogen gas. The reference electrode was Ag/AgCl (aqueous 3 M KCl), and the counter electrode was a platinum coil. All electrochemical experiments were commenced after degassing solutions with nitrogen for 10 min.

TEM measurements were performed with a JEOL 1010 TEM operated at an accelerating voltage of 100 kV after dropcasting the nanomaterials on to a carbon-coated TEM grid.

Results and discussion

The electrochemistry of gold has been widely studied and is often regarded as a model system consisting of an extended double layer region and a monolayer oxide formation/reduction region as shown in Fig. 1. However, numerous studies have shown that gold is not as inert as its d^{10} configuration suggests which accounts for its pronounced catalytic and electrocatalytic activity [6, 17–24]. This has been attributed to active sites on the surface that consists of atoms or clusters of atoms that have low co-ordination number and have the ability to partake in electrocatalytic reactions [21, 22, 25, 26]. Recent work by Scholz has demonstrated in the case of gold that active or defect sites are located on the asperities of an electrode surface which are the loci of partially filled d orbitals that can stabilise free radical intermediates [27, 28]. These sites have been shown to dictate electrocatalytic reaction rates for inner sphere reactions that involve free radical intermediates such as oxygen reduction and hydrogen evolution. This was concluded via treatment of gold with Fenton's reagent which dissolved surface asperities and knocked out these active sites as evidenced by a decrease in activity for the reactions mentioned. These active sites were also shown by the same group to be highly important for electrodeposition processes. In a study on platinum electrodeposition on gold electrodes, it was demonstrated that gold

electrodes treated with OH$^{\bullet}$ radicals resulted in a reduction in the number of active sites for Pt nucleation [29]. Prior to this work, Kolb investigated the cyclic voltametric behaviour of single crystal gold surfaces where it was found that there was a significant peak prior to the monolayer oxide formation process that could be attributed to the level of surface defects which were step edges on Au(111) electrodes [30, 31]. This peak was labelled as OA1 which can also be seen for the bulk polycrystalline electrode shown in Fig. 1. Burke et al. have shown that this peak can be enhanced, as well as other processes well within the double layer region, by thermal and electrochemical activation of bulk gold electrodes [32–34]. This creates a disordered surface which is prone to oxidation at potentials below that for monolayer oxide formation.

To investigate if this type of cyclic voltammetric behaviour is observed at gold nanomaterials, commercially available gold spheres and rods were utilised (Fig. 2). The spheres are capped with carboxylic acid and the rods with cetyl trimethyl ammonium bromide (CTAB). The surfactants were used to prevent aggregation and in the latter case, provide directional growth. The size and shape of these nanomaterials were confirmed by TEM imaging (insets of Fig. 2). It is immediately apparent that the cyclic voltammetric responses for the spheres and rods are significantly different to each other as well as that for a bulk gold electrode (Fig. 1). A distinct OA1 peak occurs prior to monolayer oxide formation on the spherical gold nanoparticles of 10-nm diameter (Fig. 2a). The OA1 process is even more pronounced at the 25-nm-diameter nanoparticles (Fig. 2c) in relation to the charge passed during oxide formation. As discussed, this peak has been attributed to the presence of surface defect sites when single crystal electrodes were studied but was recently confirmed by Compton [35] at electrodeposited gold nanoparticles which were devoid of capping agents. The oxidation product is speculative in nature but Burke has attributed it to the formation of hydrous oxide type species [26]. Later, Bard demonstrated that incipient oxides are formed on well-polished bulk gold surfaces

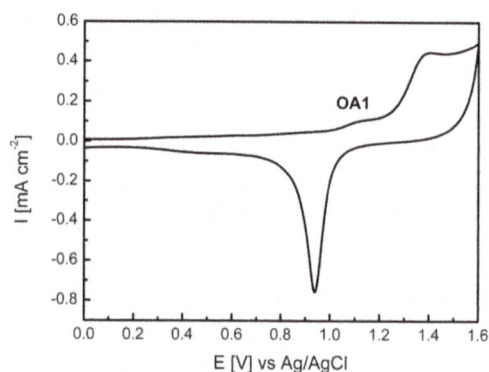

Fig. 1 Cyclic voltammogram recorded in 1 M H_2SO_4 at 100 mV s^{-1} for a bulk gold electrode (d=1.6 mm)

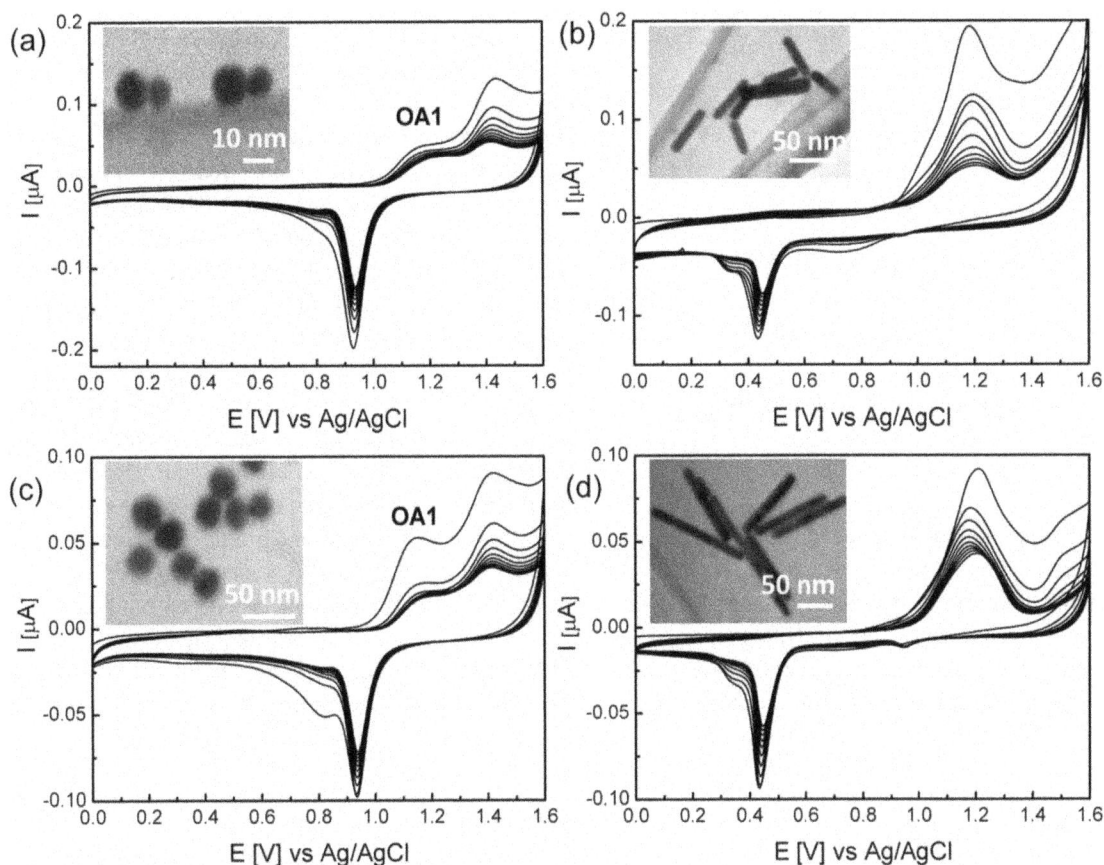

Fig. 2 Cyclic voltammograms recorded in 1 M H₂SO₄ at 100 mV s⁻¹ for **a** 10-nm Au spheres, **b** 10×50 nm Au rods, **c** 25-nm Au spheres and **d** 25× 256 nm Au rods. The *insets* are TEM images of the nanomaterials

with a coverage of 0.2 of a monolayer at potentials within the double layer region [36]. There are several minor features in the double layer region at ca. 0.50 and 0.85 V for the 10-nm-diameter nanospheres (Fig. 2a) which are more clearly evident from the magnified double layer region shown in Fig. 3. These features have been attributed to the oxidation of adatoms or clusters of adatoms on the surface with low co-ordination number. They were also confirmed to be Faradaic in nature by using large-amplitude Fourier-transformed ac voltammetry and involved in electrocatalytic reactions [25]. These features and the OA1 process are stable upon repetitive cycling for 10 and 25-nm nanoparticles but the overall response does decrease indicating a reduced surface area. However, this demonstrates that the carboxylic acid functional group does not block access to the gold surface.

A dramatic difference in behaviour, however, is observed at the gold nanorods. The onset potential for surface oxidation is similar to the spheres; however, it has a distinctly different shape. It is more reminiscent of a stripping process associated with anodic stripping voltammetry than surface oxide formation which proceeds via OH⁻ ion adsorption followed by a place exchange reaction and further oxidation to a Au(III) oxide. The major reduction peak is also shifted by a significant amount to ca. 0.42 V for the nanorods compared to the spheres

where the oxide reduction peak occurred at 0.92 V. Therefore, we believe that the reduction peak seen for the nanorods is not related to surface oxide reduction but an entirely different process. CTAB contains Br⁻, and it is known that gold in the presence of halides oxidises at potentials below that seen for oxide formation [37]. Therefore, the large peak seen on the forward sweep is not due to the OA1 process but rather the electrodissolution of gold. On the reverse sweep, this soluble species which will be present at the electrode surface on the

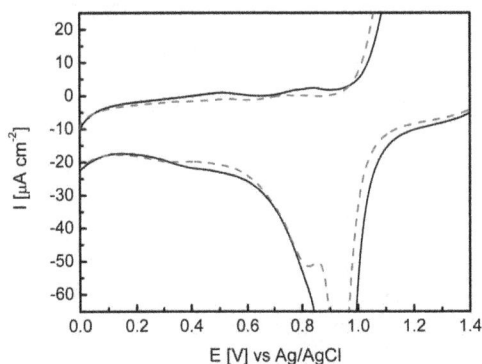

Fig. 3 Cyclic voltammograms recorded in 1 M H₂SO₄ at 100 mV s⁻¹ for 10-nm Au spheres (*black line*) and 25-nm Au spheres (*red dash lines*) showing the double layer region only

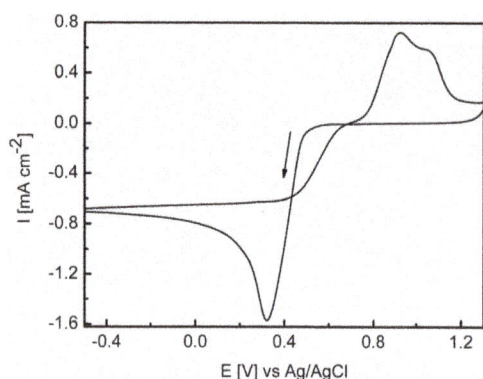

Fig. 4 Cyclic voltammogram obtained at a GC electrode in 10 mM KAuBr$_4$ recorded at 10 mV s^{-1}

timescale of this experiment will be redeposited on the electrode at potentials <0.60 V. Remarkably, this electrodissolution/re-deposition process is quite stable with only a slight decrease in signal intensity with cycling. The same phenomenon is seen with the higher aspect ratio nanorods, and therefore, the difference in the oxidation onset

potential compared to spheres is unlikely to be a size effect. Recent studies have shown that a gold electrode cycled in an aqueous solution of CTAB also undergoes an electrodissolution process where Au$^+$ ions are stabilised by bromide in CTAB which are then redeposited on the reverse sweep at ca. 0.45 V vs AgCl [38]. This observation is also consistent with the oxidation of citrate-capped gold nanoparticles in the presence of KBr which occurred at potentials below that in the absence of bromide in the electrolyte [37]. To investigate this further, gold was electrodeposited from an aqueous solution of 1 mM KAuBr$_4$ on to a GC electrode and is shown in Fig. 4. The peak potential for the deposition of AuBr$_4^-$ to Au^0 is at 0.33 V which is quite close to that observed in Fig. 2b, d. On the subsequent positive sweep, the electrodissolution of gold can be observed with two peaks at 0.92 and 1.06 V. The overall profile of this response is quite similar to that seen for the oxidation of gold nanorods in H$_2$SO$_4$ and supports the conclusion that regular monolayer oxide formation does not occur to a significant extent on the CTAB-stabilised nanorods. However, it should be noted that a minor process can be observed on the negative sweep at

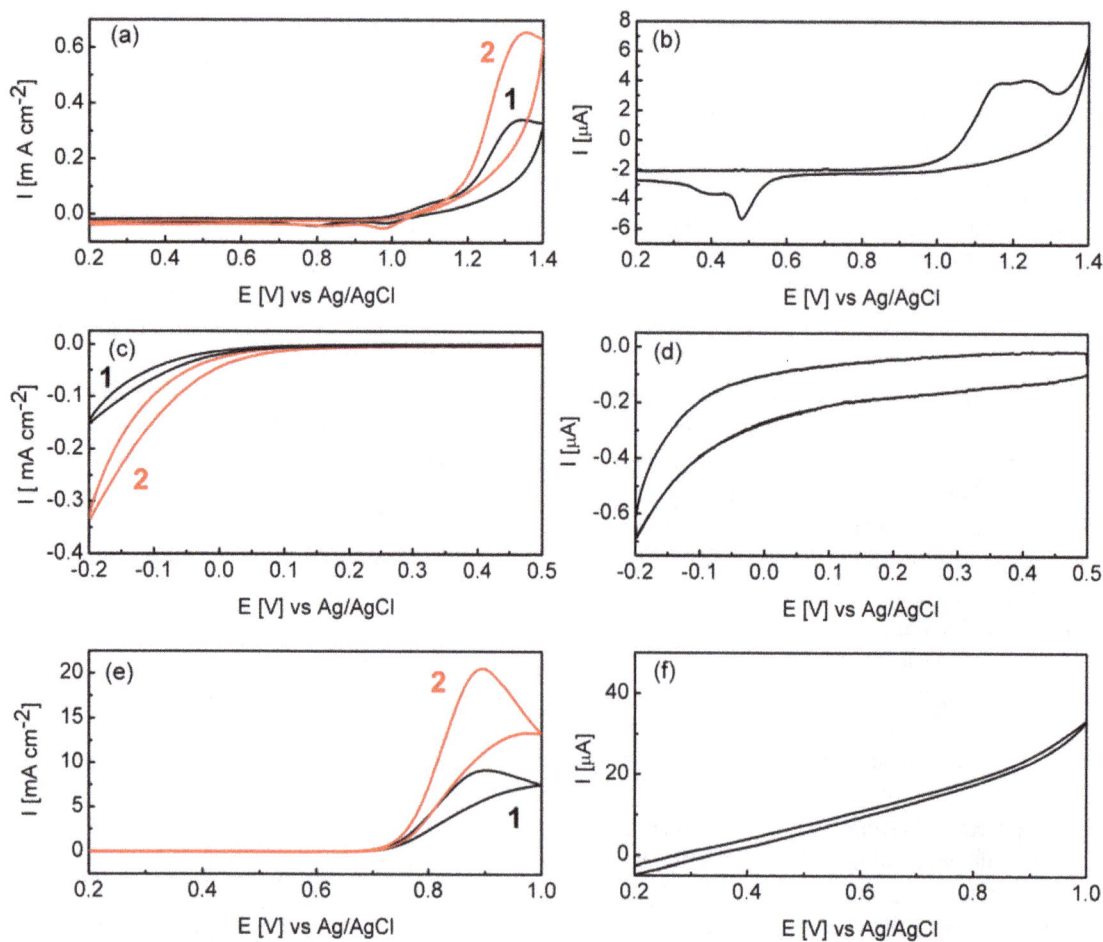

Fig. 5 Cyclic voltammograms in 1 M H$_2$SO$_4$ containing **a–d** 20 mM H$_2$O$_2$ recorded at 20 mV s^{-1} and **e–f** 50 mM hydrazine recorded at 50 mV s^{-1} for 10-nm (1) and 25-nm Au spheres (2). **b, d, f** Data for 25×256 nm rods

Fig. 6 Cyclic voltammograms recorded in 1 M H_2SO_4 and 20 mM H_2O_2 at 20 mV s^{-1} at **a** 25-nm Au spheres and **b** 25× 256 nm rods that are pristine (1) and electrochemically cleaned (2)

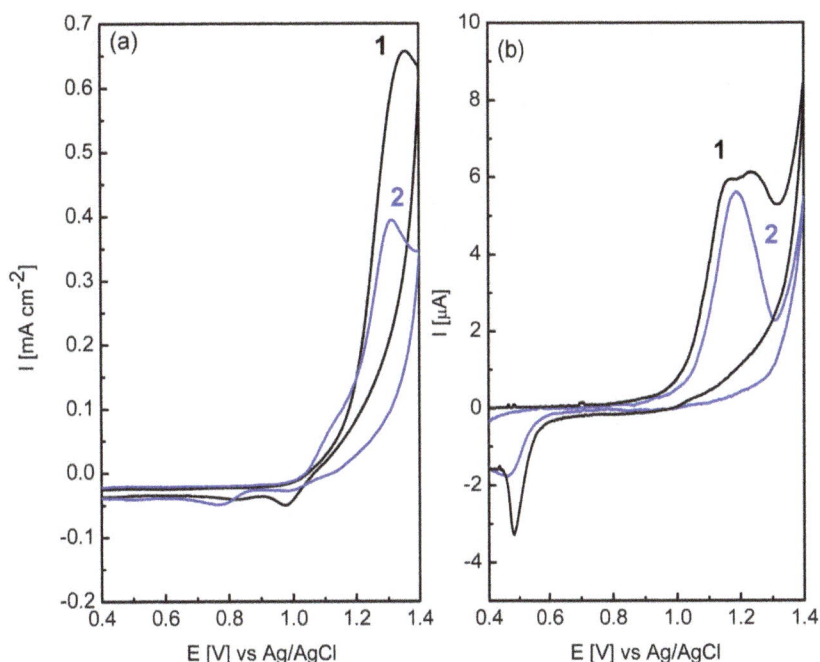

0.94 V at the higher aspect ratio nanorods (25×256 nm) (Fig. 2d) indicating some access to the underlying gold surface.

The oxidation of H_2O_2 was chosen to probe the effect of both the capping agent and the role of the OA1 process in electrocatalysis (Fig. 5a, b) where it is clear that activity at the gold nanorods is completely suppressed. A cathodic peak is present at 0.48 V (Fig. 5b) which suggests that the electrodissolution/re-deposition process described for Fig. 2c still occurs. Koper [39] reported that PVP-stabilised Pt could be cleaned in a solution of H_2O_2/H_2SO_4 where Pt catalyses the decomposition of H_2O_2 into water and O_2. It was suggested that PVP was physically removed by the generated oxygen bubbles. The electrooxidation of peroxide on gold also generates oxygen; however, in this system, it appears that this is insufficient to remove CTAB from the gold rods. The incipient hydrous oxide adatom mediator (IHOAM) model [22, 40] of electrocatalysis predicts that reactions are mediated by a surface-confined redox couple based on the oxidation of active surface sites (M*) to hydrous oxide species (OA1 process). The gold spheres show a clear oxidation process indicating that the carboxylic acid group allows H_2O_2 to interact with the surface. In Fig. 5a, current density is presented which indicates the larger nanoparticles have the highest specific activity. This may be related to the intensity of the OA1 process seen in the blank electrolyte and correlates with the IHOAM model which indicates that the formation of a hydrous oxide species at surface defect sites mediates electrocatalytic oxidation reactions. The potential at which this occurs is also consistent with previous studies on electrochemically activated bulk gold electrodes [34].

The electrocatalytic reduction of H_2O_2 (Fig. 5c, d) and oxidation of hydrazine (Fig. 5e, f) were also investigated, and the same trend was observed, i.e. activity is completely shut down in the case of the nanorods due to the presence of CTAB on the surface and that the specific activity of the larger gold spheres was higher. Even though all reactions investigated occur at distinctly different onset potentials, the outcome is the same in that the nanoparticles with the most distinctive OA1 process (in particular on the first sweep) is the most active. This suggests that even though other M*/hydrous oxide transitions are not clearly visible by conventional dc voltammetry in the double layer region, which require more sophisticated experiments such as large-amplitude Fourier-transformed ac voltammetry to be observed [25], that the OA1 process may in general be indicative of a highly active surface.

The effect of a repetitive cycling cleaning procedure was then investigated (Fig. 6). The procedure involved cycling between 0.0 and 1.60 V in 1 M H_2SO_4 at 100 mV s^{-1} for 20 cycles. For the case of the 25-nm-diameter nanoparticles, there is a decrease in specific activity after the cleaning procedure (note the reduced surface area after cycling is accounted for) which is consistent with the work of Compton who showed a decrease in the activity of electrodeposited gold nanoparticles for the oxygen reduction reaction after a similar procedure [35]. Recent work by Mayrhofer has shown that if a polycrystalline gold electrode is cycled in an acidic solution to potentials where oxide formation occurs, as is the case here, then the electrodissolution of gold occurs [41]. In this study, the lower activity can be correlated with the decrease in intensity of the OA1 process upon repetitive cycling

(Fig. 2c) as the reduction in surface area via this electrodissolution process has been accounted for. This demonstrates that the density of defect sites which are responsible for the electrooxidation of hydrogen peroxide decreases. The same cleaning procedure was applied to the gold nanorods which showed that this approach is totally ineffective in generating an electrocatalytically active surface (Fig. 6b).

Conclusions

The electrochemical behaviour of commercially available gold spheres and rods in acidic electrolyte is distinctly different due to the capping agent employed on the surface. The presence of CTAB resulted in an electrodissolution/redeposition process whereas the presence of a carboxylic group allowed surface oxide processes to be observed which indicated the presence of a high level of defect sites. Significantly, a repetitive potential cycling cleaning protocol did not improve the electrocatalytic performance of the nanorods and also led to a slight deactivation of the gold spheres due to a decrease in the number of surface defect sites.

Acknowledgment AOM gratefully acknowledges funding from the Asian Office of Aerospace Research and Development (FA2386-13-1-4073).

References

1. O'Mullane AP (2014) From single crystal surfaces to single atoms: investigating active sites in electrocatalysis. Nanoscale 6:4012–4026
2. Kleijn SEF, Lai SCS, Koper MTM, Unwin PR (2014) Electrochemistry of nanoparticles. Angew Chem Int Ed 53:3558–3586
3. Campbell CT (2013) The energetics of supported metal nanoparticles: relationships to sintering rates and catalytic activity. Acc Chem Res 46:1712–1719
4. Saha K, Agasti SS, Kim C, Li X, Rotello VM (2012) Gold nanoparticles in chemical and biological sensing. Chem Rev 112:2739–2779
5. Herves P, Perez-Lorenzo M, Liz-Marzan LM, Dzubiella J, Lu Y, Ballauff M (2012) Catalysis by metallic nanoparticles in aqueous solution: model reactions. Chem Soc Rev 41:5577–5587
6. Saint-Lager MC, Laoufi I, Bailly A, Robach O, Garaudee S, Dolle P (2011) Catalytic properties of supported gold nanoparticles: new insights into the size-activity relationship gained from in operando measurements. Faraday Discuss 152:253–265
7. Plowman BJ, Bhargava SK, O'Mullane AP (2011) Electrochemical fabrication of metallic nanostructured electrodes for electroanalytical applications. Analyst 136:5107–5119
8. Sardar R, Funston AM, Mulvaney P, Murray RW (2009) Gold nanoparticles: past, present, and future. Langmuir 25:13840–13851
9. Perez-Alonso FJ, McCarthy DN, Nierhoff A, Hernandez-Fernandez P, Strebel C, Stephens IEL, Nielsen JH, Chorkendorff I (2012) The effect of size on the oxygen electroreduction activity of mass-selected platinum nanoparticles. Angew Chem Int Ed 51:4641–4643
10. Rhee CK, Kim B-J, Ham C, Kim Y-J, Song K, Kwon K (2009) Size effect of Pt nanoparticle on catalytic activity in oxidation of methanol and formic acid: comparison to Pt(111), Pt(100), and polycrystalline Pt electrodes. Langmuir 25:7140–7147
11. Chen W, Chen S (2009) Oxygen electroreduction catalyzed by gold nanoclusters: strong core size effects. Angew Chem Int Ed 48:4386–4389
12. Wang C, Daimon H, Onodera T, Koda T, Sun S (2008) A general approach to the size- and shape-controlled synthesis of platinum nanoparticles and their catalytic reduction of oxygen. Angew Chem Int Ed 47:3588–3591
13. Hebié S, Cornu L, Nappom TW, Rousseau J, Kokoh BK (2013) Insight on the surface structure effect of free gold nanorods on glucose electrooxidation. J Phys Chem C 117:9872–9880
14. El-Deab MS, Sotomura T, Ohsaka T (2005) Oxygen reduction at electrochemically deposited crystallographically oriented Au(100)-like gold nanoparticles. Electrochem Commun 7:29–34
15. Fischer LM, Tenje M, Heiskanen AR, Masuda N, Castillo J, Bentien A, Émneus J, Jakobsen MH, Boisen A (2009) Gold cleaning methods for electrochemical detection applications. Microelectron Eng 86:1282–1285
16. Carvalhal RF, Sanches Freire R, Kubota LT (2005) Polycrystalline gold electrodes: a comparative study of pretreatment procedures used for cleaning and thiol self-assembly monolayer formation. Electroanalysis 17:1251–1259
17. Shim JH, Kim J, Lee C, Lee Y (2011) Electrocatalytic activity of gold and gold nanoparticles improved by electrochemical pretreatment. J Phys Chem C 115:305–309
18. Lai SCS, Dudin PV, Macpherson JV, Unwin PR (2011) Visualizing zeptomole (electro)catalysis at single nanoparticles within an ensemble. J Am Chem Soc 133:10744–10747
19. Zeng J, Zhang Q, Chen J, Xia Y (2010) A comparison study of the catalytic properties of Au-based nanocages, nanoboxes, and nanoparticles. Nano Lett 10:30–35
20. Zhou X, Xu W, Liu G, Panda D, Chen P (2009) Size-dependent catalytic activity and dynamics of gold nanoparticles at the single-molecule level. J Am Chem Soc 132:138–146
21. Burke LD (2004) Scope for new applications for gold arising from the electrocatalytic behavior of its metastable surface states. Gold Bull 37:125–135
22. Burke LD, Nugent PF (1998) The electrochemistry of gold. II The electrocatalytic behavior of the metal in aqueous media. Gold Bull 31:39–50
23. Plowman BJ, O'Mullane AP, Bhargava SK (2011) The active site behaviour of electrochemically synthesised gold nanomaterials. Faraday Discuss 152:43–62
24. Plowman BJ, Field MR, Bhargava SK, O'Mullane AP (2013) Exploiting the facile oxidation of evaporated gold films to drive electroless silver deposition for the creation of bimetallic Au/Ag surfaces. ChemElectroChem 1:76–82
25. Lertanantawong B, O'Mullane AP, Surareungchai W, Somasundrum M, Burke LD, Bond AM (2008) Study of the underlying electrochemistry of polycrystalline gold electrodes in aqueous solution and electrocatalysis by large amplitude fourier transformed alternating current voltammetry. Langmuir 24:2856–2868
26. Burke LD, Nugent PF (1997) The electrochemistry of gold: I. The redox behavior of the metal in aqueous media. Gold Bull 30:43–53
27. Nowicka AM, Hasse U, Sievers G, Donten M, Stojek Z, Fletcher S, Scholz F (2010) Selective knockout of gold active sites. Angew Chem Int Ed 49:3006–3009
28. Nowicka A, Hasse U, Donten M, Hermes M, Stojek Z, Scholz F (2011) The treatment of Ag, Pd, Au and Pt electrodes with OH• radicals reveals information on the nature of the electrocatalytic centers. J Solid State Electrochem 15:2141–2147

29. Sievers G, Hasse U, Scholz F (2012) The effects of pretreatment of polycrystalline gold with OH˙ radicals on the electrochemical nucleation and growth of platinum. J Solid State Electrochem 16:1663–1673

30. Schneeweiss MA, Kolb DM, Liu D, Mandler D (1997) Anodic oxidation of Au(111). Can J Chem 75:1703–1709

31. Kolb DM (2000) Structure studies of metal electrodes by in-situ scanning tunneling microscopy. Electrochim Acta 45:2387–2402

32. Burke LD, Ahern AJ, O'Mullane AP (2002) High energy states of gold and their importance in electrocatalytic processes at surfaces and interfaces. Gold Bull 35:3–10

33. Burke LD, Hurley LM, Lodge VE, Mooney MB (2001) The effect of severe thermal pretreatment on the redox behavior of gold in aqueous acid solution. J Solid State Electrochem 5:250–260

34. Burke LD, O'Mullane AP (2000) Generation of active surface states of gold and the role of such states in electrocatalysis. J Solid State Electrochem 4:285–297

35. Wang Y, Laborda E, Crossley A, Compton RG (2013) Surface oxidation of gold nanoparticles supported on a glassy carbon electrode in sulphuric acid medium: contrasts with the behaviour of 'macro' gold. Phys Chem Chem Phys 15:3133–3136

36. Rodriguez-Lopez J, Alpuche-Aviles MA, Bard AJ (2008) Interrogation of surfaces for the quantification of adsorbed species on electrodes: oxygen on gold and platinum in neutral Media. J Am Chem Soc 130:16985

37. Masitas RA, Zamborini FP (2012) Oxidation of highly unstable <4 nm diameter gold nanoparticles 850 mV negative of the bulk oxidation potential. J Am Chem Soc 134:5014–5017

38. Nambiar SR, Aneesh PK, Rao TP (2014) Studies on deposition of gold atomic clusters on to polycrystalline gold electrode from aqueous cetyl trimethyl ammonium bromide solutions. J Electroanal Chem 722–723:60–67

39. Monzó J, Koper MTM, Rodriguez P (2012) Removing polyvinyl-pyrrolidone from catalytic Pt nanoparticles without modification of superficial order. ChemPhysChem 13:709–715

40. Burke LD, Casey JK, Morrissey JA, Murphy MM (1991) Incipient hydrous oxide/adatom mediator (IHOAM) model of electrocatalysis. Bull Electrochem 7:506–511

41. Cherevko S, Topalov AA, Katsounaros I, Mayrhofer KJJ (2013) Electrochemical dissolution of gold in acidic medium. Electrochem Commun 28:44–46

19

Alumina hollow microspheres supported gold catalysts for low-temperature CO oxidation: effect of the pretreatment atmospheres on the catalytic activity and stability

Yu-Xin Miao · Lei Shi · Li-Na Cai · Wen-Cui Li

Abstract Hierarchically organized γ-Al_2O_3 hollow microspheres were prepared via a hydrothermal method using potassium aluminum sulfate and urea as reactants. The corresponding Au/Al_2O_3 catalysts were obtained using a deposition-precipitation (DP) method. The effect of the pretreatment under different atmospheres (N_2, air, and H_2) on the activity and stability of the Au/Al_2O_3 catalysts in CO oxidation was investigated. The results showed that the pretreatment under H_2 atmosphere improved the low-temperature CO oxidation activity. Furthermore, a 50 h long-term test at 30 °C showed no significant deactivation for the H_2-pretreated catalyst. Moreover, the catalytic activity was promoted by H_2O vapor in all cases, and the H_2-pretreated catalyst exhibited a good tolerance in the co-presence of CO_2 and H_2O. Finally, oxygen temperature-programmed desorption (O_2-TPD) and in situ diffuse reflectance infrared Fourier transform spectra (DRIFTS) revealed that the reductive atmosphere pretreatment greatly improved the CO adsorption capacity and facilitated the oxygen activation.

Keywords Pretreatment atmosphere · CO oxidation · Gold catalyst · Stability · DRIFTS

Introduction

Supported gold nanoparticles (Au NPs) have been intensively studied in catalysis [1], particularly for the low-temperature CO oxidation [2–6], water-gas shift (WGS) reaction [7], VOC removal [8], and selective oxidation of organic compounds

Y.-X. Miao · L. Shi · L.-N. Cai · W.-C. Li (✉)
State Key Laboratory of Fine Chemicals, School of Chemical Engineering, Dalian University of Technology, Dalian 116024, People's Republic of China
e-mail: wencuili@dlut.edu.cn

[9]. It is generally considered that the catalytic activity of Au NPs is strongly dependent on the size (<5 nm) and morphology [10–12]. The Au NPs supported on reducible oxides such as Fe_2O_3 [13], TiO_2 [14], and CeO_2 [15] may have smaller sizes and present higher catalytic performance compared to non-reducible oxide materials (SiO_2 [16], Al_2O_3 [17]). Generally, the strong interaction between Au NPs and the reducible oxide supports helps to stabilize small Au NPs and also increases the catalytic activity [18]. However, we reported that the alumina nanosheets with rough surface can efficiently stabilize the Au NPs and thus contribute a high activity for CO oxidation [19].

Besides the nature of supports [20, 21], it is believed that the pretreatment process of the Au catalyst can strengthen the interaction between Au and the supports, and so improve the CO oxidation activity [22, 23]. Wang et al. [24] prepared a series of Au/α-Mn_2O_3 catalysts by a deposition-precipitation method and found that the best activity was obtained when the catalyst was pretreated with O_2 because a specific formed oxygen-enriched interface gave enhanced metal-support synergy. On the contrary, the Au/α-Mn_2O_3 catalysts pretreated with He and H_2 had inferior catalytic activities on account of severe deactivation and over-reduction of the surface of support, respectively. In the work of Xu et al. [25], a highly active "NiO-on-Au" nanocatalyst was synthesized using a two-step method. They suggested that the catalyst pretreated with H_2 had a better catalytic performance since the formed NiO-Au boundaries can provide dual sites for O_2 activation and CO adsorption. Our recent work also confirmed that the pretreatment atmospheres significantly influence the catalytic activity of the Au/CeO_2 catalyst for CO oxidation [26]. The characterization results showed that pretreatment can significantly change the surface interaction between Au species and CeO_2 support.

However, concerning the influence of the pretreatment atmospheres on the catalytic activity, current studies mainly revolve around the reducible-oxide-supported Au catalysts. There were few studies in terms of the non-reducible-oxide-supported Au NPs. In the present work, we take the home-made highly active Au/Al$_2$O$_3$ catalyst as an example and study the effect of the pretreatment atmospheres on the activity towards CO oxidation. In addition, the stability of the pretreated Au/Al$_2$O$_3$ catalyst under CO$_2$ and H$_2$O steam was also conducted. The oxygen temperature-programmed desorption (O$_2$-TPD) and in situ diffuse reflectance infrared Fourier transform spectra (DRIFTS) characterization were performed to investigate the surface characteristics of the pretreated Au/Al$_2$O$_3$ catalyst with the aim of correlating with the catalytic behavior.

Experimental

Catalyst preparation

Alumina hollow microspheres were prepared using a hydro-thermal method [27]. In detail, 5 mmol of KAl(SO$_4$)$_2$·12H$_2$O was dissolved in 50 mL of deionized water, and then 10 mmol of CO(NH$_2$)$_2$ dissolved in 50 mL of deionized water was added into the KAl(SO$_4$)$_2$·12H$_2$O solution under vigorous stirring at room temperature for 0.5 h. The mixture was transferred into a Teflon-lined stainless steel autoclave and heated at 180 °C for 3 h. After thorough washing and centrifugation, the solid was dried at 80 °C and calcined at 600 °C in a muffle oven for 2 h.

Gold was deposited onto the surface of γ-Al$_2$O$_3$ hollow microspheres by a deposition-precipitation (DP) method using (NH$_4$)$_2$CO$_3$ as precipitant and HAuCl$_4$ solution (7.888 g/L) as the gold precursor, similar to the procedure used in our previous work [28]. In a typical preparation, HAuCl$_4$ was added dropwise to an aqueous suspension of Al$_2$O$_3$, and the pH of the suspension was adjusted to 8–9 by addition of 0.5 M (NH$_4$)$_2$CO$_3$ solution at 60 °C for 2 h. Afterwards, the product was washed several times with deionized water until it was clear of Cl$^-$ (tested by AgNO$_3$), followed by centrifugal separation and drying under vacuum. The theoretical content of Au is 1 wt%. Prior to the catalytic test, the samples were pretreated under flowing N$_2$, 14 vol% H$_2$/N$_2$ and air atmospheres at 250 °C for 2 h, and the corresponding samples were named Au/Al$_2$O$_3$-N$_2$, Au/Al$_2$O$_3$-H$_2$, and Au/Al$_2$O$_3$-air, respectively.

Catalyst characterization

The morphology of the Al$_2$O$_3$ sample was observed using a Hitachi S4800 scanning electron microscope (SEM) operated at 20 kV. Transmission electron microscope (TEM) images were obtained with a FEI Tecnai G220 S-Twin microscope. The X-ray diffraction (XRD) pattern was collected on a Siemens D/Max 2400 X-ray powder diffractometer (Cu K$_\alpha$ radiation, λ=1.54056 Å) with a working voltage of 40 kV and a current of 100 mA. The Brunauer-Emmett-Teller (BET) surface area of the Al$_2$O$_3$ sample was measured by N$_2$ adsorption at −196 °C on a Micromeritics Tristar 3000 instrument. The samples were degassed at 200 °C for 4 h prior to analysis. In situ DRIFTS were recorded by a Nicolet 6700 spectrometer equipped with MCT detector and KBr window. The Au/Al$_2$O$_3$ catalyst was heated to 200 °C for 2 h under vacuum prior to the test. The background spectrum was collected in a flowing He atmosphere at RT (30 °C), and in situ DRIFTS were collected in 5 vol% CO/N$_2$ or 1 vol% CO/air atmosphere for 20 min. O$_2$-TPD experiments were conducted on a Micromeritics Autochem II 2920 apparatus. The Au/Al$_2$O$_3$ catalyst was first treated in an Ar flow at 200 °C for 2 h. After cooling to 30 °C, the pretreated sample was exposed to O$_2$ for 1 h and heated to 800 °C, with a heating rate of 10 °C/min. The actual loading of Au was determined using an inductively coupled plasma atomic emission spectrometer (ICP-AES) on the Optima 2000 DV.

Catalytic test

The activity of the Au/Al$_2$O$_3$ catalyst for CO oxidation was evaluated with a fixed-bed flow quartz reactor (8 mm i.d.). A typical flow rate was 1 vol% CO and 20 vol% O$_2$ in N$_2$ (79 vol%), giving a total flow rate of 67 mL/min. The catalyst sample was 100 mg, and the corresponding space velocity was 40,000 mL/h g$_{cat}$. The composition of the effluent gas was analyzed using an online GC-7890 gas chromatograph equipped with a thermal conductivity detector (TCD) and a 5A molecular sieve column (T=80 °C, H$_2$ as the carrier gas).

Results and discussion

Alumina hollow microspheres

In Fig. 1a–c, the SEM and TEM images of the obtained sample display hollow microsphere structures with a diameter of 4–6 μm and a shell thickness of 600–700 nm. The spheres consist of closely packed nanoflakes. The main diffraction peaks of alumina are present at 2θ=31.9° (220), 37.3° (311), 45.7° (400), and 66.9° (440) (Fig. 1d), which can be assigned to the γ-alumina crystalline phase (JCPDS card 10-0425). The BET specific surface area and the pore volume of Al$_2$O$_3$ were calculated as 209 m^2/g and 0.66 cm^3/g, respectively. These results are similar to our recent work [27]. The actual Au content is 0.53 wt% by ICP measurement.

Fig. 1 a, b SEM images, c TEM
image, and d XRD patterns of the
Al_2O_3 sample

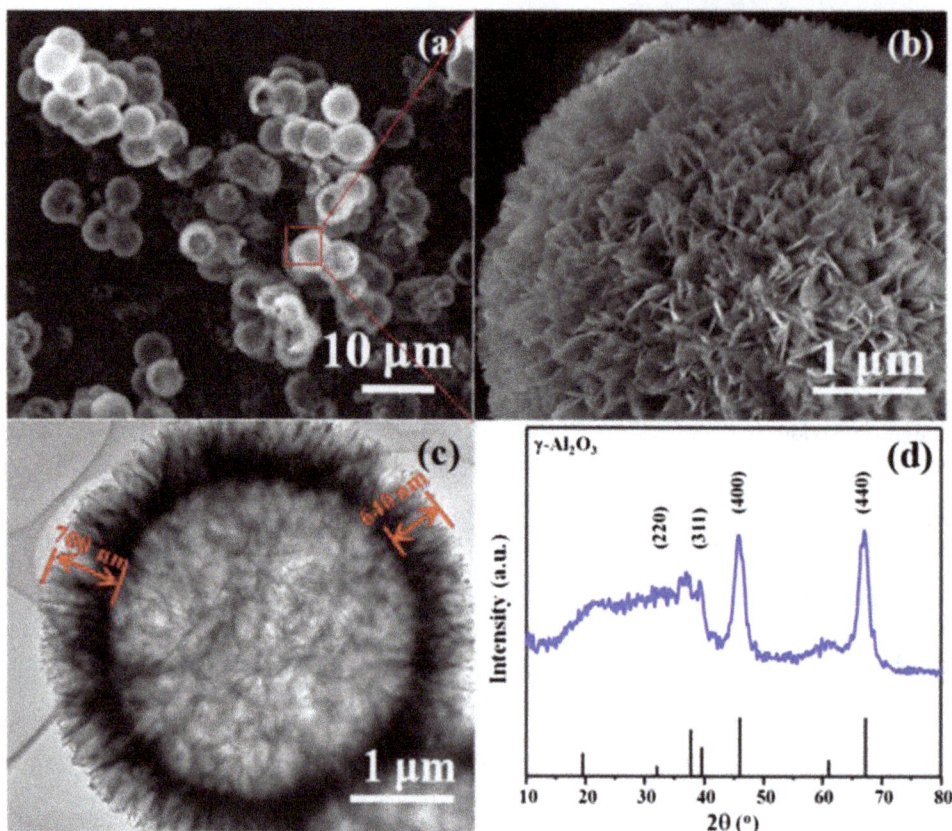

Fig. 1 a, b SEM images, c TEM image, and d XRD patterns of the Al_2O_3 sample

Catalytic activities and stabilities

Figure 2 gives the CO oxidation performance over the Au/Al_2O_3 catalysts pretreated under different atmospheres. The initial CO conversion of Au/Al_2O_3-H_2 is 62 % at 30 °C, which is significantly higher than those of Au/Al_2O_3-air and Au/Al_2O_3-N_2 catalysts under the same reaction conditions. Moreover, a complete conversion can be obtained for the Au/Al_2O_3-H_2 at a low temperature of 60 °C. In contrast, the Au/Al_2O_3-air catalyst showed a slightly increased initial

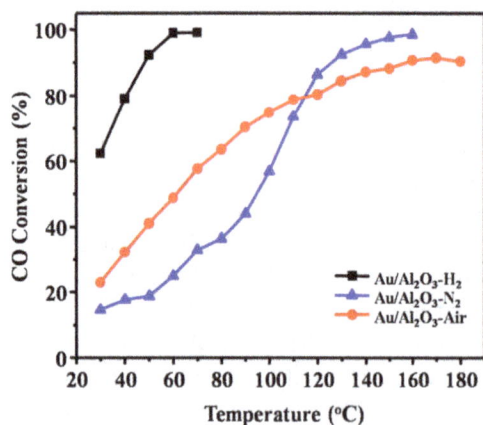

Fig. 2 CO conversions as a function of reaction temperature over Au/Al_2O_3 catalysts with different pretreatment atmospheres

activity, but a lower activity at high temperature compared to the Au/Al_2O_3-N_2 catalyst. In addition, based on the best results from Fig. 2, the reactive rate at 30 °C for the Au/Al_2O_3-H_2 catalyst was calculated as 2.109 mol/h g_{Au}, which is comparable and even higher than the results in the literature [19, 28–30] (Table 1).

The stabilities of the Au/Al_2O_3 catalysts pretreated under different atmospheres in CO oxidation were measured at 30 °C. As shown in Fig. 3a, the conversion of CO over the Au/Al_2O_3-H_2 catalyst increased from 62 % in the initial time to 75 % in 10 h and then maintained an excellent stability over 50 h on stream. The Au/Al_2O_3-air and Au/Al_2O_3-N_2 catalysts also exhibit stable catalytic performances, but the CO conversions were only 27 and 7 %, respectively. It is evident that the CO oxidation activities and stabilities over the Au/Al_2O_3 catalysts are sensitive to the pretreatment atmospheres. The possible explanation is supplied in the following by means of in situ DRIFTS characterization.

Furthermore, the effects of the H_2O vapor and CO_2 with varied concentration on the stability of the Au/Al_2O_3-H_2 catalyst were also tested. As expected, it can be observed in Fig. 3b that the CO conversion increased when the concentration of H_2O vapor was raised. The CO conversion of the Au/Al_2O_3-H_2 catalyst under 500 ppm of H_2O vapor at 30 °C is ~70 % and increased to ~87 % under 5000 ppm. As a result, complete CO conversion is achieved at lower temperatures

Table 1 Comparison of catalytic performances with other reported Au NPs

Catalysts	Au loading (wt%)[a]	Composition of feed gas (vol%)		Space velocity	$T_{100\%}$ (°C)[b]	T (°C)[c]	TOF (/s)[d]	Rate (mol/h g$_{Au}$)	Stability test (h)	Reference
		CO	O$_2$							
Au/Al$_2$O$_3$	0.53	1	20	40,000 mL/h g$_{cat}$	60	30	0.096	2.109	50	This work
Au/Al$_2$O$_3$	2.4	1	20	80,000 mL/h g$_{cat}$	2	0	0.141	1.433	–	[19]
Au/Al$_2$O$_3$	2.08	1	20	80,000 mL/h g$_{cat}$	10	20	0.0878	1.604	–	[28]
Au/Al$_2$O$_3$	1.0	1	20	15,000/h	40	25	0.25	1.62	–	[29]
Au/Al$_2$O$_3$	0.065	1	0.5	20,000 mL/h g$_{cat}$	24	24	–	1.803	0.5	[30]
Au/Fe-La -Al$_2$O$_3$	1	1	20	16,000 mL/h g$_{cat}$	−10	25	–	0.48	30	[31]
Au/FeO$_x$/Al$_2$O$_3$	1.35	1	10	45,000/h	30	–	–	–	30	[32]
Au-Rh/Al$_2$O$_3$	0.94	0.2	20	54,000/h	0	–	–	–	33	[33]

[a] The actual loadings of Au were determined by ICP technique

[b] $T_{100\%}$ represents the temperatures for 100 % CO conversion

[c] The temperature of the corresponding reactive rate and TOF of Au catalysts were calculated

[d] Turnover frequency (TOF) was calculated based on the number of supported Au atoms

(30 °C) under 10,000 ppm of H$_2$O vapor. However, with the addition of CO$_2$ (Fig. 2c), the catalyst suffers from a rapid decrease in the activities of CO oxidation; for example, CO conversion is approximately 40 % with the addition of 15 and 20 % CO$_2$. A possible explanation is that the increase in CO$_2$ concentration leads to the formation of carbonate-like species and/or the competitive adsorption of CO$_2$ on the active sites

which inhibit the oxygen mobility. Figure 2d gives the corresponding CO conversion curves in the co-presence of H$_2$O and CO$_2$ at 30 °C. It is seen that the Au/Al$_2$O$_3$-air and Au/Al$_2$O$_3$-H$_2$ catalysts gave transient 100 % CO conversion after which the catalytic activities gradually decreased and reached a stable activity of 90 %. In contrast, low catalytic activities occurred with the Au/Al$_2$O$_3$-N$_2$ catalyst under the same

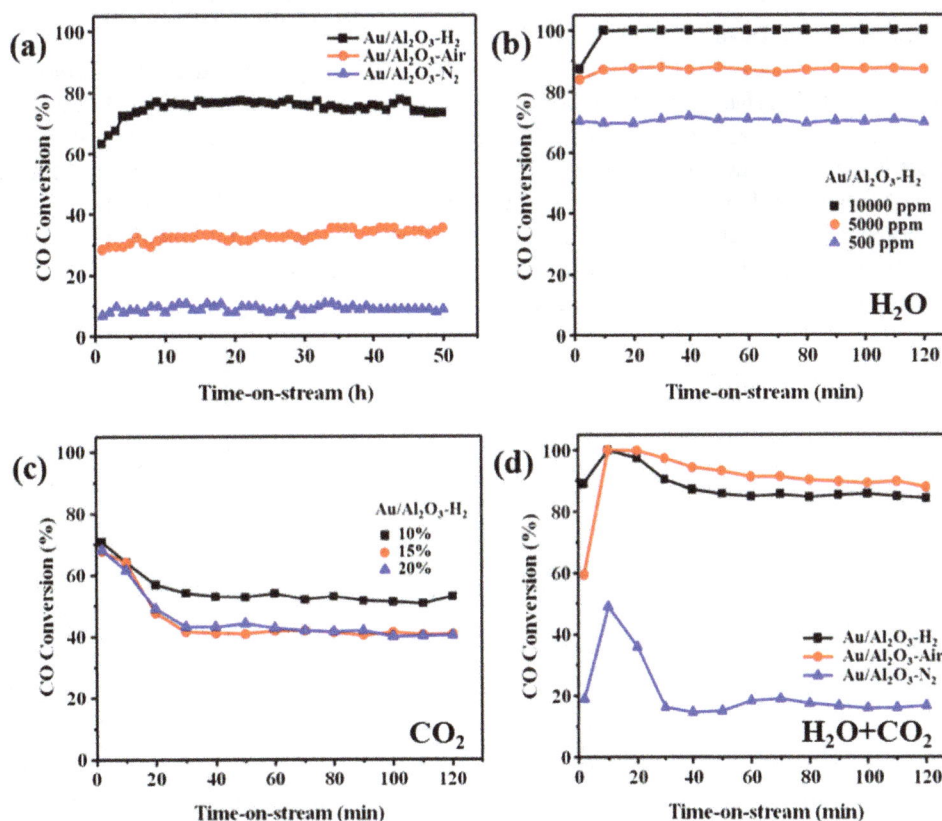

Fig. 3 a The stabilities of various Au/Al$_2$O$_3$ catalysts for the CO oxidation at 30 °C. **b** Au/Al$_2$O$_3$-H$_2$ catalyst in the presence of H$_2$O vapor at 30 °C. **c** Au/Al$_2$O$_3$-H$_2$ catalyst in the presence of CO$_2$ at 30 °C. **d** Au/Al$_2$O$_3$ catalysts for the CO oxidation in the co-presence of 10,000 ppm H$_2$O and 15 % CO$_2$ at 30 °C (reaction conditions: 1 vol% CO, 20 vol% O$_2$, 15 vol% CO$_2$, 1 vol% H$_2$O, and balance N$_2$. WHSV, 40,000 mL/h g$_{cat}$)

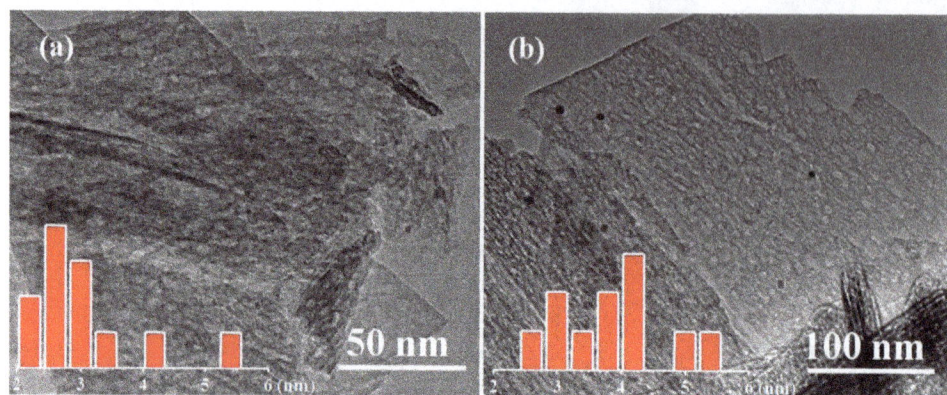

Fig. 4 a TEM image of a fresh Au/Al$_2$O$_3$-H$_2$ catalyst. **b** TEM image of a used Au/Al$_2$O$_3$-H$_2$ catalyst

reaction conditions. The noticeable promotion of initial activity for Au/Al$_2$O$_3$-air and Au/Al$_2$O$_3$-N$_2$ catalysts may be considered to be moisture-assisted oxygen activation [34]. These results indicate a promoting effect of reductive or oxidative atmosphere pretreatment on the activity of the Au/Al$_2$O$_3$ catalysts under the co-presence of H$_2$O vapor and CO$_2$.

In order to investigate the stability of Au NPs after treatment, one representative Au/Al$_2$O$_3$-H$_2$ catalyst was characterized by TEM (Fig. 4a, b). One can see that the gold particle sizes of the used Au/Al$_2$O$_3$-H$_2$ catalyst (3.9±0.4 nm) are similar to those of the fresh sample (3.0±0.4 nm). The outstanding stability could be relating to the novel surface structure of our Al$_2$O$_3$ support, which contributes to a solid stabilization of Au NPs [19, 28].

CO adsorption on the Au/Al$_2$O$_3$ catalyst

To study the initial surface property of various Au/Al$_2$O$_3$ catalysts, we chose CO as a probe molecule and employed in situ DRIFTS to study the adsorption on the catalyst surface at 30 °C. As shown in Fig. 5, when the Au/Al$_2$O$_3$ catalysts were exposed to CO for 20 min at 30 °C, several bands at 1440, 1560, 1641, 2056, 2114, and 2171 cm^{-1} were observed. Concomitant with the CO adsorption, the peaks in the 1400–

1800 cm^{-1} region are related to the vibration of carbonate-like species as suggested by other studies [35, 36]. The absorption band at 2056 cm^{-1} in our study was also observed by Liu et al. [37] at 2048 cm^{-1}, which is assigned to negatively charge gold carbonyls [38]. At the same time, one weak absorption band at 2114 cm^{-1} can be assigned to Au^0–CO. On the other hand, the band at 2171 cm^{-1} can be attributed to Au$^{\delta+}$–CO [39]. There are no striking differences between the peak position of Au$^{\delta+}$/Au^0–CO for Au/Al$_2$O$_3$ catalysts with different pretreatment atmospheres. This result indicates that the co-presence of Au$^{\delta+}$/Au^0 on the surface of Au/Al$_2$O$_3$ catalysts [27] and the pretreatment atmospheres had little effect on CO adsorption on the surface of the Au/Al$_2$O$_3$ catalyst.

Oxygen temperature-programmed desorption

To gain the adsorption/desorption capacity of oxygen on the catalyst surface, O$_2$-TPD measurements were carried out. The O$_2$-TPD profiles of the Au/Al$_2$O$_3$ catalysts pretreated under different atmospheres are shown in Fig. 6. The oxygen desorption peaks of Au/Al$_2$O$_3$-N$_2$, Au/Al$_2$O$_3$-air, and Au/Al$_2$O$_3$-H$_2$ are 71, 69, and 59 °C, respectively, which can be ascribed to the decomposition of Au$_2$O$_3$ or surface oxygen species

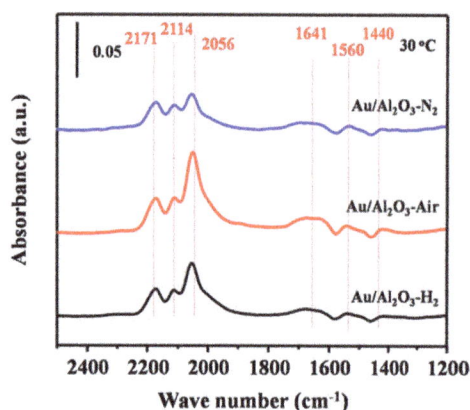

Fig. 5 In situ DRIFTS of various Au/Al$_2$O$_3$ catalysts after CO adsorption for 20 min

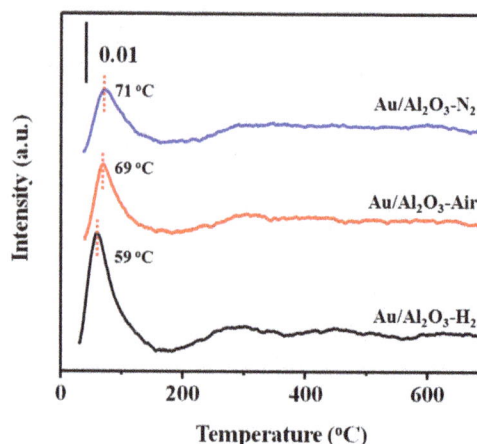

Fig. 6 The O$_2$-TPD profiles of Au/Al$_2$O$_3$ catalysts with different pretreatment atmospheres

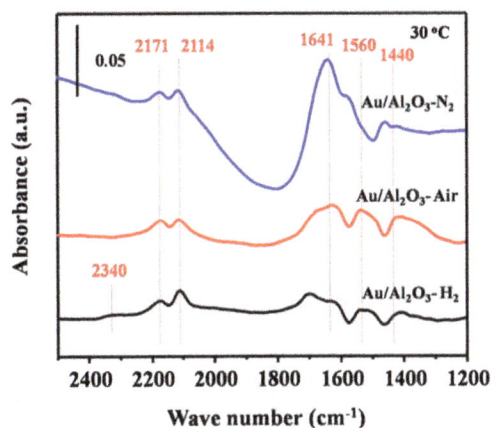

Fig. 7 In situ DRIFTS of various Au/Al_2O_3 catalysts after CO and O_2 co-adsorption for 20 min

weakly interacting with Au particles. For the Au/Al_2O_3-H_2 sample, the area of oxygen desorption was found to be higher than those of Au/Al_2O_3-air and Au/Al_2O_3-N_2 catalysts, suggesting that the Au/Al_2O_3-H_2 catalyst had more active oxygen species. It is known that the oxygen desorption behavior depends on the amount and strength of chemisorbed oxygen

species which are easily desorbed at low temperature [40, 41]. A shift of the oxygen desorption peak to the lower temperature indicates an effective way to achieve lower reaction energy for CO oxidation and higher catalytic activity. The above results are in agreement with the catalytic activities (Fig. 2).

In situ DRIFTS analysis of the Au/Al_2O_3 catalyst

In order to explain the difference between catalytic behaviors of various Au/Al_2O_3 catalysts, in situ DRIFTS analysis was tested under CO oxidation conditions at 30 °C. As shown in Fig. 7, the surfaces of Au/Al_2O_3-N_2 and Au/Al_2O_3-air catalysts were mainly covered by carbonate-like species (1400–1800 cm^{-1}) with CO and O_2 co-adsorption for 20 min. This implies that carbonate-like species were easily accumulated on the surface of these catalysts at 30 °C. In contrast, few carbonate-like species were accumulated on the Au/Al_2O_3-H_2 catalyst surface. Simultaneously, a new band appeared at 2340 cm^{-1} which could be ascribed to CO_2 adsorbed on the Au/Al_2O_3-H_2 catalyst [42, 43], suggesting that CO can be oxidized into CO_2 products at 30 °C on the H_2-pretreated

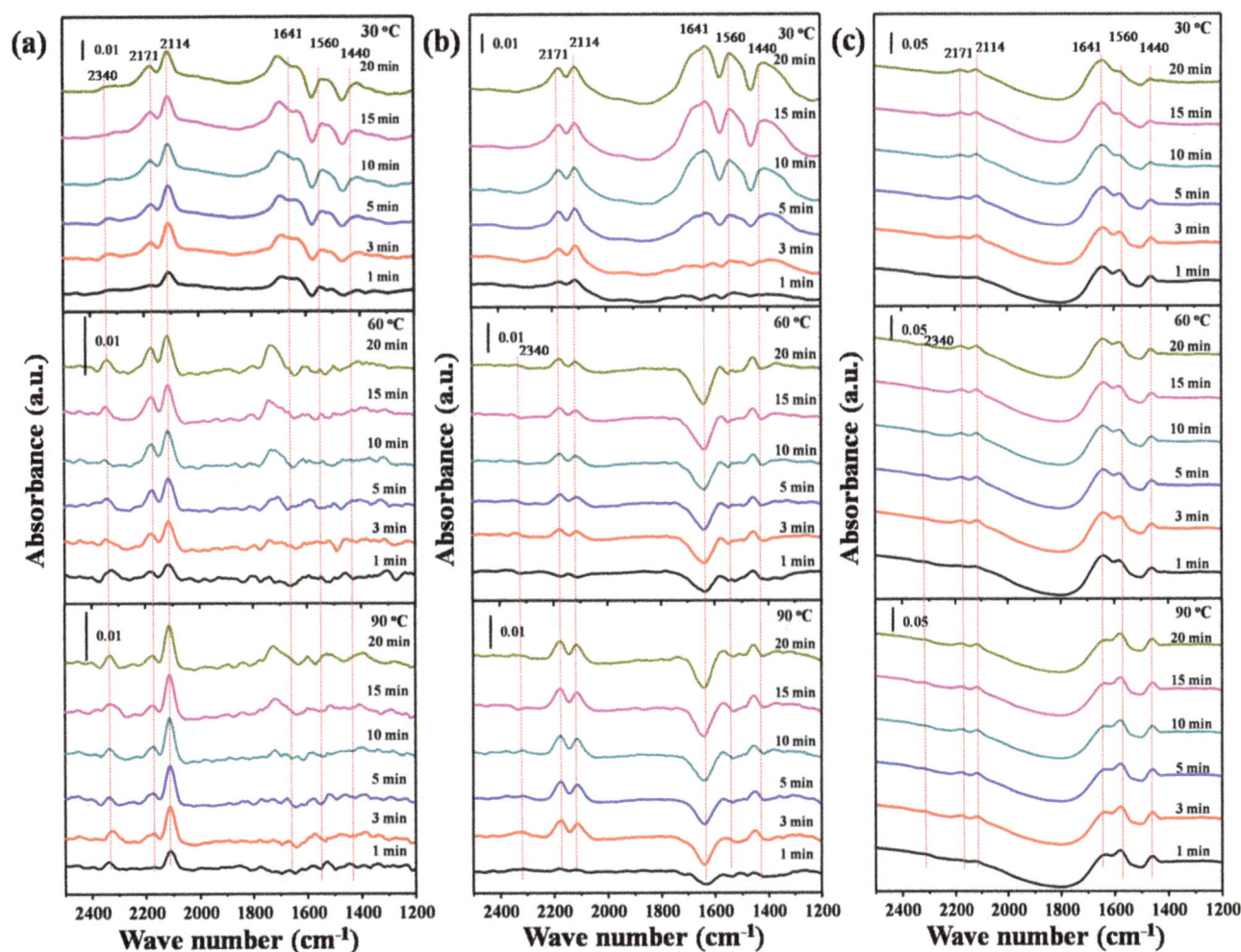

Fig. 8 In situ DRIFTS of the Au/Al_2O_3 catalysts for CO reaction at 30, 60, and 90 °C. **a** Au/Al_2O_3-H_2. **b** Au/Al_2O_3-air. **c** Au/Al_2O_3-N_2

catalyst. However, the CO_2 adsorption bands were not obvious for Au/Al$_2$O$_3$-N$_2$ and Au/Al$_2$O$_3$-air catalysts.

At 30 °C under CO oxidation conditions, the Au/Al$_2$O$_3$ catalyst surface is covered with CO, CO_2, and carbonate-like species. Comparing with the CO-DRIFTS in Fig. 5, the intensity of CO adsorption bands at 2171 and 2114 cm^{-1} decreased and the bands at 2056 cm^{-1} disappeared in the CO and O_2 co-adsorbed DRIFTS. A similar adsorption process was also reported by Liu et al. [44] with the Au-Cu/SBA-15 catalyst and was interpreted as the O_2 participating in the CO oxidation reaction, suggesting that the CO adsorbed on Au^0 readily reacts with O_2 even at a low temperature. This result indicates that the O_2 adsorption competes with CO on the catalyst surface. The intensity of peaks at 2114 cm^{-1} on various Au/Al$_2$O$_3$ catalysts with different pretreatment atmospheres follows the order of Au/Al$_2$O$_3$-H$_2$>Au/Al$_2$O$_3$-air>Au/Al$_2$O$_3$-N$_2$, which is consistent with their catalytic activities (Fig. 2).

To investigate the accumulation of adsorbed surface species during the reaction, we recorded the in situ DRIFTS experiments at high temperature. As shown in Fig. 8, the CO-derived species formed at 60 and 90 °C on the surface of the Au/Al$_2$O$_3$ catalysts are similar to those formed at 30 °C. Furthermore, the coverage of the Au$^{\delta+}$/Au^0–CO (the bands at 2171 and 2114 cm^{-1}) species decreased remarkably with temperature increasing. The intensity of the main carbonyl band (Au$^{\delta+}$/Au^0–CO) increases with time. Figure 8a shows that the Au/Al$_2$O$_3$-H$_2$ catalyst reached equilibrium of CO adsorption/desorption after 10 min. However, the Au/Al$_2$O$_3$-air and Au/Al$_2$O$_3$-N$_2$ catalysts reached equilibrium for CO adsorption/desorption after 15 min (Fig. 8b, c). These results suggest that the CO adsorption rate is higher on the surface of the Au/Al$_2$O$_3$-H$_2$ catalyst and thus contributes an excellent catalytic activity.

Additionally, Fig. 8 shows that the intensity of the adsorption band assigned to Au$^{\delta+}$/Au^0–CO (2171 and 2114 cm^{-1}) decreased when the CO oxidation temperature was raised from 30 to 90 °C. Meanwhile, the intensity of the CO_2 peak increased. This result indicates that all Au/Al$_2$O$_3$ catalysts are reactive in the CO oxidation at 60 and 90 °C. In our work, the Au/Al$_2$O$_3$-H$_2$ catalyst is more active as shown by the stronger intensity of the CO_2 absorption band at 2340 cm^{-1} (Fig. 8). On the other hand, the peak intensity of carbonate-like species over the Au/Al$_2$O$_3$-H$_2$ and Au/Al$_2$O$_3$-air catalysts was lower at 60 and 90 °C, indicating that the adsorption on the catalyst was reversible at this temperature (Fig. 8a, b). In contrast, the inferior activity and rapid deactivation of the Au/Al$_2$O$_3$-N$_2$ catalyst can be well understood by the continuous accumulation of carbonate-like species with CO and O_2 co-adsorption tested at 90 °C (Fig. 8c). This is in good agreement with the catalytic activities. The presence of water vapor can enhance the catalytic activities due to the promotion of the decomposition of carbonate-like species (Fig. 3b), which is consistent with the literature conclusions [6]. These results in both O_2-

TPD and in situ DRIFTS measurements well explain why the catalyst pretreated under H$_2$ atmosphere displays a higher activity and stability in CO oxidation (Fig. 3a, d).

Conclusions

The influence of pretreatment atmospheres on the surface properties and catalytic performances of Au/Al$_2$O$_3$ catalysts was investigated. The low-temperature CO oxidation activity of various Au/Al$_2$O$_3$ catalysts was found to be Au/Al$_2$O$_3$-H$_2$> Au/Al$_2$O$_3$-air>Au/Al$_2$O$_3$-N$_2$. The catalyst pretreated under a H$_2$ atmosphere shows excellent catalytic activity and stability in the co-presence of CO_2 and H_2O at room temperature. The O_2-TPD and in situ DRIFTS results revealed that the Au/Al$_2$O$_3$-H$_2$ catalyst greatly enhanced the CO adsorption capacity and facilitated the oxygen activation. The deactivation observed on Au/Al$_2$O$_3$ catalysts was caused by adsorption of carbonate-like species. As compared with the Au/Al$_2$O$_3$-N$_2$ sample, the superior stability of the Au/Al$_2$O$_3$-H$_2$ catalyst may result from its suppressed accumulation of carbonate-like species under the co-presence of H_2O vapor and CO_2.

Acknowledgments This work was supported by the National Program on Key Basic Research Project (No. 2013CB934104).

References

1. Hashmi ASK, Hutchings GJ (2006) Gold catalysis. Angew Chem Int Ed 45:7896–7936
2. Haruta M, Yamada N, Kobayashi T, Iijima S (1989) Gold catalysts prepared by coprecipitation for low-temperature oxidation of hydrogen and of carbon monoxide. J Catal 115:301–309
3. Haruta M, Tsubota S, Kobayashi T, Kageyama H, Genet MJ, Delmon B (1993) Low-temperature oxidation of CO over gold supported on TiO$_2$, α-Fe$_2$O$_3$, and Co$_3$O$_4$. J Catal 144:175–192
4. Okumura M, Nakamura S, Tsubota S, Nakamura T, Azuma M, Haruta M (1998) Chemical vapor deposition of gold on Al$_2$O$_3$, SiO$_2$, and TiO$_2$ for the oxidation of CO and of H$_2$. Catal Lett 51:53–58
5. Cunningham DAH, Vogel W, Haruta M (1999) Negative activation energies in CO oxidation over an icosahedral Au/Mg(OH)$_2$ catalyst. Catal Lett 63:43–47
6. Daté M, Okumura M, Tsubota S, Haruta M (2004) Vital role of moisture in the catalytic activity of supported gold nanoparticles. Angew Chem Int Ed 43:2129–2132
7. Tabakova T, Ilieva L, Ivanov I, Zanella R, Sobczak JW, Lisowski W, Kaszkur Z, Andreeva D (2013) Influence of the preparation method and dopants nature on the WGS activity of gold catalysts supported on doped by transition metals ceria. Appl Catal B Environ 136–137:70–80

8. Haruta M, Ueda A, Tsubota S, Sanchez RMT (1996) Low-temperature catalytic combustion of methanol and its decomposed derivatives over supported gold catalysts. Catal Today 29:443–447

9. Long J, Liu H, Wu SJ, Liao S, Li Y (2013) Selective oxidation of saturated hydrocarbons using Au-Pd alloy nanoparticles supported on metal-organic frameworks. ACS Catal 3:647–654

10. Valden M, Lai X, Goodman DW (1998) Onset of catalytic activity of gold clusters on titania with the appearance of nonmetallic properties. Science 281:1647–1650

11. Hughes MD, Xu YJ, Jenkins P, McMorn P, Landon P, Enache DI, Carley AF, Attard GA, Hutchings GJ, King F, Stitt EH, Johnston P, Griffin K, Kiely CJ (2005) Tunable gold catalysts for selective hydrocarbon oxidation under mild conditions. Nature 437:1132–1135

12. Corma A, Serna P (2006) Chemoselective hydrogenation of nitro compounds with supported gold catalysts. Science 313:332–334

13. Li L, Wang A, Qiao B, Lin J, Huang Y, Wang X, Zhang T (2013) Origin of the high activity of Au/FeO_x for low-temperature CO oxidation: direct evidence for a redox mechanism. J Catal 299:90–100

14. Li WC, Comotti M, Schüth F (2006) Highly reproducible syntheses of active Au/TiO_2 catalysts for CO oxidation by deposition-precipitation or impregnation. J Catal 237:190–196

15. Carrettin S, Concepción P, Corma A, Nieto JML, Puntes VF (2004) Nanocrystalline CeO_2 increases the activity of Au for CO oxidation by two orders of magnitude. Angew Chem Int Ed 43:2538–2540

16. Zhang Y, Zhaorigetu B, Jia M, Chen C, Zhao J (2013) Clay-based SiO_2 as active support of gold nanoparticles for CO oxidation catalyst: pivotal role of residual Al. Catal Commun 35:72–75

17. Wen L, Fu JK, Gu PY, Yao BX, Lin ZH, Zhou JZ (2008) Monodispersed gold nanoparticles supported on γ-Al_2O_3 for enhancement of low-temperature catalytic oxidation of CO. Appl Catal B Environ 79:402–409

18. Schubert MM, Hackenberg S, Veen ACV, Muhler M, Plzak V, Behm RJ (2001) CO oxidation over supported gold catalysts-"inert" and "active" support materials and their role for the oxygen supply during reaction. J Catal 197:113–122

19. Wang J, Lu AH, Li MR, Zhang WP, Chen YS, Tian DX, Li WC (2013) Thin porous alumina sheets as supports for stabilizing gold nanoparticles. ACS Nano 7:4902–4910

20. Comotti M, Li WC, Spliethoff B, Schüth F (2006) Support effect in high activity gold catalysts for CO oxidation. J Am Chem Soc 128:917–924

21. Wang GH, Li WC, Jia KM, Spliethoff B, Schüth F, Lu AH (2009) Shape and size controlled α-Fe_2O_3 nanoparticles as supports for gold-catalysts: synthesis and influence of support shape and size on catalytic performance. Appl Catal A Gen 364:42–47

22. Park ED, Lee JS (1999) Effects of pretreatment conditions on CO oxidation over supported Au catalysts. J Catal 186:1–11

23. Szabó EG, Tompos A, Hegedűs M, Szegedi Á, Margitfalvi JL (2007) The influence of cooling atmosphere after reduction on the catalytic properties of Au/Al_2O_3 and Au/MgO catalysts in CO oxidation. Appl Catal A Gen 320:114–121

24. Wang LC, He L, Liu YM, Cao Y, He HY, Fan KN, Zhuang JH (2009) Effect of pretreatment atmosphere on CO oxidation over α-Mn_2O_3 supported gold catalysts. J Catal 264:145–153

25. Xu X, Fu Q, Guo X, Bao X (2013) A highly active "NiO-on-Au" surface architecture for CO oxidation. ACS Catal 3:1810–1818

26. Zhang RR, Ren LH, Lu AH, Li WC (2011) Influence of pretreatment atmospheres on the activity of Au/CeO_2 catalyst for low-temperature CO oxidation. Catal Commun 13:18–21

27. Wang J, Hu ZH, Miao YX, Li WC (2014) Hollow γ-Al_2O_3 microspheres as highly "active" supports for Au nanoparticle catalysts in CO oxidation. Gold Bull 47:95–101

28. An AF, Lu AH, Sun Q, Wang J, Li WC (2011) Gold nanoparticles stabilized by a flake-like Al_2O_3 support. Gold Bull 44:217–222

29. Han YF, Zhong ZY, Ramesh K, Chen FX, Chen LW (2007) Effects of different types of γ-Al_2O_3 on the activity of gold nanoparticles for CO oxidation at low-temperatures. J Phys Chem C 111:3163–3170

30. Lee SJ, Gavriilidis A (2002) Supported Au catalysts for low-temperature CO oxidation prepared by impregnation. J Catal 206:305–313

31. Qi C, Zhu S, Su H, Lin H, Guan R (2013) Stability improvement of Au/Fe-La-Al_2O_3 catalyst via incorporating with a Fe_xO_y layer in CO oxidation process. Appl Catal B Environ 138–139:104–112

32. Zou X, Xu J, Qi S, Suo Z, An L, Li F (2011) Effects of preparation conditions of $Au/FeO_x/Al_2O_3$ catalysts prepared by a modified two-step method on the stability for CO oxidation. J Nat Gas Chem 20:41–47

33. Wang X, Lu G, Guo Y, Zhang Z, Guo Y (2011) Role of Rh promoter on increasing stability of Au/Al_2O_3 catalyst for CO oxidation at low temperature. Environ Chem Lett 9:185–189

34. Shang C, Liu ZP (2010) Is transition metal oxide a must? Moisture-assisted oxygen activation in CO oxidation on gold/γ-alumina. J Phys Chem C 114:16989–16995

35. Piccolo L, Daly H, Valcarcel A, Meunier FC (2009) Promotional effect of H_2 on CO oxidation over Au/TiO_2 studied by operando infrared spectroscopy. Appl Catal B Environ 86:190–195

36. Leba A, Davran CT, Önsan ZI, Yıldırım R (2012) DRIFTS study of selective CO oxidation over Au/γ-Al_2O_3 catalyst. Catal Commun 29:6–10

37. Liu X, Liu MH, Luo YC, Mou CY, Lin SD, Cheng H, Chen JM, Lee JF, Lin TS (2012) Strong metal-support interactions between gold nanoparticles and ZnO nanorods in CO oxidation. J Am Chem Soc 134:10251–10258

38. Chakarova K, Mihaylov M, Ivanova S, Centeno MA, Hadjiivanov K (2011) Well-defined negatively charged gold carbonyls on Au/SiO_2. J Phys Chem C 115:21273–21282

39. Venkov T, Klimev H, Centeno MA, Odriozola JA, Hadjiivanov K (2006) State of gold on an Au/Al_2O_3 catalyst subjected to different pre-treatments: an FTIR study. Catal Commun 7:308–313

40. Wang YZ, Zhao YX, Gao CG, Liu DS (2008) Origin of the high activity and stability of Co_3O_4 in low-temperature CO oxidation. Catal Lett 125:134–138

41. Xu H, Li W, Shang S, Yan C (2011) Influence of MgO contents on silica supported nano-size gold catalyst for carbon monoxide total oxidation. J Nat Gas Chem 20:498–502

42. Schumacher B, Denkwitz Y, Plzak V, Kinne M, Behm RJ (2004) Kinetics, mechanism, and the influence of H_2 on the CO oxidation reaction on a Au/TiO_2 catalyst. J Catal 224:449–462

43. Denkwitz Y, Makosch M, Geserick J, Hörmann U, Selve S, Kaiser U, Hüsing N, Behm RJ (2009) Influence of the crystalline phase and surface area of the TiO_2 support on the CO oxidation activity of mesoporous Au/TiO_2 catalysts. Appl Catal B Environ 91:470–480

44. Liu X, Wang A, Li L, Zhang T, Mou CY, Lee JF (2011) Structural changes of Au-Cu bimetallic catalysts in CO oxidation: in situ XRD, EPR, XANES, and FT-IR characterizations. J Catal 278:288–296

20

Evolution and investigation of copper and gold ball bonds in extended reliability stressing

C. L. Gan · F. C. Classe · B. L. Chan · U. Hashim

Abstract This paper discusses the microstructure evolution of copper (Cu) and gold (Au) ball bonds after various extended reliability stresses such as biased highly accelerated temperature and humidity test (HAST), unbiased highly accelerated temperature and humidity test (UHAST), temperature cycling (TC), and high temperature storage life (HTSL) in BGA package. Objective of this study is to study the microstructure evolution and changes after long hours and long cycles of component reliability stressing and its predicted failure mechanisms and to determine the long-term reliability comparison with combination of bonding wires in HAST, UHAST, and TC. Secondary electron microscopy (SEM) and energy dispersive X-ray (EDX) have been carried out to understand the respective microstructure of failed samples in HAST, UHAST, TC, and HTSL long-term reliability failures. Respective failure mechanisms of copper and gold ball bonds carrion under HAST and UHAST, ball bond lifting in TC and HTSL have been analyzed and proposed. The evolution of surface morphology, including copper and gold ball bond micro cracking, gold ball bond Kirkendall microvoiding and intermetallic compound (IMC) formation, was studied in FBGA package with copper and gold ball bonds during various reliability stresses. Biased HAST, UHAST, TC, and HTSL mechanisms were proposed to explain the observed morphological changes and the resulting ball bond wear out modes after extended reliability stresses. Weibull reliability analyses have been established to compare the performance of copper and gold ball bonds under humid and dry environmental tests.

Keywords Microstructure evolution · Copper and gold ball bonds · Failure mechanisms · SEM · EDX analysis · Extended reliability · Weibull plot

Introduction

Gold and copper wire bondings are two most common bonding techniques used in microelectronic packaging in semiconductor industry. Recently, copper wire bonding appears to be the alternate materials and various engineering studies on copper wire development have been reported [1]. Technical barriers and reliability challenges of Cu wire bonding in microelectronics packaging are well identified [2–11]. Au-Al microstructure evolution and intermetallic compound (IMC) formation is widely studied by Karpel et al. [12]. Two types of failures occurred during annealing: crack formation at the bond periphery due to an increase in volume during intermetallic growth and the formation of stresses; and oxidation of the $AlAu_4$ phase adjacent to the Au ball, which resulted in the formation of continuous cracks between the Au ball and the intermetallic region [10]. Drozdov et al. [13, 14] evaluated CuAl IMC formation on as-bonded stage and post annealing to study the interface composition and morphology of copper wire bonds heat-treated at 175 °C for 2, 24, 96, and 200 h in argon. The main intermetallic phase was Al_2Cu, which was found to grow via solid state diffusion. In specimens heat-treated for 96 and 200 h, the Al_4Cu_9 phase was also detected. Void formation at the Al–Cu bonds heat-treated up to 200 h was not found to be a source of bond failure. Xu C et al. [15] studied oxidation behavior of two types of bulk gold aluminides, $AuAl_2$, and Au_4Al, using thermogravimetry. Xu H et al. [16, 17] characterized behavior of aluminum oxide, intermetallics, and voids in

C. L. Gan (✉) · F. C. Classe · B. L. Chan
Spansion (Penang) Sdn Bhd, Phase II Free Industrial Zone,
Penang 11900 Bayan Lepas, Malaysia
e-mail: chong-leong.gan@spansion.com

C. L. Gan · U. Hashim
Institute of Nanoelectronic Engineering (INEE), Universiti Malaysia
Perlis, Perlis 01000, Kangar, Malaysia

Cu–Al wire bonds. Zeng Y et al. [18] plotted Pourbaix (Eh–pH) diagrams of Al–Cu alloys (instead of pure metals) are generated on the basis of critical assessment of thermodynamic data. The Eh–pH diagram is used to conduct comprehensive studies on the thermodynamic equilibrium of Cu–Al bonding at humid environment with chlorine.

Cu ball bond is more susceptible to moisture corrosion compared to gold ball bonds and undergo different corrosion mechanisms in microelectronic packaging [19–21]. There are different ball bonds corrosion mechanisms of Au and Cu ball bond under humid reliability test. Uno T [21] reported CuAl IMC interfacial corrosion under HAST environmental test. Yamaji Y et al. [22] and Su P et al. [23] reported similar CuAl IMC interfacial corrosion post HAST and UHAST tests and effects of pH of molding compounds on HAST failure rates. Lu YH et al. [24] observed the growth rates of IMCs in Pd-coated Cu wire bonds are very sensitive to temperature, but the sequence of IMC formation remains the same for temperature below 350 °C. Pd atoms in the Pd-coated Cu wire do not participate in the interfacial reaction, and have no marked effect on the growth rate of IMCs. Gan et al. conducted studies on effects of bonding wires on UHAST and TC reliability and found Au with better UHAST reliability compare to Cu wire [25–31]. Au ball bond is well known with its Au atomic diffusion into Al metallization and caused resistive ball bonds with non-optimized bonding parameter [1, 25]. The interdiffusion between Au and Al across a thermally exposed Au-Al ball bond causes the movement of the void line towards the Au bump and shows that Au interdiffuses faster than Al. Although the movement of the void line appears to be associated with movement of the Au_4Al or Au_8Al_3 interface, it is actually analogous to the Kirkendall microvoiding. Yu CH et al. [32] studied HTSL failure mechanism of Cu ball bond after aging at 205 °C in air from 0 h to 2,000 h. The cracks grew towards the ball bond center with an increase in the aging time, and the Cl ions diffused through the crack into the ball center. This diffusion caused a corrosion reaction between the Cl ions and the Cu–Al intermetallic phases, which in turn caused copper wire bonding damage [32].

Experimental procedures

Specimen preparation

Copper wires were bonded on top of silicon wafers coated with thermally grown SiO_2 and covered by uniform aluminum (Al) metallization. The Al metallization consisted of 0.5 wt% Cu and 1 wt% Si. Materials used include 0.8 mil Pd-coated Cu wire (Cu) and 4 N (99.99 % purity) gold (Au) wire, 90 nm and 110 nm flash devices packaged into fortified fine-pitch BGA packages, with green (<20 ppm chloride in content) molding compound and substrate. Thermosonic ball bonding of each

Si die was performed at 175 °C for an approximate time of 18 s per device with a pre-heat and post-heat of 18 s at 150 °C. The bonding parameters were optimized to ensure zero pad peals, which is an essential condition for successful copper wire bonding. In order to ensure uniform and symmetric bonds, free air balls at the Cu wire tip were formed by melting the tips of the Cu wires in a reducing atmosphere (95 % N_2, 5 % H_2) prior to the bonding stage.

The corresponding stress tests and its conditions are tabulated in Table 1. Extended reliability stresses include biased HAST (130 °C, 85 %RH, 3.60 V biasing voltage), unbiased HAST (130 °C, 85 %RH), temperature cycling (−40 °C to 150 °C), and HTSL (150 °C and 200 °C). All direct material used in this evaluation study for the 90 nm and 110 nm, flash device (with top Al metallization bondpad) for packaging purpose. Forty-five units of Au and Pd-coated Cu wire bonded on fine-pitch 64-ball BGA packages are subjected for 150 °C aging temperature. Electrical testing was conducted after each hours and cycles of stress to check Au and Cu ball bond integrity in terms of its high temperature ball bonds reliability with various aging conditions. The package construction of test vehicle, FBGA 64 as depicted in Fig. 1, is assembled with gold or copper wires.

Prior to biased HAST, UHAST and TC stresses, the electronic packages were subjected to preconditioning (30 °C, 60 %RH) for 192 h in a temperature and humidity chamber, followed by three cycles of reflow at 260 °C by using reflow chamber as per JEDEC IPC-STD 020 [35]. After preconditioning and electrical test, samples were loaded in respective HAST, UHAST, and TC chambers according to the stress conditions as tabulated in Table 1. After each read point, electrical opens, shorts, and device datasheet functionality was verified by using a commercial electrical tester. Cu and Au ball bonds microstructure analysis were measured by using secondary electron microscopy (SEM) to understand the morphology and microstructure evolution. Composition analysis on failed samples were determined by using energy dispersive X-ray (EDX) on FBGA 64 package with different sets of extended reliability stresses. Wear out reliability tests were conducted on HAST, UHAST, and TC package reliability stresses to predict its reliability margins of Au and Cu wires used in FBGA package. The time and cycles-to-failures reliability modeling can be predicted based on the stress-to-failures in respective reliability stresses [33].

Table 1 Summary of extended reliability matrix (for Au and Cu wires)

Extended reliability stresses	Test conditions	Sample size
Biased HAST	85 %RH, 110 °C, 3.6 V	80
Unbiased HAST	85 %RH, 130 °C	80
TC	−40 °C to 150 °C	80
HTSL	150 °C, 200 °C	45

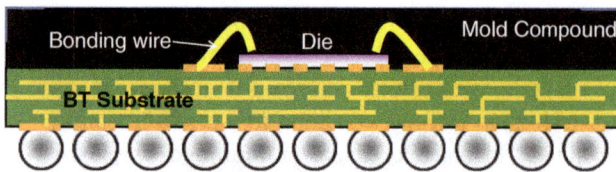

Fig. 1 FBGA 64 package constructions with Cu and Au wires used in reliability sample preparation

Result and discussion

Ball bond corrosion under highly humidity and temperature test

Biased HAST test

In order to obtain information regarding the Al–Cu interface composition after HAST or UHAST tests, CuAl and AuAl samples were studied using EDX. EDX analysis was conducted using FEI Tecnai microscopes, following the standard procedure. In order to ensure acquisition of the signal from a selected region, the specimen was tilted 15° towards the EDX detector. Typical CuAl IMC microcracking is found for the HAST 2,000 h electrical open failure (see Fig. 2).

Figure 3 shows representative SEM cross section and EDX analysis of failed Cu ball bond. EDX analysis on the micro crack at the edge of Cu ball bond indicates presence of O and Cl peaks. This proves the hydrolysis of CuAl IMC under UHAST moist conditions and Cl peak is originated from AlCl$_3$. The trace Cl$^-$ is usually found in epoxy mold compound. Tables 2 and 3 tabulates the summary of EDX analysis of Au and Cu ball bonds.

Figure 4 reveals SEM micrograph of HAST 2,000 h open found on Au ball bond and presents SEM micrographs taken the periphery of a bond subjected for HAST 2,000 h. The intermetallic coverage of the Au–Cu interface is not complete. Regions at the interface with no intermetallics can still be found. The intermetallics found at the bond periphery are less uniform than those found along the whole gold ball bonds. The intermetallics found at the left side of gold ball bond are more continuous. Voiding between the gold ball and the intermetallics located at the bond periphery can be detected. EDX analysis (in Table 2) confirmed presence of O peak and this might be induced by hydrolysis of Au$_4$Al into Al$_2$O$_3$ (see Eq. 3).

Unbiased HAST test

UHAST test is pretty similar to biased HAST test except is without biasing condition (85 %RH, 130 °C). Corrosion is typically found after long hours of UHAST test. Figure 5 presents SEM micrograph of failed Cu ball bond (electrical open) after 3,000 h of UHAST stress. EDX analysis on failing Cu ball bond shows higher percentage of O and Cl peaks compared to good ball bond (see Table 2). The source of Cl$^-$ could be originating from the non-green molding compound used in assembly of FBGA 64 and corroded the Cu ball bond under highly temperature and humidity stresses such as UHAST or HAST.

Fig. 2 Cu ball bond corrosion found after biased HAST 2,000 h. CuAl IMC interface microcracking found beneath Cu ball bond in extended hours of HAST failure

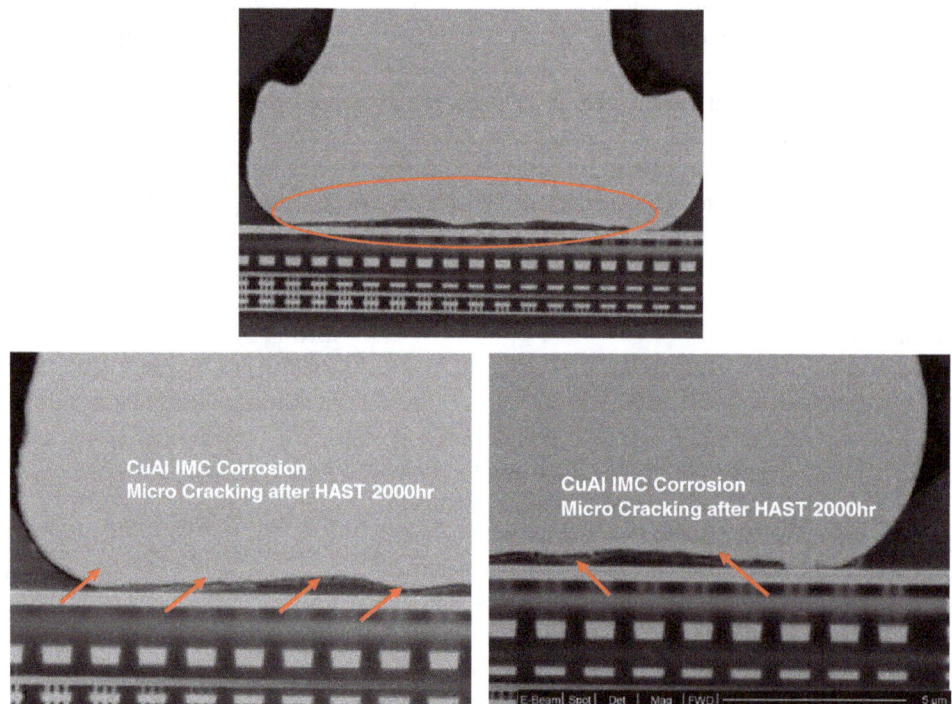

Fig. 3 EDX analysis shows presence of Cl⁻ content at the CuAl IMC interfacial microcracking region. Cu ball bond corrosion found after biased HAST 2,000 h

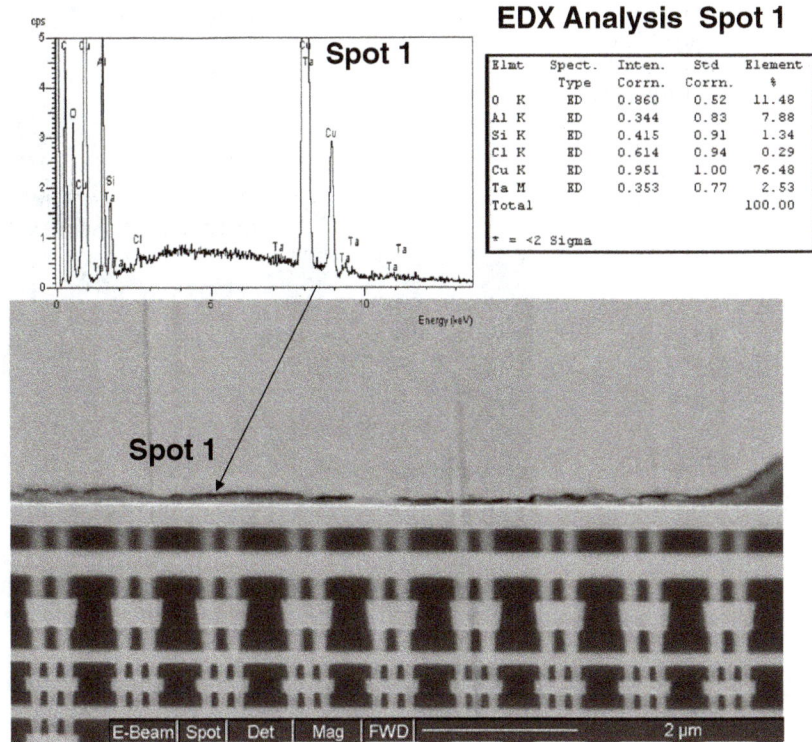

EDX Analysis Spot 1

Elmt	Spect. Type	Inten. Corrn.	Std Corrn.	Element %
O K	ED	0.860	0.52	11.48
Al K	ED	0.344	0.83	7.88
Si K	ED	0.415	0.91	1.34
Cl K	ED	0.614	0.94	0.29
Cu K	ED	0.951	1.00	76.48
Ta M	ED	0.353	0.77	2.53
Total				100.00

* = <2 Sigma

Ball bond microcracking under dry environmental stress

Temperature cycling test

Figure 6a presents SEM micrograph of CuAl IMC microcracking after undergoing extended temperature cycling (TC) stress of 9,500 cycles of −40 °C to 150 °C. Obvious molding compound to die passivation delamination is observed across the row of failing Cu ball bonds. This indicates the mismatches of coefficient of thermal expansion (CTE) between molding compound and silicon die which induced the CuAl IMC microcracking after extended cycles of TC. Figure 6b shows closed up failed Cu ball bond after TC 9,500 cycles and full separation of CuAl IMC beneath Cu ball bond. EDX analysis on site 1 and site 2 near the CuAl IMC interfacial microcracking shows presence of C, Cu, O, Si, and Al elements without Cl⁻ element (see Fig. 6b). Table 4 tabulates the EDX analysis comparing failing Au and Cu ball bond after TC 9,500 cycles.

High temperature storage life test

Specimens heat-treated in nitrogen for more than 3,000 h at 150 °C Figure 7a, b presents SEM micrographs of failed Cu ball bonds after aged for 3,500 h at 150 °C. We found CuAl IMC full separation and microcracking along beneath Cu ball bonds at the edge and also center regions of CuAl IMC. EDX area scan reveals presence of O, Al, Si, and Cu elements at the edge of failed Cu ball bond. No signature of halide element such as Cl (see Fig. 7c) since HTSL is conducted in a dry environmental chamber with nitrogen purging gas.

Figure 8 indicates the uneven AuAl IMC layer formation after aging for 3,500 h at 150 °C. The SEM micrographs show the Au HTSL 3,500 h opens. Variation of thicknesses of AuAl IMCs is noted due to the faster Au atom diffusion into Al metallization on Al bondpad. Hence, thicker AuAl IMC will be found in HTSL stress, and Kirkendall microvoiding is observed after long aging time in Au ball bonds (see Fig. 8).

Representative EDX analysis of failed Cu and Au ball bonds after 3,500 h of HTSL at 150°C is shown in Table 5

Table 2 EDX analysis of failed Cu and Au ball bonds after HAST 2,000 h open failures

Sample	Element (atomic %)						
	Au	Cu	O	Al	Si	Ta	Cl
Au ball	32.15	–	24.76	43.15	–	–	–
Cu ball	–	76.48	11.48	7.88	1.34	2.53	0.29

Table 3 EDX analysis of failed Cu and good Cu ball bonds after UHAST 3,000 h

Sample	Element (atomic %)						
	Au	Cu	O	Al	Si	Ta	Cl
Good Cu ball	–	83.14	1.48	11.02	0.98	3.38	–
Failed Cu ball	–	78.28	8.64	8.17	1.89	2.40	0.68

Fig. 4 Au ball bond corrosion found after biased HAST 2,000 h. Thicker AuAl IMC is formed unevenly beneath Au ball bond. Microcracking is observed between Au ball bond and AuAl IMC. EDX analysis reveals presence of Au, O, and Al elements

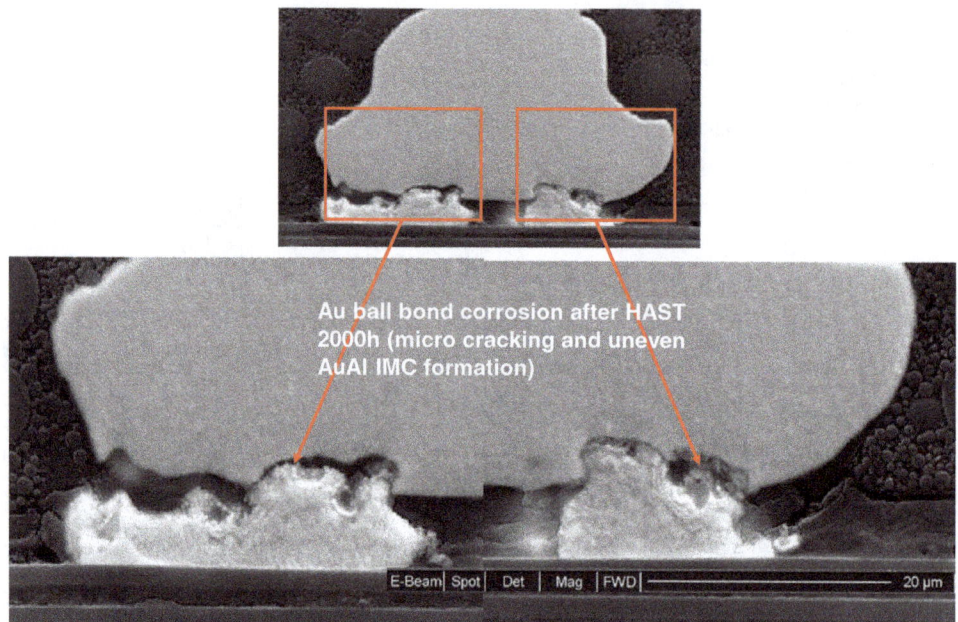

Au ball bond corrosion after HAST 2000h (micro cracking and uneven AuAl IMC formation)

whereby no presence of Cl$^-$ ion in both failing Au and Cu ball bonds. This is noted as HTSL test is conducted in a dry environment condition at 150°C. As noted in our previous report [34, 35], discontinuous intermetallics were found in aged HTSL specimens on Au and Cu ball bonds. The predicted HTSL wear out mechanisms are illustrated in Fig. 11.

Cu and Au ball bond corrosion (biased HAST and unbiased HAST)

CuAl IMC growth mechanism is slightly different from AuAl IMC in microelectronic packages. The IMC between Cu wire and Al pad can be distinguished into five types, as in the case of Au wire. However, only two IMC, Cu$_9$Al$_4$ and CuAl$_2$, can be typically observed because the CuAl IMC forms very slowly and it is very thin in Cu ball bond IMC growth. CuAl IMC will be formed thicker at the edge of Cu ball bond compared to the center of Cu ball bond. This is mainly due to the thermo-

compression effect during capillary compression onto Al bondpad during wire bonding induced by capillary compression.

Moisture in HAST or UHAST chamber will attack Cu ball bond at both edges of Cu ball bonds. Trace Cl$^-$ ion from molding compound will corrode the thin CuAl IMC layer beneath Cu ball bond and hydrolysis of CuAl IMC will occur (see Eq. 1). CuAl IMC microcracking will occur as a result of hydrogen outgassing or embrittlement (as in Eq. 2). Hydrolysis of CuAl IMC will form a brittle IMC, still conductive in Cu ball bond but resistive and will reach wear out opens (lifted ball bond after corrosion) after extended reliability stressing under HAST or UHAST (see Fig. 9).

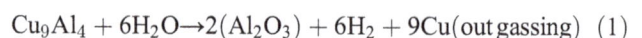

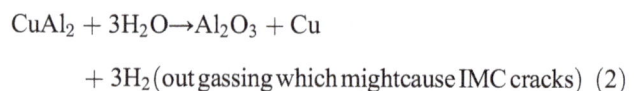

$$Cu_9Al_4 + 6H_2O \rightarrow 2(Al_2O_3) + 6H_2 + 9Cu(out gassing) \quad (1)$$

$$CuAl_2 + 3H_2O \rightarrow Al_2O_3 + Cu$$
$$+ 3H_2(out gassing which might cause IMC cracks) \quad (2)$$

Fig. 5 Cu ball bond corrosion found after unbiased HAST 3,000 h. Microcracking is observed at the edge of Cu ball bond region. EDX analysis reveals presence of Cu, O, Cl, and Al elements

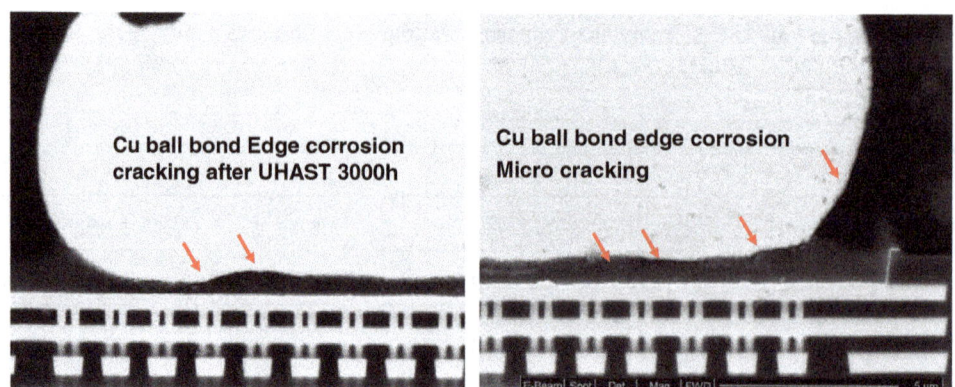

Cu ball bond Edge corrosion cracking after UHAST 3000h

Cu ball bond edge corrosion Micro cracking

Fig. 6 Cu ball bond microcracking is observed after TC 9,500 cycles (**a**). Mold compound to die interfacial delamination is found and might be the factor induced CuAl IMC microcracking. EDX analyses reveal presence of C, Cu, O, Si, and Al elements without Cl⁻ element (**b**)

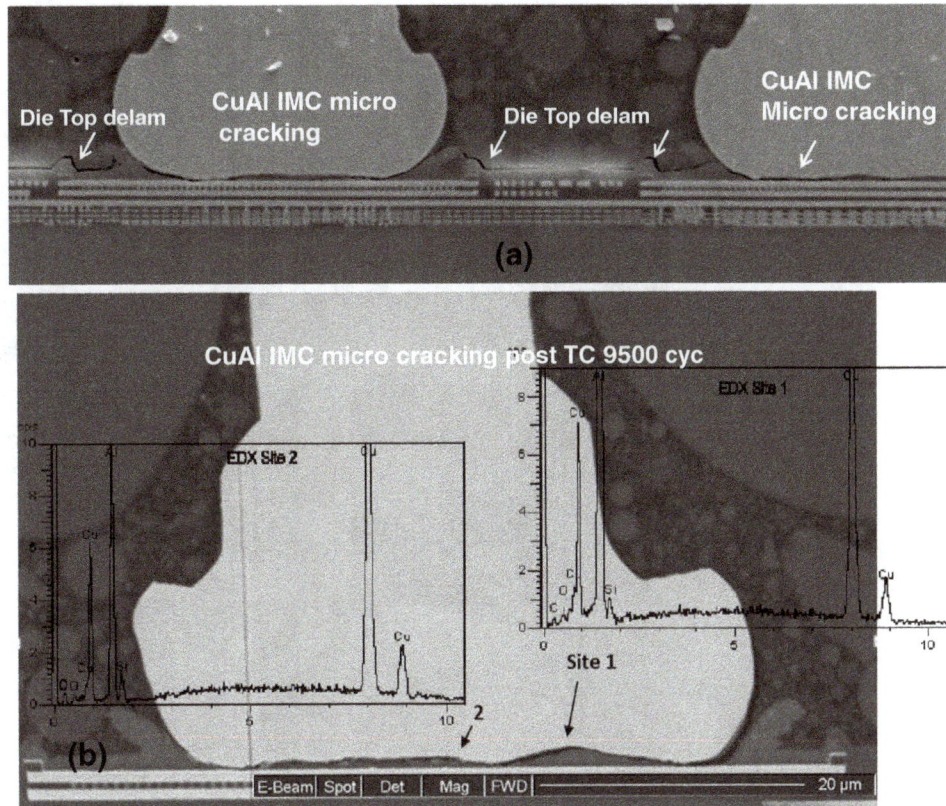

Au ball bond undergoes a slightly different corrosion mechanism in HAST or UHAST reliability stressing. Moisture in HAST or UHAST chamber will penetrate from the edge of Au ball bond after long hours of HAST and UHAST stressing. AuAl IMC will react with moisture and form Al_2O_3 and hydrogen outgassing (as shown in Eq. 3). Hydrogen gas evolution due to moisture in contact with intermetallics has been extensively documented and is one of the known causes of embrittlement [10]. Stress-induced microcracking will cause AuAl microcracking together with the formation of Kirkendall microvoiding which results in uneven AuAl IMC formation (refer Fig. 4). Lifted Au ball bond will occur after long hours of HAST or UHAST stress and this is typical wear out failure (as shown in Fig. 10).

$$2Au_4Al + 3H_2O \rightarrow Al_2O_3 + 8Au + 6H \qquad (3)$$

Table 4 EDX analysis of failed Cu and Au ball bonds after TC 9,500 cycles

Location	Element (atomic %)						
	Au	Cu	O	Al	Si	C	Cl
Au ball	38.25	–	1.76	59.99	–	–	–
Cu ball	–	77.58	2.11	13.74	3.74	2.83	

Au and Cu ball bond microcracking in HTSL

Au ball bond is found with higher IMC growth rate at least 5× compared to Cu ball bond in HTSL aging test [1, 25]. Hence, there is slightly different HTSL failure mechanism after long duration of aging stress in Au and Cu ball bonds. Both CuAl and AuAl IMCs are formed in long hours of HTSL test except more uniform AuAl IMC formation compared to CuAl. Thicker CuAl IMC is formed at the edge of Cu ball bond (see Figs. 2 and 7c). AuAl IMC is observed with more uniform and thicker (as indicated in Fig. 8) but Kirkendall microvoiding will occur which might induce AuAl IMC microcracking (Fig. 11b). HTSL wear out opens occur as lifted ball bonds for both Au and Cu balls except with Kirkendall microvoiding in Au ball bonds (Fig. 11a, b, respectively).

Au and Cu ball bond microcracking in TC

Table 6 tabulates coefficient of thermal expansion (CTE) for materials used in package bills of materials in FBGA 64. The mismatch in CTE between Cu (17.8 ppm/°C) and Au ball bond (14.2 ppm/°C) to the silicon die (3.0 ppm/°C) induced different thermal expansions and contraction rates in the temperature cycling test. The CTE mismatch between Au and Cu ball

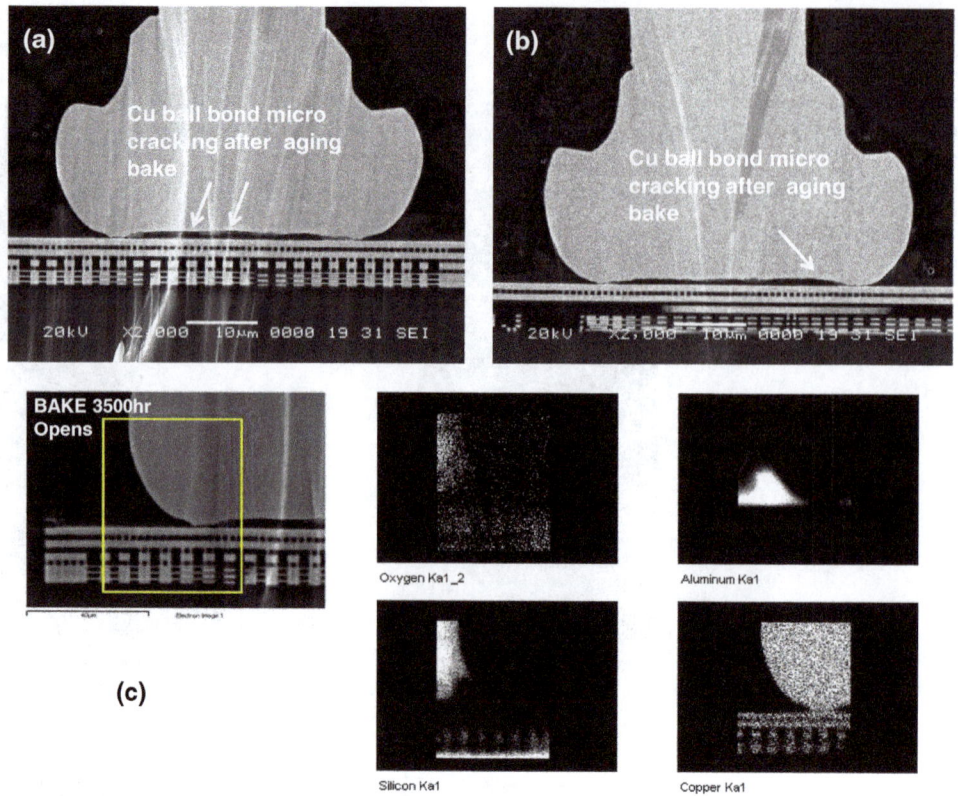

Fig. 7 Cu ball bond IMC microcracking is found after HTSL 3,500 h of stressing (a, b). EDX area scan reveals presence of O, Al, Si, and Cu elements at the edge of failed Cu ball bond. No signature of halide element such as Cl (c)

bonds with Al bondpad of silicon die will impose different thermal expansion rates during hot cycles (150 °C) and contraction rates during cold cycles (−40 °C). IMC formation initiated at the edge of the ball bond (due to the ball bond pressing force by bonding capillary) and

microcracking will be induced after long cycles of thermal cycling effects. The microcracking occurs in between ball bond IMC (as shown in Fig. 12). This predicted TC mechanism also correlated to SEM images as indicated in Fig. 12.

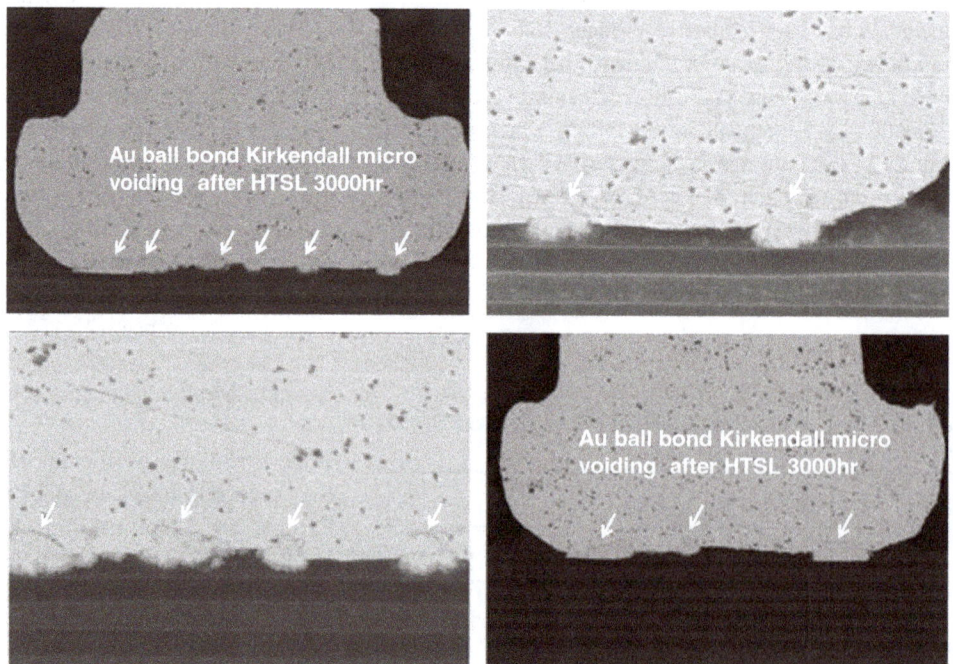

Fig. 8 Au ball bond with uneven AuAl IMC formation and Kirkendall microvoiding are found along the Au ball bond after HTSL 3,000 h at 150 °C aging condition. Hairline cracking is observed along the Kirkendall microvoiding region

Table 5 EDX analysis of failed Cu and Au ball bonds after HTSL 3,500 h

Location	Element (atomic %)					
	Au	Cu	O	Al	Si	Cl
Au ball	48.65	–	0.97	50.38	–	–
Cu ball	–	77.58	2.11	13.74	3.74	–

Extended reliability analysis of Au and Cu ball bonds

Humidity reliability analysis (UHAST)

All package reliability plots belong to wear out reliability mode in bathtub curve since its shape parameter (ß) is more than 1.0. Au ball bonds show better UHAST package reliability performance with higher mean-time-to failure hours (t_{50}) and characteristics life ($t_{63.2}$, η) in UHAST reliability plot (fitted to Weibull distribution) compared to Cu ball bonds. Figure 13 illustrates a Cu ball bond with lower package reliability margin and usually more susceptible to Cu moisture corrosion test under UHAST condition. This has been reported in our previous literature works [8, 25–27, 31]. Cu ball bond has a layer of Pd coated on low-corrosive resistance Cu wire which inhibits moisture ball bond corrosion in UHAST conditions (130 °C, 85 %RH). However, Au wire is well known with corrosion resistant material and shows higher hours-to-failure in UHAST wear out reliability plot (Fig. 13).

Dry environmental reliability analysis (TC)

Cu ball bonds is found with higher mean-time-to failure hours (t_{50}) and characteristics life ($t_{63.2}$) in TC reliability plot (fitted to Weibull distribution), as indicated in Fig. 14. Apparently, Cu performs well under dry conditions in the TC cycling test (−40 °C to 150 °C). In this case, we observe Cu ball bonds withstand higher cycle-to-failure compared to Au ball bonds in TC stress test conditions.

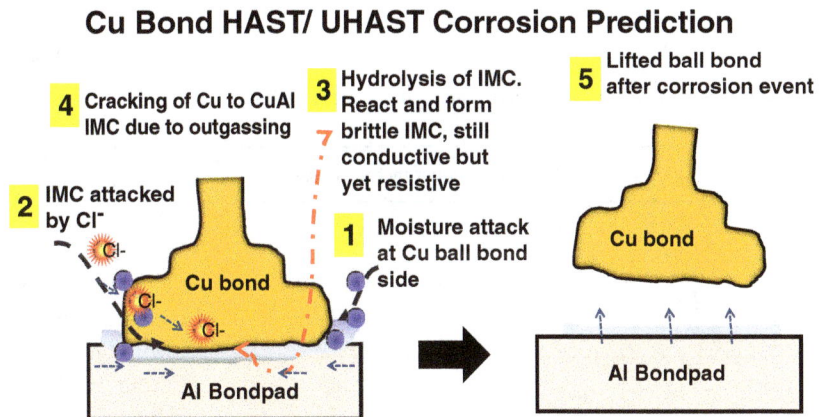

Fig. 9 Proposed CuAl IMC corrosion mechanism on 110 nm device FBGA 64 package after extended hours of biased HAST or unbiased HAST stressing [31]

Cu Bond HAST/ UHAST Corrosion Prediction

4 Cracking of Cu to CuAl IMC due to outgassing

3 Hydrolysis of IMC. React and form brittle IMC, still conductive but yet resistive

5 Lifted ball bond after corrosion event

2 IMC attacked by Cl⁻

1 Moisture attack at Cu ball bond side

Cu bond

Al Bondpad

Cu bond

Al Bondpad

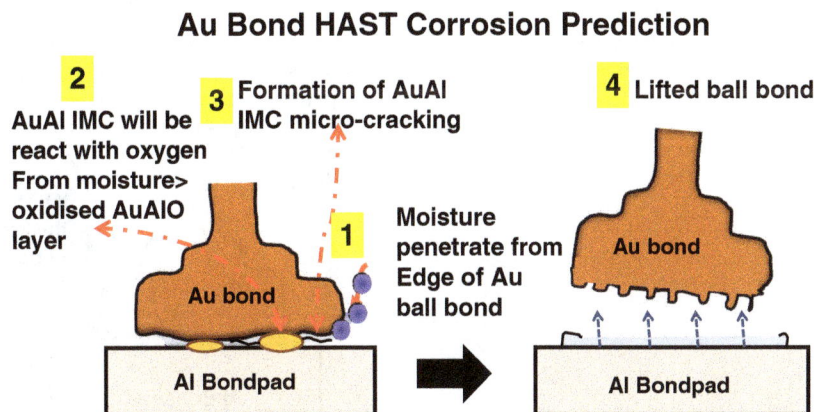

Fig. 10 Proposed AuAl IMC Kirkendall microvoiding mechanism and induced opens after long aging hours on 110 nm device FBGA 64 package [31]

Au Bond HAST Corrosion Prediction

2 AuAl IMC will be react with oxygen From moisture> oxidised AuAlO layer

3 Formation of AuAl IMC micro-cracking

4 Lifted ball bond

1 Moisture penetrate from Edge of Au ball bond

Au bond

Al Bondpad

Au bond

Al Bondpad

Fig. 11 Schematic representation of Cu ball bond microcracking after long hours of HTSL stressing (a) and Kirkendall microvoiding as a function of aging hours in Au ball bond (b)

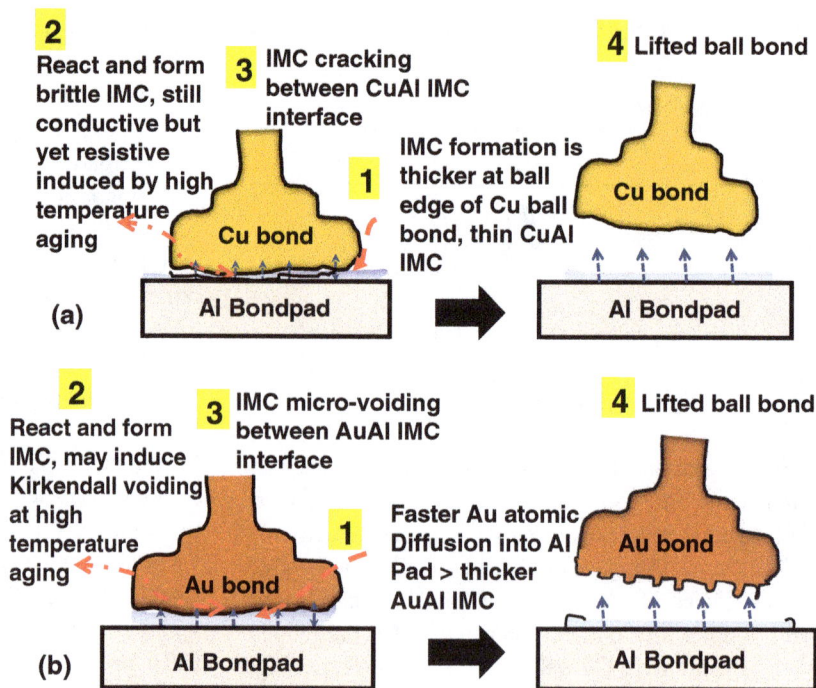

2 React and form brittle IMC, still conductive but yet resistive induced by high temperature aging

3 IMC cracking between CuAl IMC interface

1 IMC formation is thicker at ball edge of Cu ball bond, thin CuAl IMC

4 Lifted ball bond

Cu bond

Al Bondpad

(a)

2 React and form IMC, may induce Kirkendall voiding at high temperature aging

3 IMC micro-voiding between AuAl IMC interface

1 Faster Au atomic Diffusion into Al Pad > thicker AuAl IMC

4 Lifted ball bond

Au bond

Al Bondpad

(b)

Table 6 Key material characteristics of epoxy mold compound (EMC) A and B

Material	Units	CTE (coefficient of thermal expansion)
Au	ppm/°C	14.2
Cu	ppm/°C	17.8
Silicon	ppm/°C	3.0
Al	ppm/°C	22.2

Summary and conclusions

The purpose of this research is to investigate and understand the microstructural evolution of Au and Cu ball bonds in extended reliability stressing such as HAST, UHAST, TC, and HTSL. To achieve this goal, both AuAl and CuAl intermetallic growth was studied at elevated temperatures in HTSL. In our study, respective failure mechanisms of copper and gold ball bonds carrion under HAST and UHAST, ball bond lifting in TC and HTSL have been analyzed and proposed. The evolution of surface morphology, including copper and gold ball bond microcracking, gold ball bond Kirkendall

Fig. 12 Proposed Cu ball bond microcracking induced by different rate of contraction and expansion between Cu bond and silicon die after extended temperature cycling

Cu Bond TC Lifted Ball Prediction

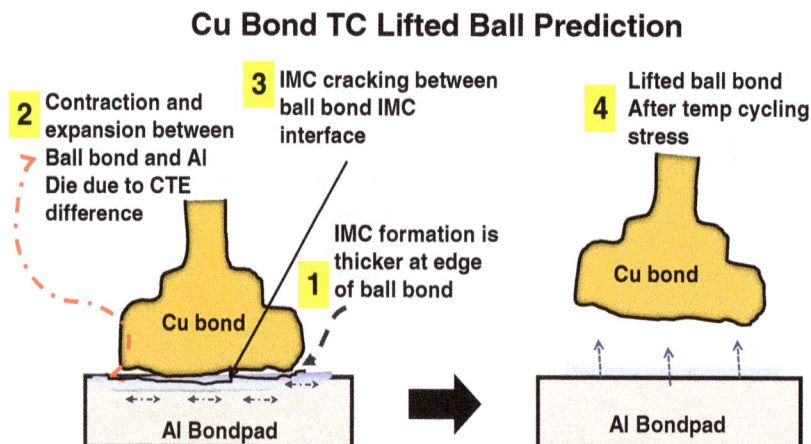

2 Contraction and expansion between Ball bond and Al Die due to CTE difference

3 IMC cracking between ball bond IMC interface

1 IMC formation is thicker at edge of ball bond

4 Lifted ball bond After temp cycling stress

Cu bond

Al Bondpad

Cu bond

Al Bondpad

Weibull Plot (UHAST Wearout Reliability)

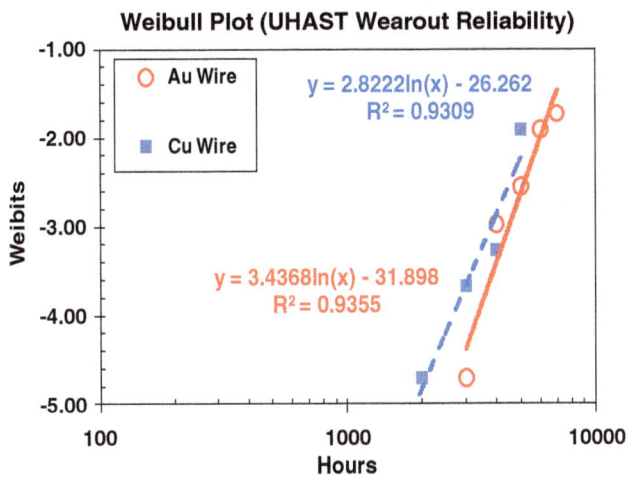

$y = 2.8222\ln(x) - 26.262$
$R^2 = 0.9309$

$y = 3.4368\ln(x) - 31.898$
$R^2 = 0.9355$

Fig. 13 Obtained Weibull plotting of Au and Cu ball bonds in unbiased HAST reliability stress with different epoxy mold compounds (EMC) [26, 31]

microvoiding and IMC formation, was studied in FBGA package with copper and gold ball bonds during various reliability stresses. Au ball bonds show superior extended UHAST reliability than Cu ball bonds for both mold compounds A and B (see Fig. 13). This could be due to Au is more stable and higher corrosion resistance under moisture UHAST conditions compared to Cu ball bonds. We observed Cu ball bonds with higher TC extended reliability performance (higher t_{first}, t_{50}, and $t_{63.2}$) compared to Au ball bonds in FBGA 64 package of both mold compounds A and B (Fig. 14). The effect of wire type is not the key factor affecting the TC reliability performance but we observed higher extended reliability performance in Cu ball bonds compared to Au ball bonds.

Weibull Plot (TC Wearout Reliability)

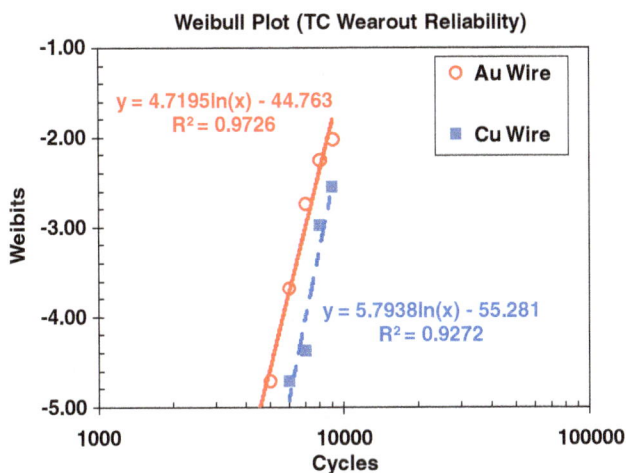

$y = 4.7195\ln(x) - 44.763$
$R^2 = 0.9726$

$y = 5.7938\ln(x) - 55.281$
$R^2 = 0.9272$

Fig. 14 Obtained Weibull plotting of Au and Cu ball bonds in extended TC reliability stress with different epoxy mold compounds (EMC) [26, 31]

Acknowledgement The authors would like to take this opportunity to thank Spansion management for their management support for the paper publication.

References

1. Harman GG (1999) Wirebonding in microelectronic: materials, processes, reliability and yield, 2nd edn. McGraw Hill, New York
2. Chauhan P, Zhong ZW, Pecht M (2014) Copper wire bonding, 1st edn. Springer, New York
3. Chauhan P, Zhong ZW, Pecht M (2013) Copper wire bonding concerns and best practices. J Electron Mater.
4. Schneider-Ramelow M, Geißler U, Schmitz S et al (2013) Development and status of Cu ball/wedge bonding in 2012. J Electron Mater 42:558–595.
5. Appelt BK, Tseng A, Chen C-H, Lai Y-S (2011) Fine pitch copper wire bonding in high volume production. Microelectron Reliab 51:13–20
6. Gan CL, Ng EK, Chan BL, Hashim U (2012) Technical barriers and development of cu wirebonding in nanoelectronics device packaging. J Nanomater 2012:1–7.
7. Gan CL, Ng EK, Chan BL, T Kwuanjai, S Jakarin, Hashim U (2012) Wearout reliability study of cu and au wires used in flash memory fineline BGA package, 2012 I.E. 7th Int. Microsystems, Packag. Technol. Conf. pp 494–497
8. Gan CL, Toong TT, Lim CP, Ng CY (2010) Environmental friendly package development by using copper wirebonding. 34th IEEE CPMT IEMT, Malacca, 2010, pp. 1–5
9. Zhong ZW (2009) Wire bonding using copper wire. Microelectron Int 26:10–16
10. Breach CD (2010) What is the future of bonding wire? Will copper entirely replace gold? Gold Bull 43:150–168
11. Murali S, Srikanth N, Wong YM, Vath CJ (2006) J Mater Sci 42:615–623
12. Karpel A, Gur G, Atzmon Z, Kaplan WD (2007) Microstructural evolution of gold–aluminum wire-bonds. J Mater Sci 42:2347–2357
13. Drozdov M, Gur G, Atzmon Z, Kaplan WD (2008) Detailed investigation of ultrasonic Al–Cu wire-bonds: I Intermetallic formation in the as-bonded state. J Mater Sci 43:6029–6037
14. Drozdov M, Gur G, Atzmon Z, Kaplan WD (2008) Detailed investigation of ultrasonic Al–Cu wire-bonds: II. Microstructural evolution during annealing. J Mater Sci 43:6038–6048
15. Xu C, Breach CD, Sritharan T et al (2004) Oxidation of bulk Au–Al intermetallics. Thin Solid Films 462–463:357–362
16. Xu H, Liu C, Silberschmidt VV et al (2009) A re-examination of the mechanism of thermosonic copper ball bonding on aluminium metallization pads. Scr Mater 61:165–168
17. Xu H, Liu C, Silberschmidt VV et al (2011) Behavior of aluminum oxide, intermetallics and voids in Cu–Al wire bonds. Acta Mater 59:5661–5673
18. Zeng Y, Bai K, Jin H (2013) Thermodynamic study on the corrosion mechanism of copper wire bonding. Microelectron Reliab. In Press
19. Gan CL, Ng EK, Chan BL, Hashim U (2012) Reliability challenges of Cu wire deployment in flash memory packaging. 2012 I.E. 7th Int. Microsystems, Packag. Technol. Conf. pp 498–501
20. Tan C, Daud A, Yarmo M (2002) Corrosion study at Cu–Al interface in microelectronics packaging. Appl Surf Sci 191:67–73

21. Uno T (2011) Microelectron reliab 51:148–156
22. Yamaji Y, Hori M, Ikenosako H, et al. (2011) IMC study on Cu wirebond failures under high humidity conditions. 2011 I.E. 13th Electron. Packag. Technol. Conf. IEEE, pp 480–485
23. Su P, Seki H, Ping C, et al. (2013) Effects of reliability testing methods on microstructure and strength at the Cu wire-Al pad interface. 2013 I.E. 63rd Electron. Components Technol. Conf., pp 179–185
24. Lu YH, Wang YW, Appelt BK, Lai YS, Kao CR (2011) Growth of CuAl intermetallic compounds in Cu and Cu (Pd) wire bonding. 2011 I.E. Electron. Components Technol. Conf., pp 1481–1488
25. Gan CL, Ng EK, Chan BL et al (2012) Wearout reliability and intermetallic compound diffusion kinetics of Au and PdCu wires used in nanoscale device packaging. J Nanomater 2012:1–9
26. Gan CL, Hashim U (2013) Reliability assessment and mechanical characterization of Cu and Au ball bonds in BGA package. J Mater Sci Mater Electron 24:2803–2811
27. Gan CL, Francis C, Chan BL, Hashim U (2013) Extended reliability of gold and copper ball bonds in microelectronic packaging. Gold Bull 46:103–115
28. Gan CL, Hashim U (2013) Superior performance and reliability of copper wire ball bonding in laminate substrate based ball grid array. Microelectron Int 30:169–175
29. Gan CL, Hashim U (2013) Comparative reliability studies and analysis of Au, Pd-coated Cu and Pd-doped Cu wire in microelectronics packaging. PLoS One 8:1–8
30. Gan CL, Francis C, Chan BL, Hashim U (2014) Future and technical considerations of gold wirebonding in semiconductor packaging—a technical review. Microelectron. Int. 31. In Press
31. Gan CL, Francis C, Chan BL, Hashim U (2013) Effects of wire type and mold compound on wearout reliability of semiconductor flash fineline BGA package. IEEE 8th Int. Microsystems, Packag. Assem. Circuits Technol. Conf. pp 297–301
32. Yu C-F, Chan C-M, Chan L-C, Hsieh K-C (2011) Cu wire bond microstructure analysis and failure mechanism. Microelectron Reliab 51:119–124
33. Mc Pherson JW (2013) Reliability physics and engineering: time-to-failure modeling, 2nd edn. Springer, New York
34. Gan CL, Hashim U (2013) Reliability assessment and activation energy study of Au and Pd-coated Cu wires post high temperature aging in nanoscale semiconductor packaging. J Electron Packag 135: 021010
35. JEDEC IPC STD 020 (2008) Moisture/reflow sensitivity classification for nonhermetic solid state surface mount devices

21

Microstructure and biocompatibility of gold–lanthanum strips

Rebeka Rudolf · Sergej Tomić · Ivan Anžel ·
Tjaša Zupančič Hartner · Miodrag Čolić

Abstract Microalloying of pure gold, which has the highest biocompatibility but relatively low yield strength and poor wear resistance, might improve its applicability as adornments and biomedical implants. The objective of this study was to analyse the microstructure and biocompatibility of gold–lanthanum (Au–0.5 wt% La) microstrips as a potential biomaterial in dentistry or medicine. We found that microalloying of Au with La produced very fine nanosized grains homogeneously dispersed through the entire volume of the rapidly solidified (RS) alloys. This initiates the formation of Au_6La phase which increases strength and hardness of the alloy significantly. By RS, large reduction of grains and microsegregation increases the strength of the alloy additionally. Our results suggest that Au–La microstrips, although non-cytotoxic for L929 cells, rat thymocytes, rat peritoneal macrophages (PMØ) and human peripheral blood mononuclear cells (PBMNCs), can activate immune cells. Namely, RS Au–La microstrips stimulated the production of nitric oxide (NO) by PMØ. Using a model of phytohemaglutinine (PHA)-stimulated human PBMNCs, we found that RS Au–La strips increased the proliferation of these cells and stimulated the production of Th1, Th17 cytokines, and immunoregulatory cytokine IL-10. Our results suggest that RS Au–La microstrips are biocompatible, but they can modulate the immune response. Therefore, their use as potential implants should be considered carefully.

Keywords Microalloying · Gold–Lanthanum · Rapid solidification · Biocompatibility · Immunomodulation

Introduction

Gold alloys are used in dentistry, not only for their preferred golden colour but also because they maintain an extremely high chemical stability in the mouth. They also possess several desirable mechanical properties, such as high strength, ductility and elasticity [1]. When considering the formulations of gold-based high noble alloys for porcelain bonding and other applications in dentistry, it is important to ensure all the required biomechanical properties, including not only their easy cast into thin sections but also their good biocompatibility. Pure gold, which has the highest biocompatibility, has relatively low yield strength and poor wear resistance, which limits its applicability in adornment and other biomedical applications. To improve these properties, the alloying elements such as silver, palladium, zinc and platinum are usually added [1, 2]. Dental alloys with high gold content were shown to have good biocompatibility due to the high corrosion resistance of gold. However, in vitro studies have demonstrated the release of alloying elements in culture media, artificial saliva or distilled water. Some of them could reach the levels that cause detectable cytotoxic effect [2, 3]. Metal corrosion depends on an alloy's composition and microstructure, biomechanical conditions, mode of casting and polishing, composition and electrolyte characteristics of solutions used for

R. Rudolf (✉) · I. Anžel
Faculty of Mechanical Engineering, University of Maribor,
2000 Maribor, Slovenia
e-mail: rebeka.rudolf@um.si

R. Rudolf
Zlatarna Celje d.d., Celje, Slovenia

S. Tomić · M. Čolić
Medical Faculty of the Military Medical Academy, University of
Defence, 11000 Belgrade, Serbia

T. Z. Hartner
Ortotip d.o.o., Murska Sobota, Slovenia

M. Čolić
Medical Faculty, University of Niš, 18000 Niš, Serbia

alloy conditioning, size of the alloy's surface area exposed to solutions, duration of incubation time and other factors [4, 2].

The mechanical properties of pure gold can also be improved significantly by microalloying [5–7]. In recent years, significant effort has been made to increase the hardness of 24-carat gold alloys with a minimum of 99.5 wt% of Au [8], but such alloys exhibit only slightly higher hardness than pure gold in an annealed state. The usual alloying elements in microalloyed gold are used mainly as a grain refiner. In general, intensive work hardening can be achieved in the metals with decreasing grain size. However, in the fine-grained alloys with low stacking fault energy, deformation with twinning can prevail as the main operative deformation mode. This decreases the effect of work hardening. Therefore, in gold, which is a metal with low stacking fault energy, no obvious strengthening can be expected by plastic deformation and grain size reduction. Therefore, it is necessary to introduce some other mechanisms of alloy strengthening, such as introduction of alloying elements, which are very different to gold in atomic radius and have lower density, or the use of elements with very low solubility in gold matrix. One such approach is the microalloying of gold with rare earth elements as suggested by theoretical analysis [5] and in our study we decided for lanthanum. Microalloying with lanthanum can serve as deoxidants to facilitate bonding between the alloy and the ceramic or to enhance the strength and colour of the alloy [9]. Consequently, lanthanum can improve the possibility to prepare thin strips by melt spinning. Therefore, Au–La alloys, due to their specific characteristics, might be used as a potential biomaterial for various applications in medicine and dentistry. However, their biocompatibility and immunomodulatory properties are still unknown. Some studies suggested that lanthanum could be a toxic element [10], but little is known about its behaviour, corrosion stability and biocompatibility within gold–lanthanum alloy.

In this research, 24-carat gold was microalloyed with 0.5 wt% La, which usually have low solubility in gold. Our aim was to characterise Au–La alloy and then to investigate the alloy's biocompatibility using different in vitro models.

Experimental

Preparation and characterisation of gold–lanthanum strips and gold platelets

The investigated alloy Au–0.5 wt% La was made from pure gold 99.99 wt% Au and pure lanthanum 99.9 wt% La. The alloy was prepared by melting and casting in an induction vacuum furnace at Zlatarna Celje under argon (Ar 5.0) inert atmosphere. The starting materials were then heated up to 1,200 °C. Melting was performed in a crucible with ceramic insert. A ceramic insert made from Al_2O_3 with the addition of

Cr_2O_3 is suitable for casting precious metals and has high temperature resistance. From this graphite crucible, the melt was cast into a graphite mould with diameter Ø 18–20 mm.

Rapid solidification was performed by the chill-block melt spinning technique. The alloy was re-melted under an argon overpressure. The stream dropped on the wheel at a 90° angle from a 1.7-mm-thick nozzle positioned 0.8 mm above the wheel rotating at 21 m/s. It was estimated that this speed is optimal for the selected alloy cast at 1,230 °C.

Continuous ribbons about 2 mm in width and 0.5–10 μm in thickness were produced. The strips were cut into pieces 2 mm×4 mm, cleaned ultrasonically and sterilised by 70 % ethanol for 5 min. Control gold platelets (5 mm×5 mm× 1 mm) were polished and prepared by using the same procedure. The samples were then used for characterisation and biocompatibility and immunological studies.

The microstructural properties and differences between cast and rapidly solidified samples were generally observed with a light microscope (Nikon Epiphot 300). To accomplish more detailed information of the microstructure constituents and morphology, we used scanning electron microscopy by Sirion NC 400. Specimens for optical microscopy were obtained as vertical cross sections, prepared with standard metallographic methods and etched chemically in a mixture of 60 vol.% HCl and 40 vol.% HNO_3 (for optical microscopy) and a solution of CrO_2 in HCl (for electron microscopy).

Additionally, imaging with transmission electron microscope (TEM) Philips CM12 was performed on submicron particles inside the ribbons using the acceleration voltage 120 kV. Chemical composition was analysed by X-ray fluorescence (XRF). After the preparation of samples and their microstructural analysis, microhardness was measured by Zwick 3212 using a 50-g load.

Cells

The following cells were used as targets for biocompatibility and immunological studies: rat thymocytes; rat peritoneal macrophages (PMØ); L929 mouse fibroblasts; and human peripheral blood mononuclear cells (PBMNCs). Thymocytes were isolated from thymuses of 10-week-old, male, AO rats. The viability of the cells, as determined by 1 % Trypan Blue staining, was higher than 95 %. PMØ were collected by aspiration of peritoneum of anaesthetised rats. The viability of peritoneal cells, of which 80 % were macrophages, was higher than 95 %. L929, a mouse fibroblast cell line, was obtained from ATCC (Washington DC, USA). PBMNCs from eight healthy volunteers, who signed consent forms, were isolated from buffy coats by density centrifugation on Lymphoprep (Nycomed, Oslo, Norway). All studies on animal and human cells were approved by the ethics committee of the Military Medical Academy, Belgrade, Serbia.

The cells were resuspended in RPMI 1640 medium supplemented with 10 % foetal calf serum (FCS, Sigma) and 2-mercaptoethanol (ME, Sigma), 2 mM L-glutamine (Sigma) and antibiotics (Galenika, Zemun, Serbia) including gentamycin (10 μg/ml), penicillin (100 units/ml) and streptomycin (125 μg/ml) (complete RPMI medium) and cultivated at 37 °C with 5 % CO_2 for the indicated periods of time.

Cytotoxicity assays

The cytotoxicity of Au–La strips or control gold platelets was tested using 3-[4,5-dimethylthiazol-2-lyl-2,5 diphenyl tetrazolium bromide (MTT) test, a standard method for the assessment of mitochondrial succinic dehydrogenase (SDH) activity. L929 cells (5×10^4/ml), thymocytes (5×10^6/ml), PMØ (1×10^6/ml) and human PBMNCs (2×10^5/ml) were cultivated with Au–La strips or control Au platelets in flat-bottom 96-well plates (ICN, Costa Mesa, CA) (250 μl/well) in an incubator in complete RPMI medium for 24 h and 3 days. The surface area/volume of medium was 2.8 cm^2/ml. The cells cultivated without Au–La strips, and Au platelets served as experimental controls.

After incubation of the cells for the indicated period of times, the medium was carefully removed and the wells were filled with 100 μl of MTT (Sigma, Munich, Germany) (1 mg/mL), dissolved in the complete RPMI medium. The wells with Au–La or Au platelets filled with 100 μl of MTT, but without cells, served as controls. In addition, wells with 100 μl of MTT solution served as blank controls. After a 3-h incubation period (37 °C, 5 % CO2), 100 μl/well of 10 % sodium-dodecyl sulphate (SDS)–0.1 N HCl (Serva, Heidelberg, Germany) was added to solubilise intracellularly stored formazan. The plates were incubated overnight at room temperature. The optical density of the colour was then measured at 570 nm in a spectrophotometer (Behring ELISA Processor II, Heidelberg, Germany). The results were expressed as the percentage of optical density (metabolic activity) compared to the control (cultures without Au–La samples), used as 100 % as follows: metabolic activity (%)=(optical density (O.D.) of cells cultivated with Au–La (or Au) samples−O.D. of Au–La (or Au) samples without cells)/(O.D. of cells cultivated alone−O.D. of Au–La (or Au) samples cultivated alone)× 100.

Cell death was determined by staining the cells from cultures with 1 % Trypan Blue. The labelled cells, identified by light microscopy, were considered as dead, predominantly necrotic cells. The percentages of dead cells were determined on the basis of at least 500 total cells from one well. The percentage of viable cells was calculated as 100 % of total cells−% of dead cells. All results were expressed as a mean of triplicates. Additionally, necrosis was confirmed using a staining protocol with propidium iodide (PI) (Sigma) without cell permeabilization by flow cytometry.

Apoptosis was assessed based on the detection of DNA fragmentation. For this purpose, the cultivated cells were

collected and then washed with PBS, followed by incubation with PI (10 μg/ml) dissolved in a hypotonic solution (0.1 % sodium citrate+0.1 % Triton-X solution in distilled water). After incubation with PI, the cells were analysed by flow cytometry. L929 cells were detached from the plastic surface by using 0.25 % trypsin (Serva, Heidelberg, Germany) before staining with PI.

For morphological evaluation of apoptosis, the cells were stained with Turk solution. The solution fixes and stains the nuclei, enabling a clear distinction between the chromatin structure in viable and apoptotic cells. At least 500 cells were examined in each sample, and the results are expressed as percentages of apoptotic cells. The detection of different stages of apoptosis and necrosis was analysed by flow cytometry using Annexin-V–fluorescein isothiocyanate (FITC)/PI staining kit (R&D), following the manufacturer's protocol.

Proliferation assays

PBMNCs (1.5×10^6/ml) were cultivated with Au–La strips or control Au platelets in flat-bottom 96-well plates (ICN, Costa Mesa, CA) (200 μl/well) in an incubator, using complete RPMI medium for 3 days. For stimulation of the cells, phytohemaglutinine (PHA, 30 μg/ml) was added to cell cultures. During the last 18 h of incubation, the cells were pulsed with 1 μCi/well [3H] thymidine (6.7 Ci/mmol, Amersham, Bucks, U.K.). Labelled cells were harvested onto glass fibre filters, and the incorporation of the radionuclide into DNA was further measured by β-scintillation counting (LKB-1219 Rackbeta, Finland). The results were expressed as counts per minute (cpm)±SD of triplicates.

NO and cytokine assays

The effect of RS Au–La strips on production of nitric oxide (NO) by PMØ was assessed after 24-h cultures by measuring the nitrite levels using Griess reaction and calculating the unknown concentrations from the standard curve. The effect of Au–La and Au platelets on cytokine production was studied using a model of PHA-stimulated PBMNCs. PHA-stimulated PBMNC cultures were incubated with Au–La and Au samples for 48 h, harvested and centrifuged, so the cell-free supernatants were collected and stored at −20 °C for the subsequent determination of cytokine levels. The levels of cytokines in PHA-stimulated PBMNC culture supernatants were determined using FlowCytomix Human Th1/Th2 11plex Kit from Bender MedSystems (Vienna, Austria), to determine interleukin (IL)-1, IL-2, IL-4, IL-5, IL-6, IL-8, IL-10, IL-12, tumour necrosis factor (TNF)-α, TNF-β and interferon (IFN)-γ levels. The levels of IL-17 in those cultures was determined using corresponding ELISA kits (R&D Systems, Minneapolis, USA), following the manufacturer's instructions.

Statistical analysis

All values are given as mean±standard deviation (SD). The number of samples was four to six. The Student t test, paired t test and ANOVA test were used for evaluating the differences between the experimental and corresponding control samples. The values $p < 0.05$ or less were considered statistically significant.

Results and discussion

Preparation and characterisation of the Au–La alloys

Upon preparation of conventionally solidified 24-carat gold microalloyed with 0.5 wt% La, which usually have low solubility in gold, we aimed to characterise the alloy. The phase diagram of Au–0.5 wt% La [11], suggests that at the Au-rich corner, eutectic reaction occurs and results in two-phase microstructure with solid solution αAu and eutectic (αAu + Au_6La).

Next, we compared the microstructure of conventionally solidified Au–0.5 wt% La alloy with pure Au (Fig. 1). Pure

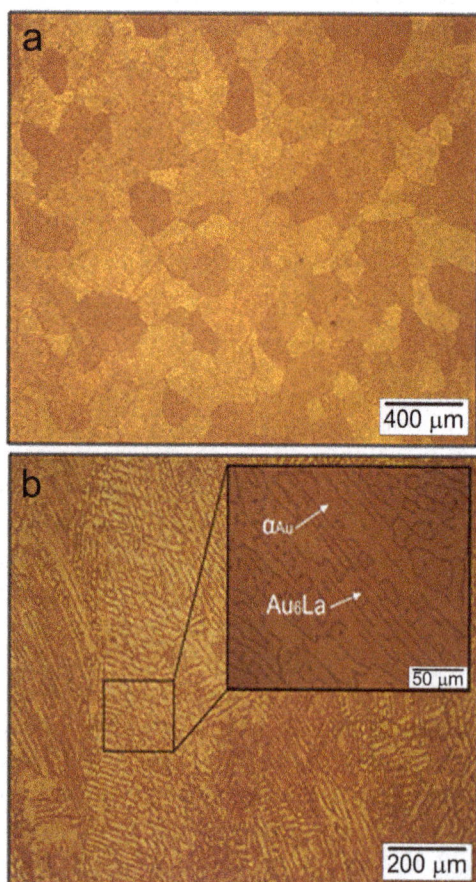

Fig. 2 Analysis of RS Au–La strips prepared by rapid solidification. Cross section of melt spun ribbon Au–La was observed by **a** optical microscopy or **b** TEM

gold consists of equiaxed grains approximately 120 µm in size. On the other hand, the microstructure of Au microalloyed by La consists of two phases. It contains mainly primary dendrites with some amount of eutectic phase in the interdendritic space. According to the phase diagram [11], the primary phase is a solid Au solution, and the eutectic is a combined αAu and intermetallic phase Au_6La. Au dissolves an infinitively small quantity of La, so the obtained solid solution in stable conditions is almost pure Au. As a result of the relatively fine distribution of eutectic, it can be predicted that the mechanical properties could be improved significantly compared to pure Au already in cast condition.

Rapid discharge of heat from the melt increases undercooling and, consequently, increases nucleation rate. As a result, grain refinement and metastable microstructure are achieved. Rapid solidification was performed by the chill-block melt spinning technique [3]. Figure 2a presents the microstructure of RS Au–0.5 wt% La ribbons 0.5–10-µm

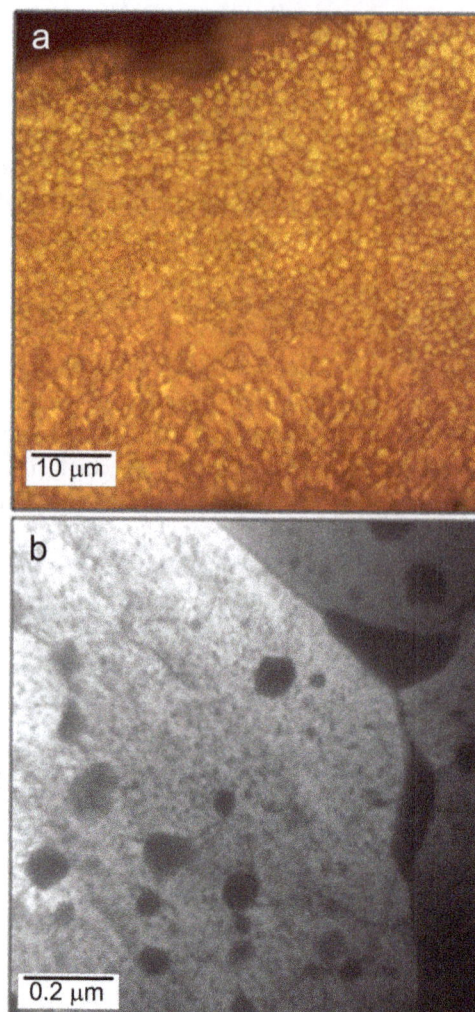

Fig. 1 Analysis of RS Au–La strips prepared by conventional solidification. Optical micrographs of **a** pure Au and **b** RS Au–La strips, in as cast condition are shown. αAu and Au_6La phases are labelled by *white arrows* on the magnified inset

thick in transverse cross section. Figure 2b suggests that the wheel side of the ribbons has a smooth surface and edge, and at the top side of the ribbons, the surface is wavy as a consequence of the streaming of argon gas above the melt in the travelling direction of the solidifying ribbons. The characteristic texture of the microstructure exhibits the direction of solidification. The wheel side of the ribbon, where solidification is initiated, is at the bottom. Close to this surface, a thin layer of fine equiaxed grains (outer equiaxed zone) is formed. These grains transit into columnar grains and, approximately at the centre of ribbon, into coarse equiaxed grains. The transition from the outer equiaxed zone to the columnar zone can be understood in terms of anisotropic growth effects [12, 13].

With rapid solidification, grain size was reduced to a few microns and less [13]. Because of unstable conditions during solidification and cooling, besides the reduction of grain size, an increased concentration of thermodynamically stable and unstable microstructure defects is present [14]. We can expect much higher concentrations of vacancies and substitution elements, in this case lanthanum; so the solid solution of gold could be oversaturated and, consequently, there should be smaller fraction of intermetallic phases [15]. With these unstable conditions, microsegregation on primary dendrites can be caused. In our case, coarser intermetallic particles appeared in the interdendritic space. Its degree is increased in the transition from columnar and equiaxed grains [16].

Rapid solidification in such a system can also have an effect on the fraction of eutectic and its phase composition [17]. If, in such condition, gold solutes over 0.5 wt% of La at eutectic temperature, eutectic reaction could be bypassed. Consequently, we could get a solid solution decorated only by coarser particles. Better interpretation of the formatted microstructure could be obtained by TEM electron images. Figure 2b presents the microstructure of the rapidly solidified ribbons (coarse equiaxed zone), obtained with TEM. Coarser particles can be observed inside the grains and on grain boundaries. They are the result of microsegregation during rapid solidification. Inside the grains, small evenly dispersed particles also formed and the participations present nucleated from the oversaturated solid solution during cooling.

Finally, cast and melt spun samples were analysed individually to observe any possible deviations of chemical composition during heat treatment (re-melting). The data obtained from XRF analysis is presented in Table 1. One can see that there is a small loss of La after rapid solidification by melt spinning from the initial alloy, possibly because of La lost during the manufacturing process (manipulation, more working phases, etc.) and because La oxidises more easily compared to Au.

Microhardness of Au–La alloys

Next, we investigated the mechanical properties of the obtained alloys by measuring microhardness. In Fig. 3, the microhardness of pure Au, slowly solidified Au–La and rapidly solidified Au–La is presented. From these results, the high strengthening effect achieved by La and additional strengthening by rapid solidification can be determined as is predicted, besides a small loss of La during rapid solidification by melt spinning [17]. Pure Au has almost three times lower hardness in comparison to Au–0.5 wt% La, which reached up to 79 HV. This is due to participation of a hard secondary phase containing the intermetallic phase Au_6La. Rapid solidification of this alloy increased hardness further to a maximum value of approximately 99 HV.

More detailed research was made on rapidly solidified samples. Microhardness was measured in the columnar and coarse equiaxed grain zones. In the columnar zone, much higher microhardness was achieved than in the coarse zone. These results cannot give us exact values, only relative predictions, because of the very small thickness of the ribbons. Namely, if the smaller indentations were chosen, measurements would not give us the hardness of material but hardness of the phases. The microstructural gradient in the cross section of ribbon, which is due to the gradient of solidification rate, leads to the large differences in material hardness. In this case, the difference between hardness close to the free surface and hardness close to the wheel surface is almost 25 %.

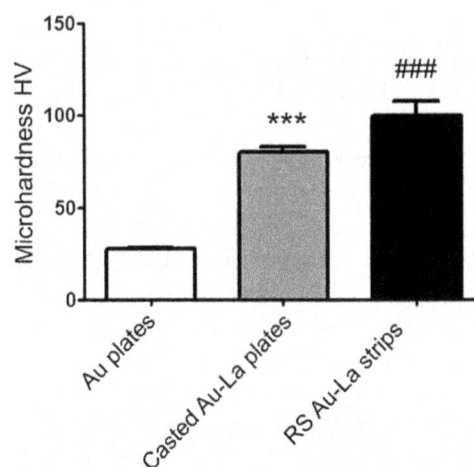

Fig. 3 Microhardness of Au–La alloys. Microhardness of Au–La alloys, prepared either by conventional solidification or rapid solidification, and microhardness of pure gold is shown. The results are presented as mean±SD of five measurements on different spots from a representative experiment, out of two with similar results. ***$p < 0.005$ compared to pure gold; ###$p < 0.005$ compared to casted Au platelets (conventionally solidified)

Table 1 Chemical composition of Au–La alloys obtained by XRF analysis	Wt% Au	Wt% La
Castings, Au–La	99.28	0.72
Ribbons, Au–La	99.43	0.57

Cytotoxicity of RS Au–La strips in animal cell cultures

After showing that microalloying of Au with La could introduce better mechanical properties to the alloy, and that these could further be improved by rapid solidification, we wondered whether the RS Au–La alloys induce cytotoxicity in vitro. First, we tested the effect of RS Au–La strips on cytotoxicity of rat thymocytes, rat PMØ and L929 cells animal cells in vitro. These three different cell types were chosen for this part of study for the following reasons: L929 cells are the most commonly used targets for biocompatibility testing due to their sensitivity to the toxic effect of different materials and their soluble products and therefore are recommended by ISO standards as a screening cell target in biocompatibility studies [18]. Our previous original studies showed that rat thymocytes are more sensitive target for testing the biocompatibility of dental and implant materials than recommended L929 cells [4]. PMØ, a part of mononuclear phagocyte system, are recommended for biocompatibility studies of implant materials because the cells are the first ones responding to foreign materials by producing pro-inflammatory mediators [19]. By using viability and apoptosis assays, it has been shown that Au–0.5 wt% La strips are not cytotoxic for the tested animal cells (Fig. 4a–c).

Lower viability of thymocytes and PMØ in both control wells and wells containing Au–La strips, compared to L929 cells, is a consequence of spontaneous cell death by apoptosis [20, 19], and such a finding was confirmed using apoptosis tests. Both morphological and flow cytometry methods showed that Au–La strips did not cause apoptosis of the examined cells. Higher viability of L929 cells followed by very low rate of apoptosis could be explained by the fact that the cells are transformed (immortalised), showing high proliferative activity and low death process in vitro. Such results are encouraging for further studies because materials which cause cytotoxic effects in short-term culture are not advisable for insertion into human body as implants [21].

Effects of RS Au–La strips on NO production by peritoneal macrophages

Our previous study has suggested that gold-based material, although not cytotoxic, could induce functional response of immune cells [22, 23]. To study the functional response of PMØ to Au–La strips, we have measured the production of NO, with or without additional stimulation by lipopolysaccharide (LPS), a known stimulator of TLR4-mediated activation of macrophages [24]. NO is an important molecule involved in different physiological and pathophysiological processes in the organism [25]. PMØ are an important source of NO, and the level of its production could serve as a sensitive parameter of PMØ activation [26]. The results presented in our study suggest that RS Au–La strips trigger the

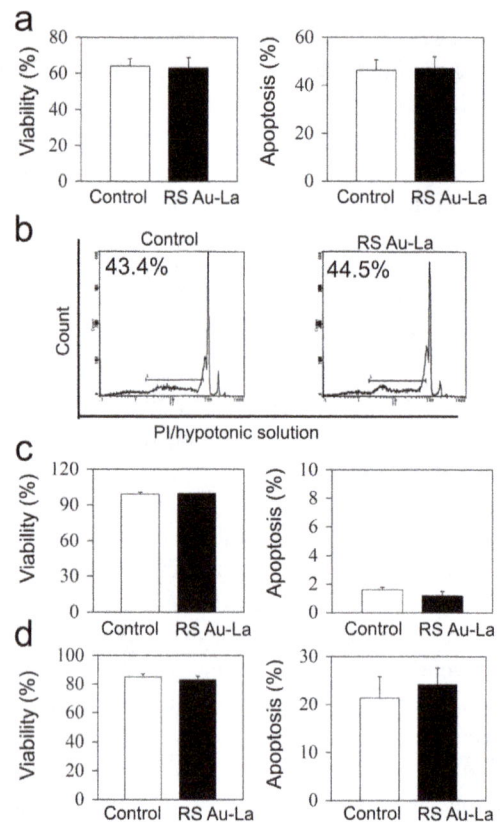

Fig. 4 Effect of RS Au–La strips on viability and apoptosis of animal cells in vitro. The effect of Au–La strips on viability and apoptosis of **a** rat thymocytes after 24-h cultures was analysed by MTT assay (viability) or staining of the cells with PI dissolved in hypotonic solution and Turk (apoptosis). The results are presented as mean±SD of three independent experiments, one of which is shown in (**b**). The effect of RS Au–La strips on viability and apoptosis of (**c**) L929 cell (**d**) rat PMØ after 24-h cultures is presented as mean±SD ($n=3$). Controls were cultures without Au–La strips

production of NO by non-stimulated PMØ, whereas the material does not significantly modify the production of NO by LPS-stimulated PMØ (Fig. 5).

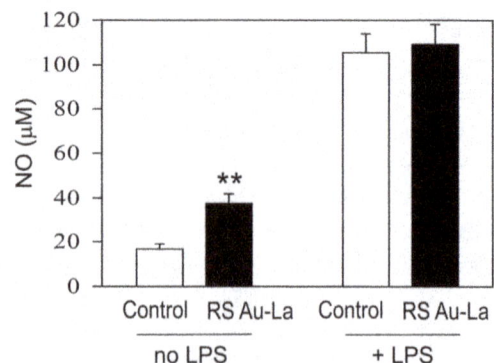

Fig. 5 Effect of Au–La strips on NO production by rat PMØ. PMØ (1×10^6/ml) were stimulated with LPS (100 ng/ml) during 24-h cultivation in the presence or absence of RS Au–La strips, after which the supernatant was collected and subjected to Griess reaction. $^{**}p<0.01$ compared to corresponding medium control

These results suggest that Au–La strips, although non-cytotoxic for PMØ, are able to activate these cells. Lack of increased NO production by PMØ treated with both LPS and RS Au–La strips could be a consequence of its over-stimulated production by LPS alone [27]. Alternatively, lanthanum may specifically interact with LPS signalling, disallowing further increase in NOS activity. In line with this, it was shown that $LaCl_3$ inhibits LPS-induced production of NO by RAW264.7 macrophages [28], by inhibiting the induction of NF-κB expression upon stimulation [29, 30]. Additionally, it was shown that small concentrations of lanthanum can inhibit the production of active oxygen free radicals by PMØ at low concentration, but it turned out contrary at high concentration [31]. Lou et al. showed that $LaCl_3$ (2.5 μM) may somewhat increase the production of NO by macrophages in the absence of LPS [28], but the effect was not significant. The much stronger effect we observed in this study could be due to a specific action of Au–La surface, in addition to the effects of ionic or surface concentrated lanthanum. To evaluate this hypothesis, an additional control was included in the following study, namely, the pure gold platelets.

Cytotoxicity of RS Au–La strips on human PBMNCs

The response of animal and human cells can differ significantly [32], including the response to a biomaterial [33]. Therefore, the second part of biocompatibility study was designed to investigate the cytotoxic and immunomodulatory effect of RS Au–La strips on human PBMNCs isolated from eight healthy volunteers. As an additional control, platelets prepared from pure gold were used.

MTT assay and Annexin–FITC/PI assay showed that neither gold–lanthanum strips nor gold platelets induce cytotoxic effects on human PBMNCs (subpanels a and b of Fig. 6, respectively). Based on the previous results on animal cells and human cells, it can be concluded that Au–La and pure Au do not induce acute cytotoxic effect. Since there are no cytotoxic effects in short-term culture, such materials are advisable for insertion into human body as implants. However, the response of the cells after prolonged exposure to RS Au–La strips remains to be evaluated before their safe biomedical application.

Immunomodulatory properties of RS Au–La strips in culture of stimulated PBMNCs

Even though the RS Au–La strips were not cytotoxic for rat PMØ, they caused the activation of these cells in culture without additional stimuli. Therefore, we aimed to investigate whether human PBMNCs could be functionally different when cultivated with RS Au–La strips.

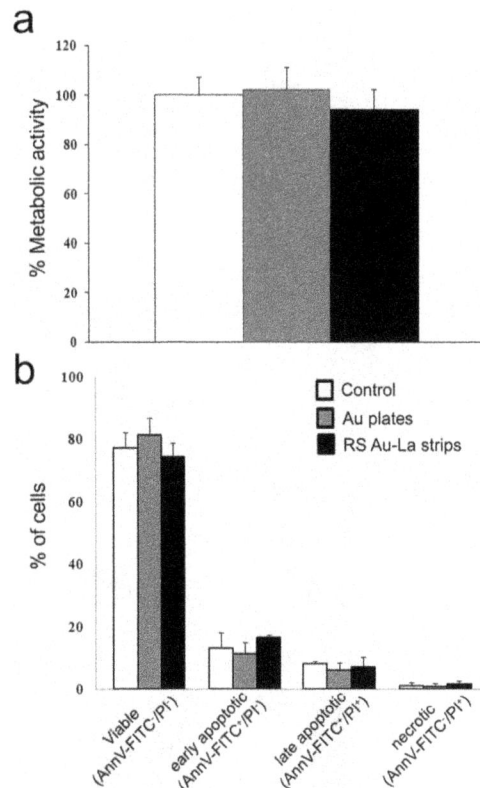

Fig. 6 Effect of RS Au–La strips on viability and apoptosis of human PBMNCs in vitro. PBMNCs were cultivated for 24 h in the presence or absence of pure Au platelets, RS Au–La strips or in medium (control). After the culture, **a** metabolic activity of the cells was determined by MTT assay or **b** Annexin V–FITC/PI staining. The results are shown as mean± SD of three independent experiments

The effect of Au–La strips and Au platelets on proliferation of PHA-stimulated PBMNC is presented in Fig. 7a. The results suggest that Au–La strips, in contrast to control Au platelets, stimulate proliferation of PHA-stimulated PBMNCs. This is a very interesting and unexpected finding which could be explained by the specific effect of La released from the surface of the strips or more probably by specific effect of Au–La surface compared to Au platelet surface. In line with the former hypothesis, it was shown that $LaCl_3$ may promote proliferation and cytoskeleton reorganisation of rat osteoblasts via activation of focal adhesion kinase (FAK) [34] and increase proliferation of 3T3-L1 cells [35]. Interestingly, the same element can inhibit growth of cancer cells, by increasing the expression of p53, p16 and p21 [36]. It is unlikely that the possible release of contaminating metal ions from Au platelets or Au–La strips is responsible for any of the observed biological effects. By using an inductively coupled plasma-atomic emission spectrometry (ICP-AES) analysis, neither of the trace elements was detected in conditioning medium prepared from these samples (data not shown).

Previous results related to the proliferation of PBMNC are confirmed by analysing the cytokine profile in supernatants of

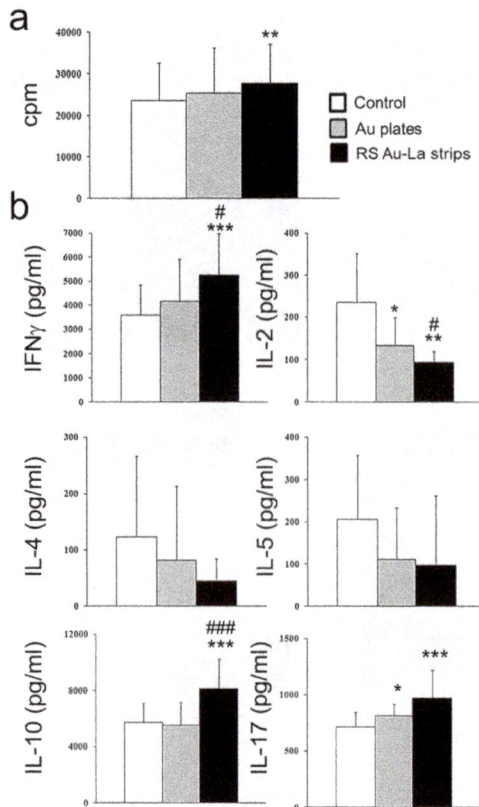

Fig. 7 Effect of RS Au–La strips on the proliferation and cytokine production of human PBMNCs. The effect of RS Au–La strips on **a** proliferation of PHA-stimulated PBMNCs was determined after 3-day cultures by ^{3}[H] labelling of the cell cultures for the last 8 h and measuring the radioactivity on β-scintillation counter. The results are shown as mean count per minute (cpm)±SD of three independent experiments, each carried out in 6-plicates. **b** The levels of IFN-γ, IL-2, IL-4, IL-5 and IL-10 were determined from 3-day culture supernatants of PHA-stimulated PBMNCs using FlowCytomix Human Th1/Th2 11plex Kit or ELISA (IL-17). $^*p<0.05$, $^{**}p<0.01$, $^{***}p<0.005$ compared to control; $^\#p<0.05$, $^{\#\#\#}p<0.005$ compared to Au platelets

these cultures. Namely, we found significantly higher levels of INF-γ in cultures with RS Au–La strips and lower levels of IL-2 (Fig. 7b).

IL-2 is a major proliferation-inducing factor of lymphocytes which they produce, and utilise via IL-2R, in an autocrine and paracrine manner [37]. Lower levels of IL-2 in culture supernatants could be a result of its increased utilisation by the proliferating cells. The levels of IL-2 were even lower in PBMNC cultures with control Au platelets, and this finding is also in accordance with slightly higher proliferative activity of these cultures compared to control PHA-stimulated cells. IFN-γ is the major cytokine produced by Th1 cells, suggesting that RS Au–La strips promote Th1 differentiation of lymphocytes.

The levels of Th2 cytokines (IL-4 and IL-5) (Fig. 7b) in cultures with Au–La strips or Au platelets were lower compared to corresponding controls, but due to significant individual variations between samples, the differences were not

statistically significant. Lower levels of Th2 cytokines could be explained by the increased production of Th1 (IFN-γ) cytokines. The IFN-γ enriched environment most probably caused the lower production of Th2 cytokines (IL-4 and IL-5) in the cultures, since it was shown that IFN-γ through the transcription factor T-bet inhibits GATA-3-dependent differentiation and proliferation of Th2 cells [38]. It is interesting that Au–La strips significantly increased the production of anti-inflammatory cytokine, IL-10, and a Th17 cytokine, IL-17 (Fig. 7b). The production of IL-17 is also triggered by pure Au. To our knowledge, such results have not been published yet, and without detailed timing of cytokine production and precise identification of their cellular sources, it is difficult to explain the physiological significance of this finding. IL-10, a key immunoregulatory cytokine, is produced by both antigen presenting cells (APCs) and T cells [39]. IL-17 is a key cytokine produced by a subset of memory T cells (Th17 cells) upon stimulation with IL-6 or IL-1β and TGF-β [40]. At the moment, it is not clear why Au–La strips trigger production of both Th1 (IFN-γ) and Th17 (IL-17) cytokines with pro-inflammatory properties, and IL-10, a cytokine with down-modulatory activity. It is possible that increased production of IL-10 is an important physiological feedback mechanism to control excessive T cell activation [41, 42]. In the context where lanthanum alone may exhibit anti-tumour effects [37], increased Th1/Th17 polarisation capacity of RS Au–La strips could be beneficial additionally for anti-tumour therapies, since such increased levels of Th1 and Th17 cytokines correlate with a good prognosis for tumour-bearing patients [43]. IFN-γ production is stimulated predominantly by IL-12 produced by APCs, and it induces a positive feedback loop for its own production [44]. Indeed, when measuring the levels of IL-12 in PHA-stimulated cultures, we found that RS Au–La strips statistically significantly increased the production of IL-12 (Fig. 8) by PBMNCs. IL-12 is a key cytokine stimulating Th1 polarisation activity of T cells [44], and our finding that the level of IL-12 correlated with increased production of IFN-γ supports this phenomenon. The levels of IL-12 in PBMNC cultures were relatively low because accessory cells, comprising up to 15 % of total PBMNCs, are its main producers.

By further analysing cytokine production by PHA-stimulated PBMNCs, we found that neither RS Au–La strips nor pure Au platelets modulate significantly the production of pro-inflammatory cytokines IL-6, IL-1β, TNF-α and TNF-β (Fig. 8). Unexpectedly, we observed that Au–La strips inhibit the production of IL-8. IL-8 plays different roles both in innate and specific immunity [45]. It is a main chemoattractant for recruitment of neutrophils to the inflammatory sites [46]. Therefore, lower production of this chemokine in the presence of Au–La, as a possible biomaterial, could be beneficial for down-regulation of non-specific inflammatory responses [45]. It is interesting that IL-17 cytokine production was increased

Fig. 8 Effect of RS Au–La strips on cytokine production of human PBMNCs. The effect of RS Au–La strips on the production of IL-12, IL-6, IL-8, IL-1β, TNF-α and TNF-β were determined from 3-day culture supernatants of PHA-stimulated PBMNCs using FlowCytomix Human Th1/Th2 11plex Kit or ELISA (IL-17). *p<0.05 compared to control; #p<0.05 compared to Au platelets

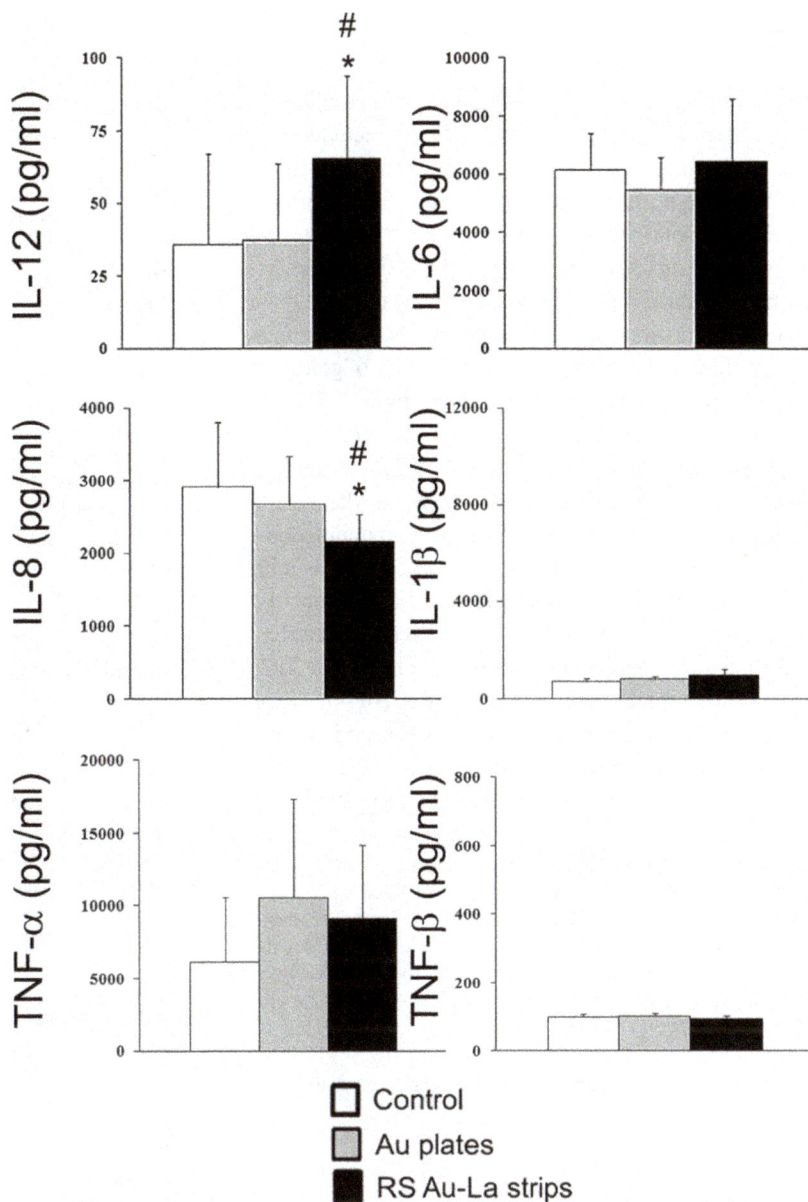

without the increase of IL-6 or IL-1β, which are important for its induction [39]. These findings suggest that other cytokines responsible for Th17 development, such as IL-23 [46], could be increased, which needs to be determined additionally.

Considering that the proliferation and cytokine production in the model of PHA-stimulated PBMNCs depends on presence of APCs within cell population, predominantly monocytes, it can be hypothesised that the effects of RS Au–La strips were mediated by functional modulation of APCs within PBMNC population. These results suggest that RS Au–La strips, although non-cytotoxic for animal cells (thymocytes, L929 cells and PMØ) and human PBMNCs, can activate them. We found an increased production of Th1 and Th17 cytokines along with increased production of IL-10, as a very

important down-regulatory cytokine in PBMNC cultures stimulated with PHA [41, 42]. It seems that Au–La strips modulate predominantly adaptive immunity, mediated by T cells, without significant influence on the components of innate immunity, as judged by measuring the levels of pro-inflammatory cytokines produced by the cells of monocyte–macrophage system. However, it remains to be tested whether and how activation of APCs in our experimental system influences this arm of immunity. The dominant effect was observed on stimulation of Th1 and Th17 and down-regulation of Th2 responses. Since all these responses are mediated by APCs, it is important to further study not only the effect of Au–La strips on Th polarisation capability of APCs but also on their immunophenotypic characteristics.

Conclusion

The microalloying of Au by La initiates the formation of Au_6La at a small alloying concentration. Such microalloying increases the strength and hardness significantly, caused by distribution of the primary phase between dendrites. Furthermore, with rapid solidification, a large reduction of grains and microsegregation increases the strength of the alloy additionally. Such mechanical properties could be outstanding for the preparation of biomedical constructs and prosthesis of various thicknesses. Since the alloy still contain 99.5 wt% of gold, it is expected to be tolerated well by the organism. Indeed, RS Au–La strips do not cause acute cytotoxicity in animal (L929 fibroblasts and rat PMØ) and human PBMNCs but can activate them. Therefore, care should be taken as to whether the ribbons will induce a desired or an adverse immunomodulatory effect in potential recipients. Namely, in conditions with the pathological Th17- or Th1-mediated inflammation, the implantation of RS Au–La strips could be adverse since they can exacerbate such conditions. On the other hand, in the Th2-dominant immunopathology, implanted RS Au–La strips could inhibit the Th2 inflammatory process and induce tolerogenicity. To prove this hypothesis, extensive in vivo studies on different animal models should be conducted.

Acknowledgments This study was supported by the EUREKA Project E!3971 BIO-SMA and the Project of the Ministry of Science and Technological Development of Serbia (Project no. 175102). The authors thank the firm Zlatarna Celje d.d. and collaborators from the Military Medical Academy, Belgrade (T. Džopalić, A. Dragićević, J. Đokić, I. Rajković, S. Vasilijić) and Faculty of Mechanical Engineering, Maribor (I. Orožim) for their helpful assistance during the experiments.

Conflict of interest The other authors also declare no conflict of interest.

References

1. Colic M, Stamenkovic D, Anzel I, Lojen G, Rudolf R (2009) The influence of the microstructure of high noble gold-platinum dental alloys on their corrosion and biocompatibility in vitro. Gold Bull 42:34–47
2. Geurtsen W (2002) Biocompatibility of dental casting alloys. Crit Rev Oral Biol Med 13:71–84
3. Tomić S, Rudolf R, Brunčko M, Anžel I, Savić V, Čolić M (2012) Response of monocyte-derived dendritic cells to rapidly solidified nickel-titanium ribbons with shape memory properties. Eur Cells Mater 23:58–81
4. Čolić M, Rudolf R, Stamenković D, Anžel I, Vučević D, Jenko M, Lazić V, Lojen G (2010) Relationship between microstructure, cytotoxicity and corrosion properties of a Cu–Al–Ni shape memory alloy. Acta Biomater 6:308–317
5. Corti CW (1999) Metallurgy of microalloyed 24 carat golds. Gold Bull 32(2):39–47
6. Corti CW (2001) Strong 24 carat golds: the metallurgy of microalloying. Gold Technol 33:27–36
7. Corti CW (2005) Microalloying of high carat gold, platinum and silver. Proceeding of the 2nd International Conference on Jewellerly Production Technology (JTF) Vicenza
8. Corti CW, Holliday RJ, Thompson DT (2002) Developing new industrial applications for gold: gold nanotechnology. Gold Bull 35:111–117
9. Shijie Z, Bingjun Z, Zhen Z, Xin J (2006) Application of lanthanum in high strength and high conductivity copper alloys. J Rare Earths 24:385–388
10. Palmer RJ, Butenhoff JL, Stevens JB (1987) Cytotoxicity of the rare earth metals cerium, lanthanum, and neodymium comparisons with cadmium in a pulmonary macrophage primary culture system. Environ Res 43:142–156
11. Okamoto H (2010) Phase diagrams for binary alloys. ASM International
12. Kurz W, Fisher D (1981) Dendrite growth at the limit of stability: tip radius and spacing. Acta Metall 29:11–20
13. Fisher D, Kurz W (1980) A theory of branching limited growth of irregular eutectics. Acta Metall 28:777–794
14. Honeycombe RWK (1975) The plastic deformation of metals. Edward Arnold, London W 1 Z 8 LL 1975, 477 p(Book)
15. Spaić S (1996) Fizikalna metalurgija I. Naravoslovnotehniška fakulteta, Oddelek za materiale in metalurgijo
16. Zupancic-Hartner T, Rudolf R, Kneissl AC, Anzel I (2009) Characterisation of the metastable microstructure of Au-La alloy. Prakt Met Sonderband 41:277
17. Ning Y (2005) Properties and applications of some gold alloys modified by rare earth additions. Gold Bull 38:3–8
18. International Standards Organisation (1997) Dentistry—preclinical evaluation of biocompatibility of medical devices used in dentistry—test methods for dental materials. ISO 7404, 1st edn. ISO, Geneva
19. Čolić M, Tomić S, Rudolf R, Anžel I, Lojen G (2010) The response of macrophages to a Cu-Al-Ni shape memory alloy. J Biomater Appl 25:269–286
20. Rinner I, Felsner P, Hofer D, Globerson A, Schauenstein K (1996) Characterization of the spontaneous apoptosis of rat thymocytes in vitro. Int Arch Allergy Immunol 111:230–237
21. Williams DF (2008) On the mechanisms of biocompatibility. Biomaterials 29:2941–2953
22. Đokić J, Rudolf R, Tomić S, Stopić S, Friedrich B, Budič B, Anžel I, Čolić M (2012) Immunomodulatory properties of nanoparticles obtained by ultrasonic spray pyrolysis from gold scrap. J Biomed Nanotechnol 8:528–538
23. Rudolf R, Friedrich B, Stopić S, Anžel I, Tomić S, Čolić M (2010) Cytotoxicity of gold nanoparticles prepared by ultrasonic spray pyrolysis. J Biomater Appl
24. Kawai T, Akira S (2010) The role of pattern-recognition receptors in innate immunity: update on Toll-like receptors. Nat Immunol 11:373–384
25. Coleman JW (2001) Nitric oxide in immunity and inflammation. Int Immunopharmacol 1:1397–1406
26. Lorsbach RB, Murphy WJ, Lowenstein CJ, Snyder SH, Russell S (1993) Expression of the nitric oxide synthase gene in mouse macrophages activated for tumor cell killing. Molecular basis for the synergy between interferon-gamma and lipopolysaccharide. J Biol Chem 268:1908–1913
27. Bredt D, Snyder S (1994) Nitric oxide: a physiologic messenger molecule. Annu Rev Biochem 63:175–195
28. Lou Y, Guo F, Wang Y, Xie A, Liu Y, Li G (2007) Inhibitory effect of lanthanum chloride on the expression of inducible nitric oxide synthase in RAW264. 7 macrophages induced by lipopolysaccharide.

Zhonghua shao shang za zhi = Zhonghua shaoshang zazhi=Chin J Burns 23:280–283

29. Guo F, He F, Xiu M, Lou Y, Xie A, Liu F, Li G (2013) Regulatory effects of lanthanum chloride on the activation of nuclear factor kappa B inhibitor kinase beta induced by tumor necrosis factor alpha. Zhonghua shao shang za zhi = Zhonghua shaoshang zazhi= Chin J Burns 29:531–536

30. Guo F, Guo X, Xie A, Lou YL, Wang Y (2011) The suppressive effects of lanthanum on the production of inflammatory mediators in mice challenged by LPS. Biol Trace Elem Res 142:693–703

31. Xue LH, Tao LX, Fen LJ, Chang LR, Kui W (2000) The effects of lanthanum, cerium, yttrium and terbium ions on respiratory burst of peritoneal macrophage (Mφ)[J]. J Beijing Med Univ 3:003

32. Mestas J, Hughes CC (2004) Of mice and not men: differences between mouse and human immunology. J Immunol 172:2731–2738

33. Pearce A, Richards R, Milz S, Schneider E, Pearce S (2007) Animal models for implant biomaterial research in bone: a review. Eur Cell Mater 13:1–10

34. Wang X, Huang J, Zhang T, Wang K (2009) Cytoskeleton reorganization and FAK phosphorylation are involved in lanthanum (III)-promoted proliferation and differentiation in rat osteoblasts. Prog Nat Sci 19:331–335

35. He M, Yang W, Hidari H, Rambeck W (2006) Effect of rare earth elements on proliferation and fatty acids accumulation of 3T3-L1 cells. Asian Australas J Anim Sci 19:119

36. Xiao B, Ji Y, Cui M (1997) Effects of lanthanum and cerium on malignant proliferation and expression of tumor-related gene. Zhonghua yu fang yi xue za zhi Chin J Prev Med 31:228–230

37. Malek TR (2008) The biology of interleukin-2. Annu Rev Immunol 26:453–479

38. Kaiko GE, Horvat JC, Beagley KW, Hansbro PM (2008) Immunological decision-making: how does the immune system decide to mount a helper T-cell response? Immunology 123:326–338

39. Steinman RM, Hawiger D, Nussenzweig MC (2003) Tolerogenic dendritic cells*. Annu Rev Immunol 21:685–711

40. McGeachy MJ, Cua DJ (2008) Th17 cell differentiation: the long and winding road. Immunity 28:445–453

41. Gabryšová L, Nicolson KS, Streeter HB, Verhagen J, Sabatos-Peyton CA, Morgan DJ, Wraith DC (2009) Negative feedback control of the autoimmune response through antigen-induced differentiation of IL-10–secreting Th1 cells. J Exp Med 206:1755–1767

42. Meyaard L, Hovenkamp E, Otto SA, Miedema F (1996) IL-12-induced IL-10 production by human T cells as a negative feedback for IL-12-induced immune responses. J Immunol 156:2776–2782

43. Qi W, Huang X, Wang J (2013) Correlation between Th17 cells and tumor microenvironment. Cell Immunol 285:18–22

44. Jouanguy E, Döffinger R, Dupuis S, Pallier A, Altare F, Casanova J-L (1999) IL-12 and IFN-γ in host defense against mycobacteria and salmonella in mice and men. Curr Opin Immunol 11: 346–351

45. Harada A, Mukaida N, Matsushima K (1996) Interleukin 8 as a novel target for intervention therapy in acute inflammatory diseases. Mol Med Today 2:482–489

46. Lukić A, Vojvodic D, Majstorović I, Čolić M (2006) Production of interleukin-8 in vitro by mononuclear cells isolated from human periapical lesions. Oral Microbiol Immunol 21:296–300

22

Gold nanoparticle synthesis using the thermophilic bacterium *Thermus scotoductus* SA-01 and the purification and characterization of its unusual gold reducing protein

Mariana Erasmus · Errol Duncan Cason ·
Jacqueline van Marwijk · Elsabé Botes ·
Mariekie Gericke · Esta van Heerden

Abstract Nanoparticles are very important materials for implementing nanotechnology in diverse areas and are abundant in nature as living organisms operate at a nanoscale. As nanoparticles exhibit interesting size- and shape-dependent physical and chemical properties, the synthesis of uniform nanoparticles with controlled sizes and shapes is of great importance. Nanoparticles are the end products of a wide variety of physical, chemical and biological processes, some of which are novel and radically different and others of which are quite commonplace. The ability to produce nanoparticles with specific shapes and controlled sizes could result in interesting new applications that can potentially be utilized in areas such as optics, electronics and the biomedical field. In the present study, we have demonstrated the ability of the thermophilic bacterium *Thermus scotoductus* SA-01 to synthesize gold nanoparticles and determined the effect of the physico-chemical parameters on particle synthesis. Furthermore, a protein purified from this bacterium is shown to be capable of reducing $HAuCl_4$ to form elemental nanoparticles in vitro. The protein was purified to homogeneity and identified through N-terminal sequencing as an ABC transporter, peptide-binding protein. It is speculated that this protein reduces Au(III) through an electron shuttle mechanism involving a cysteine disulphide bridge. Through manipulation of physico-chemical parameters, it was possible to vary nanoparticles in terms of number, shape and size. This is the first report of a transporter protein from a thermophile with the ability to produce nanoparticles in vitro thus expanding the limited knowledge around biological gold nanoparticle synthesis.

Keywords Gold nanoparticles · Thermophilic · Gold reduction · Biological synthesis · ABC transporter

M. Erasmus · E. D. Cason · J. van Marwijk · E. Botes ·
E. van Heerden (✉)
Department of Microbial, Biochemical and Food Biotechnology,
Faculty of Natural and Agricultural Sciences, University of the Free
State, Bloemfontein 9300, South Africa
e-mail: vheerde@ufs.ac.za

M. Erasmus
e-mail: erasm@ufs.ac.za

E. D. Cason
e-mail: casoned@ufs.ac.za

J. van Marwijk
e-mail: jacquelinevanmarwijk@yahoo.com

E. Botes
e-mail: elsabe75@gmail.com

M. Gericke
Biotechnology Division, Mintek, Private Bag X3015,
Randburg 2125, South Africa
e-mail: mariekieg@mintek.co.za

Introduction

Colloidal gold nanoparticles have attracted the interest of scientists for over 400 years, and the synthesis thereof includes reduction of chloroauric acid by reductants; use of these reductants produces particles of relatively uniform size, which are generally spherical [1–3]. In order to control synthesis of these particles, the methodology by which they are produced should be understood. The means by which microorganisms achieve changes in metal speciation and mobility are essential gears of the biogeochemical cycles of metals, as well as for other elements [4]. Microorganisms can interact with metals in a variety of ways [5], [6–8] and biomineralization occurs through one of two major pathways, namely biologically induced biomineralization (BIM) or boundary-organized biomineralization (BOB) [9]. When the BIM pathway is followed, the organism has little control over deposition of mineral particles and these particles are often characterized by poor crystallinity, broad particle size distribution and non-specific crystal morphology [10–12]. Alternatively, when the BOB pathway is followed, the organism could exercise control over the nucleation and growth of particles. These nanoparticles are developed under the direct regulatory control of the organism and synthesized at a specific location within the cell and only under certain conditions [11–13].

The ability to produce gold nanoparticles with predictable shapes and sizes under controlled conditions could result in interesting new applications that can potentially be utilized in areas such as optics, electronics and biomedicine [14–17]. The benefits of using a biological synthesis system range from the predictable production of uniform metal nanoparticles due to the highly structured activities of microbial cells to environmentally friendly production methods and the ability to produce nanoparticles with unique shapes and composition [18–20]. It is therefore important to gain an understanding of the mechanism of gold accumulation and reduction by bacterial cells on a molecular and cellular level. Identification and analysis of the biopolymers involved in these processes could potentially allow for a process in a cell-free environment, where the size and shape of particles can be regulated to some extent by specific proteins/enzymes [21]. This would also be advantageous from a process point of view, since it would eliminate the need to harvest the nanoparticles formed within the cells, simplifying the process and lowering operation costs.

Different hypotheses have been formulated about the mechanisms involved in gold reduction and nanoparticle synthesis, and these are all based on enzymes with metal reductase (oxidoreduction) abilities [22, 23]. However, the focus on oxidoreductases has overlooked the concept of 'metabolic promiscuity', where a single protein can have a wide range of functions [24, 25] resulting in a lack of knowledge concerning the mechanisms and capabilities of other non-reductase proteins that are still capable of gold reduction and nanoparticle formation [24, 25].

In this paper, the interaction between *Thermus scotoductus* SA-01 and gold chloride ions was evaluated on both a biomass level, where biomass refers to the whole cells of *T. scotoductus* SA-01, and by using a purified protein that was identified to be capable of regulating gold nanoparticle synthesis. The effect of the physico-chemical parameters on particle morphology was assessed, and it was determined whether these parameters can be manipulated to produce size- and shape-specific nanoparticles.

Experimental details

Microorganisms and growth

T. scotoductus SA-01 was always grown from a glycerol stock, plated out onto TYG-agar plates (5 g L^{-1} tryptone, 3 g L^{-1} yeast extract, 1 g L^{-1} glucose and 18 g L^{-1} bacteriological agar) and grown for 24 h. A 20–24-h-old culture was used to prepare the pre-inoculum (a loop full of culture in 50 ml TYG in a 250-ml Erlenmeyer flask) and grown until mid-exponential growth phase (~8 h). The inoculum was also grown to mid-exponential growth phase (~8 h) and was prepared by adding 10 ml of the pre-inoculum to 90 ml TYG (in a 500-ml Erlenmeyer flask). Five millilitres of the inoculum was added to 95 ml TYG (in a 500-ml Erlenmeyer flask) for the growth experiments. One-millilitre samples were withdrawn over time (every 2 h until 16 h, and then at 20 and 24 h), and the optical density (OD) was measured at 600 nm.

The correlation between growth phase and gold reduction was evaluated by collecting biomass at different time intervals as described before. The biomass was washed three times and resuspended in a 1:20 *w/v* ratio in 50 mM acetate buffer, pH 5.0 [21]. Au(III), to a final concentration of 2 mM, was added and the concentration determined by using a spectrophotometric method adapted from Melwanki and co-workers [26].

Biomass was obtained as described previously to be used as whole cell catalysts and exposed to varying physico-chemical parameters. The effect of pH was evaluated using a pH ranged from 3.6, 5.5 (50 mM sodium acetate buffer) and 7.4 (50 mM sodium phosphate buffer) to 9.0 (50 mM borax buffer), while gold ion concentration was evaluated for the range starting from 0 to 0.01 M. The temperature range stretched from 30 to 65 °C and the exposure time from 1 to 48 h.

Šlouf and co-workers [27] indicated that gold nanoparticles exhibit pink/purple red colours, which arises due to excitation of surface plasmon vibrations in the gold nanoparticles. UV–Vis spectroscopy can be used to record the surface plasmon band of the gold nanoparticles resulting in visual confirmation (a shift from a yellow to a pink/purple colour) indicative of gold nanoparticle production [28]. Samples were withdrawn over time and a UV–Vis spectrum (400–700 nm) was obtained using a Beckman DU-800 spectrophotometer, to evaluate the plasmon band associated with gold nanoparticles. These samples were also analysed by transmission electron microscopy (TEM), as described below.

Cell-free extraction and protein purification

T. scotoductus SA-01 was grown to the late exponential growth phase [29], biomass harvested by centrifugation (6,000×*g*; 15 min; 4 °C) and cell pellets washed three times with 50 mM sodium phosphate buffer (pH 7.4). Cell-free extracts were prepared according to adapted methods of Gaspard and co-workers (this implied that the cells were broken to release the proteins from within the cell) [30]. The soluble fraction was size fractionated and concentrated using an Amicon stirred cell fitted with a UF 30 MWCO membrane (Osmonics Inc.). The retentate was dialysed (SnakeSkin 7000 MWCO; Thermo Scientific) against 50 mM sodium phosphate buffer (pH 7.4) to remove reducing agents such as organic acids and carbohydrates. The filtrate was applied to a Super Q-Toyopearl column (TOSOH) in 50 mM sodium phosphate buffer (pH 7.4). The non-binding protein fraction, which was able to form nanoparticles in the presence of $HAuCl_4$, was pooled and dialyzed against 50 mM sodium acetate buffer (pH 5.0). The dialysate was applied to a SP-Toyopearl column (TOSOH) and eluted with a 0–1 M NaCl gradient. Active fractions were pooled, dialysed against 50 mM sodium phosphate buffer (pH 7.4) and reapplied to the SP-Toyopearl column (TOSOH). Bound proteins were eluted with a linear 0–0.3 M NaCl gradient, active fractions pooled and dialysed against 50 mM sodium phosphate buffer (pH 7.4) and applied to a Blue-Sepharose CL-6B column (Sigma-Aldrich). Proteins were eluted with a linear 0–0.6 M NaCl gradient, active fractions pooled and dialysed against 50 mM sodium phosphate buffer (pH 7.4). Ammonium sulphate was added to 1 M and applied to a Phenyl-Toyopearl column (TOSOH) pre-equilibrated with 50 mM sodium phosphate buffer (pH 7.4) containing 1 M $(NH_4)_2SO_4$. Bound proteins were eluted with a linear 1–0 M $(NH_4)_2SO_4$ gradient, active fractions pooled and dialysed against 50 mM sodium phosphate buffer (pH 7.4). The sample was concentrated to ~3 ml and applied to a Sephacryl S200HR column (Sigma-Aldrich) (2.5×65 cm). Proteins were eluted with 50 mM sodium phosphate buffer (pH 7.4) containing 50 mM NaCl at 1 ml min^{-1}. All liquid chromatography procedures were performed on an Äkta Prime Purification System (Amersham Biosciences).

Cloning and expression of protein

N-terminal sequencing of the purified protein was performed by automated Edman degradation with an Applied Biosystems 477A gas-phase sequencer (Foster City, CA) at the Protein Chemistry Facility of the Centro de Investigaciones Biológicas (CSIC; Madrid, Spain). From the obtained sequence, primers were designed and the target gene was PCR amplified with 2 mM $MgCl_2$, 0.8 mM dNTPs, 2.5 U Super-Therm polymerase, 0.2 μM of both the forward (ABC_F_Nde– 5′ CAT ATG AGA AAA GTA GGC AAG CTG GCT 3′) and reverse (ABC_R_Eco – 5′ GAA TTC TTA CTT GAC GGA AAG AGC GTA 3′) primers and 50 ng gDNA template. The gene was cloned into pET-28b(+) and the construct transformed into *Escherichia coli* Rosetta-Gami 2 (DE3) competent cells (Novagen) for expression. The transformants were inoculated into antibiotic-containing LB media (8 g L^{-1} tryptone, 5 g L^{-1} yeast extract, 5 g L^{-1} NaCl and 30 μg L^{-1} kanamycin) and cultured until an OD_{600} of 0.8 was reached before IPTG was

added to a final concentration of 1 mM to induce expression. The cells were incubated a further 4 h before being harvested, washed (50 mM sodium phosphate buffer, pH 7.4) and disrupted by ultrasonic treatment. The fractions were separated by ultra-centrifugation (100,000×g; 90 min; 4 °C). Purification of the protein was achieved with poly-histidine tag affinity chromatography using a HisTrap FF column (GE Healthcare) according to the manufacturer's instruction and an imidazole gradient ranging from 20 to 500 mM (100 ml; flow rate, 5 ml min^{-1}). Fractions were examined by Coomassie staining following SDS-PAGE [31, 32], and those containing proteins of approximately 70 kDa were pooled and further fractionated by size exclusion chromatography at a flow rate of 1 ml min^{-1} on Sephacryl S200HR (Sigma-Aldrich). Active fractions were pooled and dialyzed against 50 mM sodium phosphate buffer (pH 7.4), with Snakeskin-Pleated dialysis tubing (10000 MWCO, Thermo Scientific) at 4 °C with two 2 L buffer change. Protein concentrations were determined using the BCA^{TM} Protein Assay Kit (PIERCE–Thermo Scientific).

Gold reduction and nanoparticle synthesis using cell-free extracts and purified protein

Proteins present in cell-free extracts were routinely assayed for gold reducing ability by incubating the sample with 2 mM $HAuCl_4$ in 50 mM sodium phosphate buffer (pH 7.4) at 65 °C for 24 h. Nanoparticle synthesis using the purified protein was performed under the same reaction conditions but with 4.6 μM sodium dithionite [33] added as a reducing agent [34]. In a living cell, natural reductants are present that are capable of reducing co-factors and in this case the disulphide bridge present in the protein, but these are absent in the purified protein fractions; thus, the reductant facilitates the reduced state of the protein. The sodium dithionite reduces the disulphide bridge in the protein, therefore enabling the protein to reduce the gold ions.

To assess the influence on the size and shape of nanoparticles, reaction mixtures were prepared as follows: $HAuCl_4$ was added to 0.5, 2, 10 and 20 mM; protein concentrations were varied from 10–100 μg ml^{-1} and sodium dithionite from 0–100 μM. Incubation times ranged from 2 to 48 h, temperatures from 30–75 °C and pH ranged from 3.6, 5.5 (sodium acetate buffer) and 7.4 (sodium phosphate buffer) to 9.0 (borax buffer). Formation of a pink/purple colour indicated gold nanoparticle production [27], which was confirmed with TEM analysis, as described below.

Transmission electron microscopy

Biomass samples were prepared for TEM analysis by separating the whole cells from the liquid by centrifugation. The biomass was washed twice in 0.1 M phosphate buffer, pH 7.0 and fixed overnight in 3 % glutaraldehyde solution, prepared in sodium phosphate buffer, pH 7.0. Recovering and washing of the biomass were done by centrifugation. The biomass was enrobed in 1 % agar and dehydrated in a graded acetone-water series. Infiltration and embedding were accomplished by spurr-epoxy resin, followed by two changes of spurr. Blocks were polymerized at 70 °C for 8 h and the embedded material sectioned with an ultra-microtome, yielding sections of approximately 0.2 μm which were mounted on copper grids.

For the cell-free extracts and the purified protein, a drop of the sample was placed onto carbon-coated formvar grids; excess sample was removed after a minute using blotting paper, and the grids were air-dried before analysis. The appropriate controls were evaluated without protein or sodium dithionite to compensate for chemical reduction.

Electron micrographs were taken with either a Philips CM 100 or 200-kV Philips CM 20 TEM [35]. Elemental analysis was carried out using energy-dispersive X-ray spectroscopy (EDS) attached to the TEM.

X-ray diffraction

X-ray diffraction (XRD) analysis was conducted by the Geology Department at the University of the Free State. The data was collected at 40 kV and 30 mA with a Siemens D 8000 diffractometer using CuKα radiation.

Results and discussion

Nanoparticle formation using *T. scotoductus* SA-01 whole cells

T. scotoductus SA-01 is unique to South Africa and can reduce a number of compounds as electron acceptors through dissimilatory pathways. These compounds include Fe(III), Mn(IV), Co(III)-EDTA, Cr(VI) and U(VI) [36]. The Au(III) reducing capability of *T. scotoductus* SA-01 cells in different growth phases illustrated that maximum Au(III) removal or reduction was reached after 8 h, which corresponds to the late exponential—early stationary growth phase (Fig. 1a). When these cells were incubated in the presence of Au(III), ultra-thin section transmission electron microscopy (TEM) revealed that the gold was mainly deposited on the outer wall layer (Fig. 1b) and elemental analysis, using EDS, showed that the particles do consist of gold (Fig. 2a); also present in vast amounts are copper and carbon which can be attributed to the copper grid and the cell biomass. Since nanoparticle synthesis was associated with the biomass, it may be concluded that the cells might have a higher degree of control over the particle morphology as the BOB pathway of mineralization is observed [10–12]. XRD analysis on the biomass exposed to gold ions gave an XRD pattern of the gold precipitated in or on the bacterial cells showing distinct peaks corresponding to the (111); (200) and (220) Bragg reflections of gold [37] (Fig. 2b) which confirms the EDS results.

Effect of the physico-chemical parameters on nanoparticle formation

Various physico-chemical parameters were evaluated using whole cell biomass to determine their effect on gold reduction and nanoparticle formation. These parameters included pH, temperature, exposure time and gold ion concentration. The particle formation was confirmed visually through colour formation (Supplementary figure 1c), peaks observed on the UV–Vis (400–700 nm) spectra (Supplementary figures 1a and b) and TEM analysis. An increase in the size of nanoparticles results in a change of colour, from pink, to purple, to

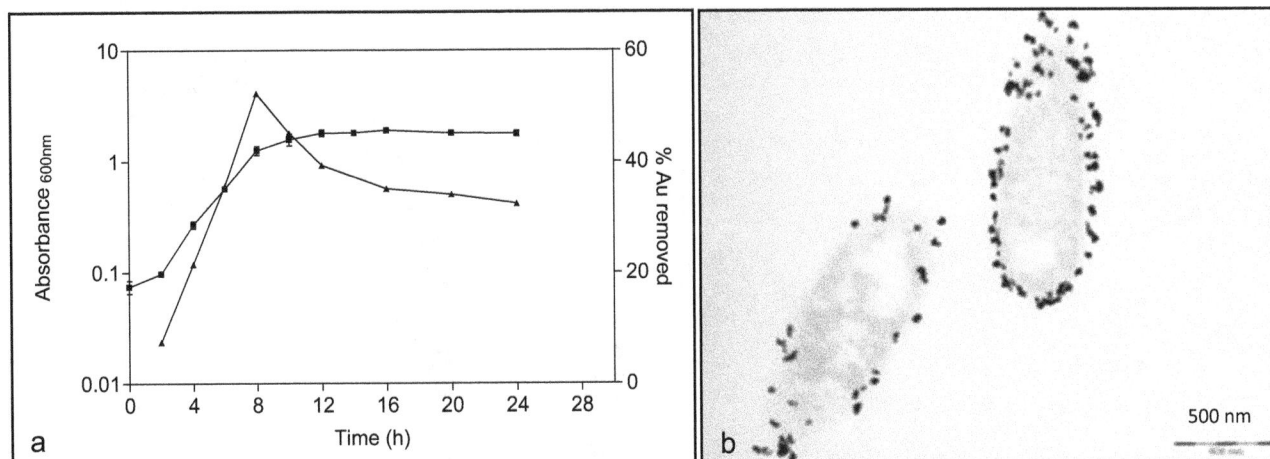

Fig. 1 Interactions of *Thermus scotoductus* SA-01 with Au(III). **a** Growth curve for *T. scotoductus* SA-01, showing the correlation between growth phase (*square*) and Au(III) removal (*triangle*). **b** Thin section TEM micrograph of *T. scotoductus* SA-01 exposed to 2 mM HAuCl4; the micrograph shows the presence of gold deposited along the outer wall layer

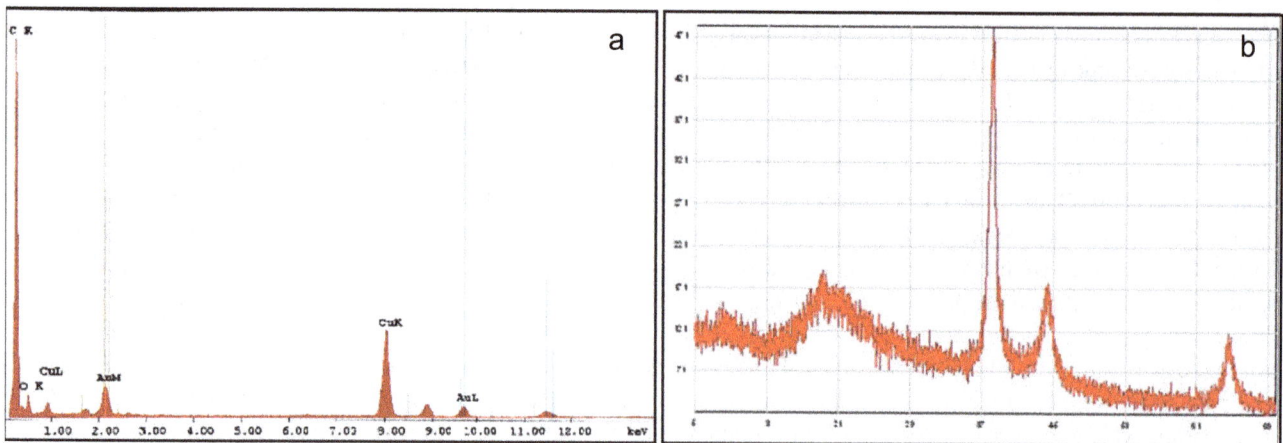

Fig. 2 **a** EDS analysis and **b** XRD pattern recorded of *T. scotoductus* SA-01 cells incubated in the presence of 2 mM Au(III)

blue, due to the surface plasmon vibrations of the gold nanoparticles [38].

The effect of pH on the production of nanoparticles indicated that a lower pH was more conducive to the formation of larger particles while a higher pH resulted in the formation of small particles (Fig. 3a, b). Although pH does affect the size, shape and location of the particles, contrary to suggestions by He and co-workers, particle shape and distribution could not be controlled at this level [21]. In Fig. 3a, the formation of octahedral gold is observed and this points to mineral diagenesis, while intracellular gold reduction leads to cell death (Supplementary figure 2) versus Fig. 3b where immobilization is focused in the cell envelope. Temperature and exposure time had no significant effect on particle morphology but influenced the amount and rate of particle formation. Lower temperatures and shorter incubation periods resulted in fewer particles formed where an increase in temperature and longer incubation periods resulted in more and slightly larger particles. These findings agree with the physico-chemical properties of nanoparticles as discussed by Gericke and Pinches [39] who stated that the rate of particle formation can be correlated with the incubation temperature where an increase in temperature will allow faster particle formation and the rate of reduction can influence the type of particle formed.

Figure 4 illustrates the effect of gold ion concentration on particle formation. At concentrations exceeding 5 mM (for example 10 mM seen in Fig. 4b), the cells are ruptured by the nanostructures formed and are released into the environment and all means of controlling particle morphology is lost. Gold ion concentrations lower than 5 mM (for example 0.5 mM as seen in Fig. 4a) result in the formation of smaller particles which are associated with the biomass. The effect of gold ion concentration on particle formation at various concentrations can be seen in supplementary Fig. 3a to e.

Nanoparticles from cell-free extracts

Incubation of soluble proteins >30 kDa with $HAuCl_4$ resulted in the formation of nanoparticles and also showed a correlation between Au concentration and types of nanoparticles synthesized (Fig. 5). Similar observations were also made by Li and co-workers [15], Shankar and co-workers [17] and Wei and co-workers [40] where it was demonstrated that gold nanosheets with triangular, hexagonal or truncated shapes were dependent on the reductant to Au ion ratio. A mixture of small amorphous particles and nebula-like formations were observed at low Au(III) concentrations (Fig. 5a), but with some particles with specific shapes at 5 mM (Supplementary

Fig. 3 TEM micrographs illustrating the effect of pH **a** 3.6 and **b** 9.5 on particle formation using whole cell biomass

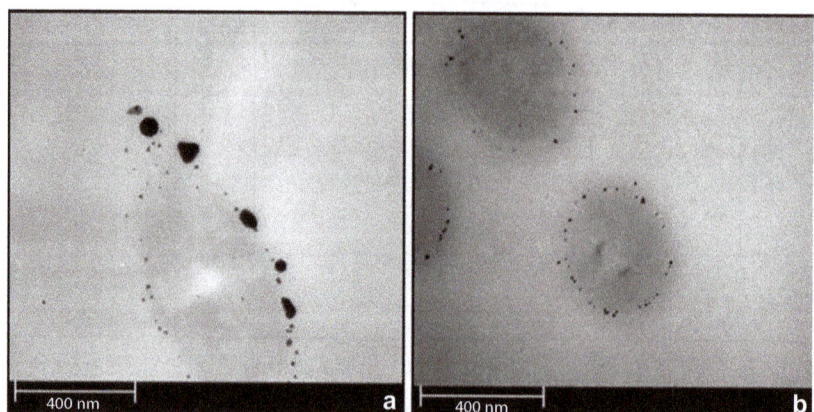

Figure 4a). Addition of 10 and 20 mM Au(III) resulted in large particles (Supplementary Figure 4b and Figure 5b), with definite crystalline surfaces. Morphologies were nanosheet-like, with shapes being mostly triangular and hexagonal. Nanosheets were crystalline with no twinning present and a prominent herring motif. The triangular nanosheets showing threefold symmetry were in the {111} orientation (data not shown), and dark lines visible through the crystals (Fig. 5b) are likely bend contours [41].

Nanoparticles from purified protein

A protein from *T. scotoductus* SA-01, able to catalyse and direct the synthesis of gold nanoparticles, was purified to homogeneity. It was determined to have a molecular mass of ~70 kDa (Supplementary Figure 5), and Edman degradation resolved the N-terminal sequence as GPQDNSLVIGAS which showed 100 % homology with the ABC transporter peptide-binding protein (YP_004203474) from *T. scotoductus* SA-01 [42] and *T. thermophilus* (AEG34049). These types of proteins are ubiquitous membrane proteins that facilitate unidirectional substrate translocation across the lipid bilayer by utilizing the energy obtained from the hydrolysis of ATP and have been found to be very diverse with respect to their physiological function and substrate unlike oxidoreductases which mainly catalyse oxidation-reduction reactions coupled to NADP/NAD$^+$ utilization [43–45].

Subsequent cloning and sequencing of the gene revealed the presence of a disulphide bond (between amino acid 337C and 481C) which confirms the hypothesis of Scott and co-workers [34], implicating an electron shuttle mechanism via a reduced disulphide bridge. This is further supported by the work done by Cason and co-workers [46] who found that reduction of the disulphide bonds resulted in electron transfer to a metal, in their case U(VI). Furthermore, a decrease in U(VI) reduction was observed when the disulphide bridge of the same ABC transporter peptide-binding protein was disrupted by mutation of the genes involved.

According to Tang and Hamley [47], nanostructures can be formed when Au(III) is gradually reduced to Au(I), and when this reduction is almost complete, the equilibrium shift towards reduction of Au(I) to Au(0). This, in turn, leads to the formation of nucleation sites that initiates nanoparticle synthesis. Nucleation sites gradually grow into more complex structures when sufficient Au(I) ions are present. The crystallization of gold is directed by the reducing as well as the stabilizing agent. It is shown that the reduced protein, purified from *T. scotoductus* SA-01, reduces Au(III) to elemental gold and acts as a stabilizing agent that directs the shape and size of the synthesized nanoparticle. To demonstrate this ability, the purified protein was used to evaluate the effects of various physico-chemical parameters on HAuCl$_4$ reduction and subsequent nanoparticle formation, further validated by no nanoparticle formation with protein-free controls.

Particle synthesis manipulations

Au(III): reductant ratio

Significant reduction of Au(III) only occurred in the presence of the proteins from cell-free extracts and once purified only when sodium dithionite was added to the reaction to keep the protein in its reduced form, thus confirming the disulphide electron shuttle mechanism. This was shown by Cason and co-workers [46] as well as Scott and co-workers [34]. Varying dithionite concentrations revealed that a stoichiometric ratio of less than 1:1 (4.6 μM sodium dithionite to protein) resulted in particles with defined edges.

Similar to the cell-free extracts, the purified protein produced small, amorphously shaped particles at low Au(III) concentrations (Fig. 6a), but exposure to higher Au(III) concentrations led to the formation of plate-like nanostructures (Fig. 6b). Varying protein concentrations confirmed the importance of Au(III) to protein (reductant) ratio, where high protein concentrations resulted in small, undefined nanoparticles and low protein concentrations in bigger particles with defined edges.

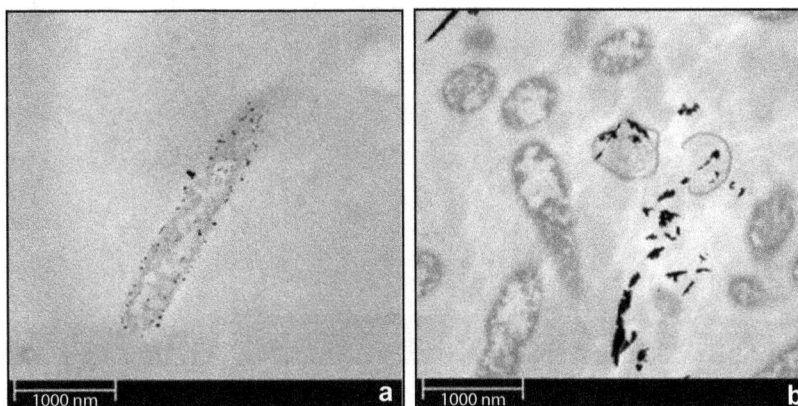

Fig. 4 TEM micrographs illustrating the effect of Au(III) concentration **a** 0.5 and (**b**) 10 mM on particle formation using whole cell biomass

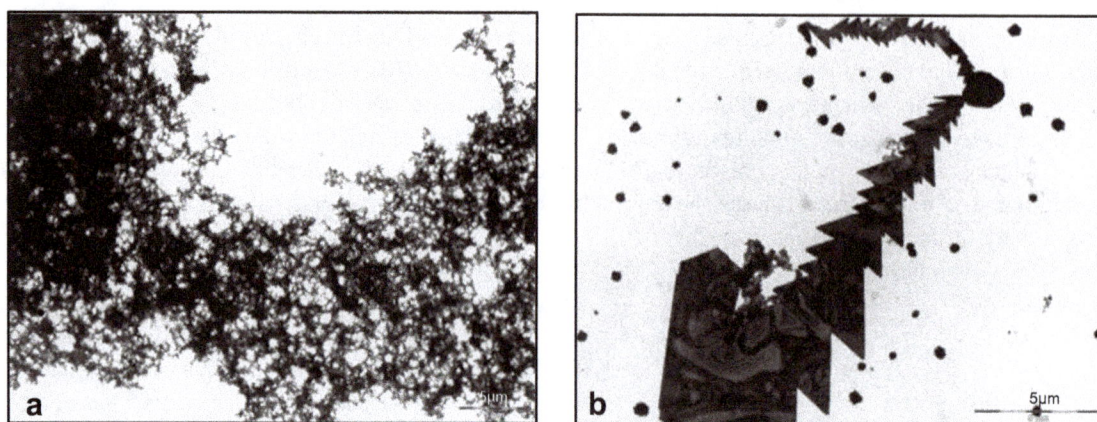

Fig. 5 Nanoparticle formation with >30 kDa soluble fraction proteins with varying HAuCl₄ concentrations. **a** 2 mM and **b** 20 mM

At high reductant to Au(III) ratios, gold ions are preferentially committed to particle initiation (nucleation) rather than particle growth [47] resulting in higher numbers of small particles. However, at lower reductant to Au(III) ratios, fewer nucleation sites are available for particle initiation which allows for bigger particles to grow as there is less competition for available gold ions [33, 48]. Thus, the less nucleation sites are available, and the bigger particles can grow as there will be less competition for gold ions to act either as part of the nucleation process or to grow the size of the particle.

Incubation time, temperature and pH

Nanoparticles were already formed after only 8 h of incubation, but longer reaction times (up to 48 h) only resulted in higher numbers of particles and did not have a significant effect on particle size. Since Au(III) reduction and eventual nanoparticle synthesis are time-dependent reactions, longer incubation times

allow for the reaction equilibrium to shift towards Au(I) reduction and formation of more nucleation sites. With longer incubation times, more nucleation sites can be occupied leading to higher numbers of particles [47, 48].

Temperature had no effect on the size of the nanoparticles, but did influence particle shape. Incubation at lower temperature resulted in fewer particles but with defined edges (Fig. 7a), while particles produced at 75 °C were mostly spherical (Fig. 7b). The overall effect of temperature is likely in influencing the nucleation seed, rather than particle growth [49].

Lower pH values were more conducive to producing nanoplate-like structures with a larger aspect ratio (Fig. 8a and Supplementary Figure 6a), while incubation at higher pH resulted in more small spherical particles (Supplementary Figure 6b). Particles with defined edges were formed at pH 7.4 (Fig. 8b). Different pH values regulate the effective proton concentration of the solution, which in turn controls

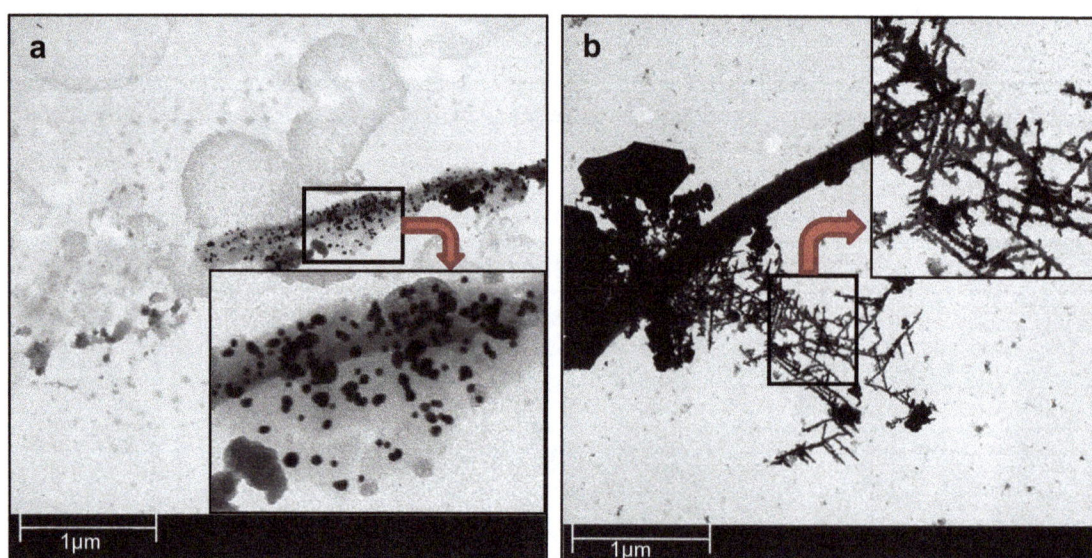

Fig. 6 TEM micrographs showing the effect of HAuCl₄ concentration **a** 0.5 mM and **b** 10 mM, on nanoparticle formation in the presence of purified protein

Fig. 7 TEM micrographs of nanoparticles formed at **a** 42 °C and **b** 75 °C with purified protein

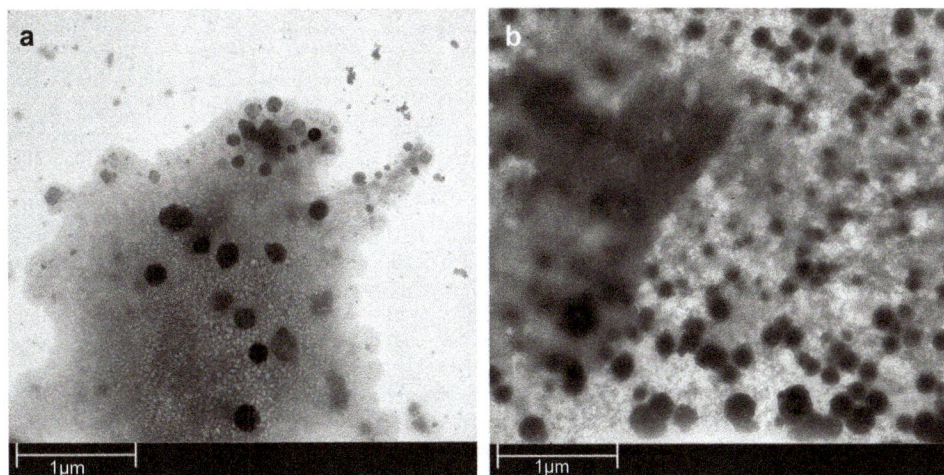

nanoparticle morphology rather than affecting nucleation and particle initiation N.

Conclusions

T. scotoductus SA-01, the thermophilic bacterium used in this study, has the ability to reduce Au(III) and produce nanoparticles, making it a suitable candidate for the production of nanoparticles.

In this study, it was found that the physico-chemical parameters have a definite influence on particle size with lower pH and higher temperatures resulting in larger particles and in contrast, higher pH and lower temperatures will produce smaller particles. However, these processes are difficult to control and a predefined morphology is still unobtainable at this time. Microorganisms are very complex structures with a variety of metabolic functions which all contribute towards the proper working of the cell. Many of these activities could be involved in the reduction of gold and the formation of nanoparticles, but a better understanding of the mechanism might be obtained if the protein/s involved in these reactions is elucidated. Gold reduction primarily occurred in the cell envelope which is strong evidence for a gold 'specific' reduction process, hinting at a location for further investigation.

An ABC transporter, peptide-binding protein of *T. scotoductus* SA-01, able to reduce and synthesize gold nanoparticles was purified to homogeneity. Even though this type of protein is not a classical oxidoreductase, a cysteine–disulphide bridge electron shuttle mechanism is likely involved in reducing Au(III). Moreover, the protein also acts as nucleation seed sites that initiate and direct nanoparticle synthesis. Through manipulation of physico-chemical parameters, it is clear that particle formation can be influenced in terms of size, shape and number of particles formed. However, since biological Au(III) reduction and nanoparticle synthesis is a complex process, manipulations of single parameters are

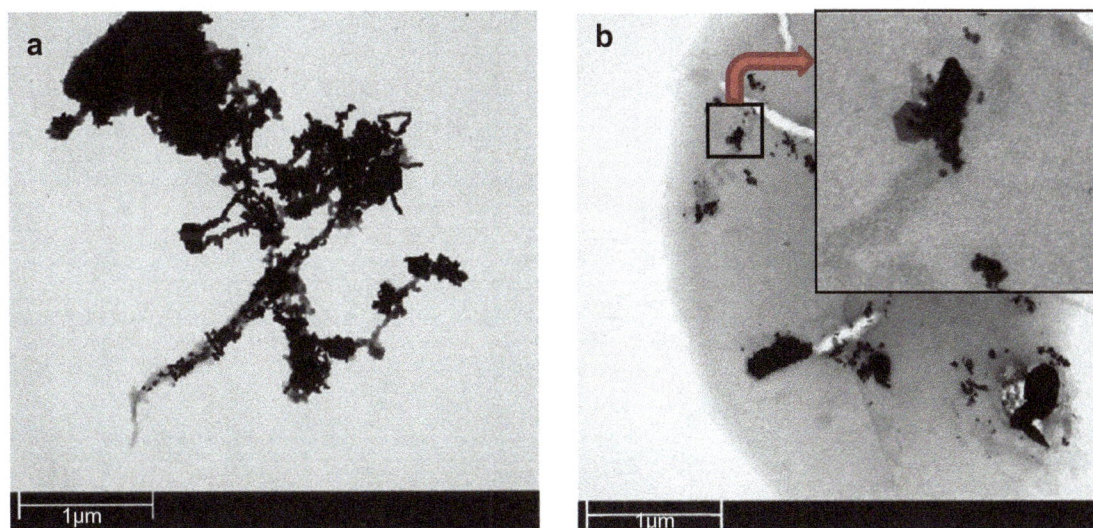

Fig. 8 TEM micrographs of nanoparticles formed at varying pH. **a** pH 5.5 and **b** pH 7.4 with purified protein

unlikely to result in the best conclusive results. This is witnessed by a lack of particle monodispersity (with the exception of small spherical particles) when evaluating any of the parameters. Varying and investigating multiple parameters simultaneously will likely shed light on the way forward to controlling and directing biological nanoparticle synthesis.

Acknowledgments This work was funded by the Project AuTek initiative and Mintek, which we gratefully acknowledge and thank for permission to publish this paper. We also thank Gordon Southam (University of Western Ontario, Canada) for his help with the TEM analysis and interpretation of the micrographs and Liesl van der Westhuizen (University of the Free State) for her assistance in the preparation of this manuscript. We also acknowledge the microscope centres at the following universities and the personnel for their help with the TEM and EDS analysis: University of the Free State, South Africa; University of Western Ontario, Canada; Nelson Mandela Metropolitan University, South Africa and the University of Ghent, Belgium.

Author contributions All work done in this manuscript is based on results from a PhD thesis by JvM with supervision by EvH. The manuscript was written by ME and EDC with input from JvM, EB and EvH. Monetary support for this work was supplied by MG from Mintek as well as NRF funding from EvH research grants.

References

1. Masala O, Seshadri R (2004) Annu Rev Mater 34:41–81
2. Handley DA (1989) Colloidal gold: principles, methods and applications. Volume 1, 13th edn. Academic, San Diego, 33
3. Sau TK, Pal A, Pal T (2001) J Phys Chem 105:9266–9272
4. Gadd GM (2001) Curr Opin Biotechnol 11:271–279
5. Ahmann D, Roberts AL, Krumholz LR, Morel FM (1994) Nature 371:750
6. Klaus-Joerger T, Joerger R, Olsson E, Granqvist CG (2001) Trends Biotechnol 19:15–20
7. Nies DH (1999) Appl Environ Biotechnol 51:730–750
8. Spain A (2003) Rev Undergrad Res 2:1–6
9. Mann S (1993) Nature 365:499–505
10. Weiner S, Dove PM (2003) Rev Mineral Geochem 54:1–29
11. Frankel RB, Bazylinski DA (2003) Rev Mineral Geochem 54:95–114
12. Frankel RB, Bazylinski DA (2003) Rev Mineral Geochem 54:217–247
13. Veis A (2003) Rev Mineral Geochem 54:249–289
14. C.W. Corti, R.J. Holliday and D.T. Thompson, First nanofabrication symposium, Ireland, (2004)
15. Li C, Cai W, Cao B, Sun F, Li Y, Kan C, Zhang L (2006) Adv Funct Mater 16:83–90
16. Kim J, Cha S, Shin K, Jho JY, Lee J (2004) Adv Mater 16:459–464
17. Shankar SS, Rai A, Ahmad A, Sastry M (2005) Chem Mat 17:566–572
18. Ahmad A, Mukherjee P, Mandal D, Senapati S, Khan MI, Kumar R, Sastry M (2002) J Am Chem Soc 124:12108–12109
19. Mukherjee P, Senapati S, Mandal D, Ahmad A, Khan MI, Kumar R, Sastry M (2002) ChemBioChem 5:461–463
20. Sastry M, Ahmad A, Khan MI, Kumar R (2003) Curr Sci 85:162–170
21. He S, Zhang Y, Guo Z, Gu N (2008) Biotechnol Prog 24:476–480
22. Duran N, Marcato PD, Alves OL, De Souza GIH, Esposito E (2005) J Nanobiotechnol 3:8–14
23. He S, Guo Z, Zhang Y, Zhang S, Wang J, Gu N (2007) Mater Lett 61:3984–3987
24. Van Heerden E, Opperman DJ, Bester PA, van Marwijk J, Cason ED, Litthauer D, Piater LA, Onstott TC (2008) Proc SPIE - Int Soc Opt Eng 7097:70970S
25. Khersonsky O, Tawfik DS (2010) Annu Rev Biochem 79:471–505
26. Melwanki MB, Masti SP, Seetharamappa J (2002) Turk J Chem 26:17–22
27. Šlouf M, Kužel R, Matej Z (2006) Krist Suppl 23:319–324
28. Lazarides AA, Schatz GC (2000) J Phys Chem B 104:460–467
29. Opperman DJ, van Heerden E (2007) J Appl Microbiol 103:1907–1913
30. Gaspard S, Vazquez F, Holliger C (1998) Appl Environ Microbiol 64:3188–3194
31. Fairbanks G, Steck TL, Wallach DFH (1971) Biochemistry 10:2606–2617
32. Lamelli UK (1970) Nature 227:680–685
33. Showe MK, DeMoss JA (1968) J Bacteriol 95:1305–1313
34. Scott D, Toney M, Muzikar M (2008) J Am Chem Soc 130:865–874
35. van Wyk PWJ, Wingfield MJ (1991) Mycologia 83:698–707
36. Kieft TL, Fredrickson JK, Onstott TC, Gorby YA, Kostandarithes HM, Bailey TJ, Kennedy DW, Li SW, Plymale A, Spadoni CM, Gray MS (1999) Appl Environ Microbiol 65:1214–1221
37. Leff DV, Brandt L, Heath JR (1996) Langmuir 12:4723–4730
38. Corti CW, Holliday RJ, Thompson DT (2002) Gold Bull 35:111–117
39. Gericke M, Pinches A (2006) Hydrometallurgy 83:132–140
40. Wei D, Qian W, Shi Y, Ding S, Xia Y (2007) Carbohydr Res 342:2494–2499
41. Rodriguez-Gonzalez B, Pastoriza-Santos I, Liz-Marzan LM (2006) J Phys Chem B 110:11796–11799
42. Gounder K, Brzuszkiewicz E, Liesegang H, Wollherr A, Daniel R, Gottschalk G, Reva O, Kumwenda B, Srivastava M, Bricio C, Berenguer J, van Heerden E, Litthauer D (2011) BMC Genomics 12:577–590
43. Braibant M, Gilot P, Content J (2000) FEMS Microbiol Rev 24:449–467
44. Locher KP, Broths E (2004) FEBS Lett 564:264–268
45. Saurin W, Hofnung M, Dassa E (1999) J Mol Evol 48:22–41
46. Cason ED, Piater LA, van Heerden E (2012) Chemosphere 86:572–577
47. Tang T, Hamley IW (2009) Colloid Surf A 336:1–7
48. Mikheenko IP, Rousset M, Dementin S, Macaskie LE (2008) Appl Environ Microbiol 74:6144–6146
49. Kawamura G, Nogami M (2009) J Cryst Growth 311:4462–4466

23

Room temperature evolution of gold nanodots deposited on silicon

C. Garozzo · A. Filetti · C. Bongiorno · A. La Magna ·
F. Simone · R. A. Puglisi

Abstract In this work, the morphological and structural evolution of gold nanodots deposited on Si substrates has been monitored for 2.4×10^3 h. Gold nanodots on Si are of great scientific interest because they can be used in numerous ways, for example as subwavelength antennas in plasmonics, as electrical contacts in nanometric devices, or as catalysts for the formation of quasi-1dimensional nanostructures. Their characteristics have been studied in a very large number of papers in literature, and among the several aspects, it is known that continuous Au films peculiarly interact with Si by interdiffusion even at room temperature. It would be expected that also small nanostructures could undergo to an interdiffusion and consequent modifications of their structure and shape after aging. Despite the cruciality of this topic, no literature papers have been found showing a detailed morphological and structural characterization of aged Au nanodots. Au nanoparticles have been deposited by sputtering on Si and stored in air at temperature between 20 and 23 °C and humidity of about 45 %, simulating the standard storage conditions of most of the fabrication labs. The morphological and structural characterizations have been performed by bright field transmission electron microscopy (TEM). A specific procedure has been used in order to avoid any modification of the material during the specimen preparation for the TEM analysis. A digital processing of the TEM images has allowed to get a large statistical analysis on the particles size distribution. Two different types of nanoparticles are found after the deposition: pure gold crystalline nanodots on the Si surface and gold amorphous nanoclusters interdiffused into the Si subsurface regions. While the nanodots preserve both morphology and structure all over the time, the amorphous agglomerates show an evolution during aging in morphology, structure, and chemical phase.

Keywords Interfaces · Gold · Silicon · Nanodots ·
Transmission electron microscopy · Aging

Introduction

Gold nanodots deposited on silicon are of significant importance because this material presents a large number of applications like as subwavelength antennas in plasmonic applications, or as catalysts for the growth of silicon nanowires (Si-NWs), or even as electrical contacts in Si nanodevices. In the first example, the gold nanoparticles are exploited to increase light absorption in solar cells, thanks to the surface plasmon resonances, when they are in contact with the active surface of semiconductor. But most of the experimental results [1–3] demonstrate that these mechanisms, and consequently the cell performance, are strongly dependent on the geometrical characteristics of the dot, such as size and shape. Another field of application for gold nanodots is in the synthesis of Si-NWs, where they can be used either as catalyst seeds in a bottom-up approach like the vapor-liquid-solid process, or as etching initiator in a top-down process like the metal chemical etching [4–8]. Also, in these cases, the main morphological characteristics of the final Si-NWs such as diameter and growth (/etching) direction are determined by the size of the catalyst and their interconnection, i.e., by the original Au dot characteristics [9–12]. The Si-Au system has been also studied from the point of view of the electrical stability in metallic contacts. Akhtari-Zavareh and

C. Garozzo · A. Filetti · C. Bongiorno · A. La Magna ·
R. A. Puglisi (✉)
Consiglio Nazionale delle Ricerche, Istituto per la Microelettronica e Microsistemi, Strada Ottava 5 Zona Industriale, 95121 Catania, Italy
e-mail: rosaria.puglisi@imm.cnr.it

F. Simone
Dipartimento di Fisica e Astronomia, Università di Catania, Via Santa Sofia 64, 95123 Catania, Italy

coworkers [13] recently studied the morphological and electrical characteristic of an Au/Si diode. They found that after air exposure there is a formation of a continuous interfacial layer due to an interaction between Au and Si. This surface or interfacial interaction leads to the degradation of the rectifying diode properties.

The evidence of an interaction between gold and silicon is shown in many works for several annealing temperatures. At $T=150$ °C, i.e., below the Au-Si eutectic point (363 °C), Rutherford backscattering spectrometry (RBS) analysis demonstrated that in a Si sample covered with an evaporated gold layer, Si atoms migrate through the Au film, accumulate on its surface, and oxidizes after air exposure [14]. Room temperature evolution of continuous Au layers evaporated on Si (111) has been investigated too [15]. The gold films were evaporated in this case at $T<50$ °C on clean silicon and characterized in situ by using electron loss spectroscopy and Auger electron spectroscopy. The interacted regions have been identified as metastable alloys or silicides [16, 17]. A critical thickness below which the gold seems inert and not able to create alloying has been identified, and, if it is less than 2 ML, gold and Si do not intermix even at temperatures as high as 800 °C [15]. For a thickness larger than 5 ML, Au reacts with silicon also at room temperature [15]. The critical thickness has been in this case estimated as the plateau exhibited by the growth rate measured by an oscillating quartz, but no morphological analysis is shown. A lower value of 1 ML for the critical thickness has also been found by Auger spectroscopy studies for gold evaporated on clean Si (111) at room temperature [16]. Experiments of minute amounts of gold deposited on NaCl demonstrated that a considerable surface mobility and coalescence is present also at room temperature. Since the diffusion coefficient of gold in silicon cannot explain this rapid interdiffusion, it was concluded that the intermixing of gold into the silicon was a chemically driven process. In the same paper, TEM analysis has been indicated as an improper technique to study the early stages of the deposition of gold on Si because early film growth can be detected only after the adatoms have formed small metal nuclei with sizes exceeding the resolving power of the technique. The top region of the interacted material has been investigated too, and in literature, a controversy is present, some papers referring to an alloy [18], some to a gold-rich material [15], others to a Si rich one [19]. The interaction can be inhibited if the Si surface is covered by SiO_2 or impurities such as carbon, water, or oxygen prior to the Au deposition [15]. The effects of the substrate orientation has been explored and the (100) Si surface exhibits higher interactions rates than the (111), due to the larger Si dangling bonds density, available for bonding to Au, present in the first [19]. The role of the deposition technique has been investigated too, and it has been found that in sputtered samples, the interaction is more active than in evaporated films due to the knock off mechanism inferred by the

energetic ions during the sputtering on the Si surface, which remove more efficiently oxide residuals and improves the Au/Si contact [19].

Despite the large amount of works present in literature on the Au/Si system and its changes with annealing process, a detailed morphological and structural characterization at the nanometer level of the room temperature evolution of a few ML of gold, i.e., a discontinuous film, deposited on Si is still missing.

In this paper, a study on gold nanodots deposited by sputtering over (100) Si substrates and after more than 2×10^3 h of aging at room temperature is presented. The samples are stored in air at temperature between 20 and 23 °C and humidity of about 40 %. The samples have been prepared to avoid any other thermal budget, included that necessary to prepare the specimen for TEM analysis, and the first characterization started just a few hours after the deposition. We used bright field TEM and diffraction analysis to get several snapshots of the changes in the shape, crystallography, and phase of the deposited and evolved nanostructures at different aging times. The kinetics have been quantified by measuring the nanostructures size and density and the Si substrate areas covered by gold. A high statistics has been guaranteed by the usage of digital analysis tools of the TEM micrographs. We demonstrate that the Au/Si phase after deposition can be modified by the specimen preparation and by the room temperature exposure in terms of shape, density, composition, thermodynamic phase, and crystallography. We also demonstrate that TEM analysis, if properly prepared, is essential to reveal the morphology of the islands and to distinguish among the different phases to which the deposited material undergoes.

Experimental

The samples have been prepared by using two different procedures depicted in Fig. 1a, b. In the first case, Fig. 1a, p-type Si substrates were subjected to a chemical etching in HF buffered solution (7 %) to remove the native oxide. Immediately after cleaning, the samples were introduced into a magnetron sputtering chamber Emitech K550X (Emitech Limited, Kent, UK), with 7×10^{-5} mbar of base pressure, 6×10^{-3} mbar of deposition pressure in Ar, for the deposition of gold at 15 mA of gun current and for 30 s. No intentional heating was used during gold deposition. The deposited equivalent gold thickness was measured by an oscillating quartz inside the chamber and resulted equal to 2 ± 0.1 nm. RBS analysis provided a measurement of the deposited total amount equal to $(1.2\pm 0.1)\times 10^{16}$ cm^{-2}. This value corresponds to about 7.9 ML of gold. After the gold deposition, the samples have been thinned for the TEM analysis by mechanical grinding from the back and ion milling polishing, which is made from the back of the sample at

3 keV with Ar. During the whole thinning procedure, the maximum temperature experienced by the specimen is 150 °C. In the second case, Fig. 1b, the p type Si substrates were first thinned as for the TEM analysis and then were subjected to the HF etch and gold deposition with the same conditions detailed above. Morphological and structural characterization were performed in both cases by TEM in planar view, by using a JEOL JEM 2010F (JEOL Ltd., Tokyo, Japan) microscope with a field emission gun operating at 200 kV accelerating voltage. The samples realized with the procedure in Fig. 1a were analyzed after the TEM preparation, while the samples prepared like in Fig. 1b were analyzed right after the gold deposition and after 330 h, 1×10^3 h, and 2.4×10^3 h. All this time, these samples have been stored in air at temperature ranging between 20 and 23 °C and humidity of about 40 %. The TEM images were digitally analyzed by a Digital Micrograph tool and were transformed in a black and white image with a procedure reported in ref. [20]. The transformed images have then been analyzed, and the corresponding nanostructure areas have been measured. In case of Au nanodots, their radius has been taken as $r_d = \sqrt{\frac{A_d}{\pi}}$, where A_d is the area of the dot. For every type of sample, a statistics of at least 100 dots has been collected.

Results and discussion

Figure 2a presents the TEM analysis in planar view of a sample sputtered with the procedure indicated in Fig. 1a, i.e., first, the HF etch plus gold deposition and then the TEM specimen preparation. Two zones can be identified:

white regions representing the Silicon substrate and black regions that represent the gold nanodots. Figure 2b, c presents TEM in planar view of the as deposited sample obtained as in Fig. 1b, i.e., the HF etch plus gold deposition performed after the TEM thinning. The analysis was performed on the sample a few hours after the deposition. Three zones can be identified: white regions representing the silicon substrate, black regions that represent gold nanodots and an intermediate zone of gray color. In Fig. 2c, d the corresponding electron diffraction patterns of the samples imaged in Fig. 2a, b, respectively, are shown. In Fig. 2c, the blue square is superimposed to the pattern of the (100) Si substrate zone axis. The indexes refer to the Si spots 004 and 022. The Au spots, indicated by the red indexes, correspond to two Au patterns rotated by 90° relative to each other with the same zone axis [110] on Si [100]. As observed, the diffraction analysis shows an in-plane alignment of the Au nanostructures. The orientation relationship is Au(110)∥Si(100) with Au[002]∥Si[022] and Au[002]∥Si[02–2]. Figure 2d shows the diffraction pattern relative to the sample prepared with the innovative procedure, i.e., by thinning the specimen before the Au sputtering. Also, in this case, the blue square indicates the Si (100) substrate zone axis. The arrows indicate the presence of rings typical of Au nanodots randomly oriented between each other and also with respect to the Si substrate. All the diffraction patterns have been taken from an area of 500×500 nm. The diffraction signal corresponding to the amorphous regions is not visible because it is covered by the ones of the randomly oriented dots.

High resolution TEM analysis (not shown here) coupled to diffraction analysis indicates that the cloudy gray region corresponds to an amorphous mixture of gold and silicon, thus

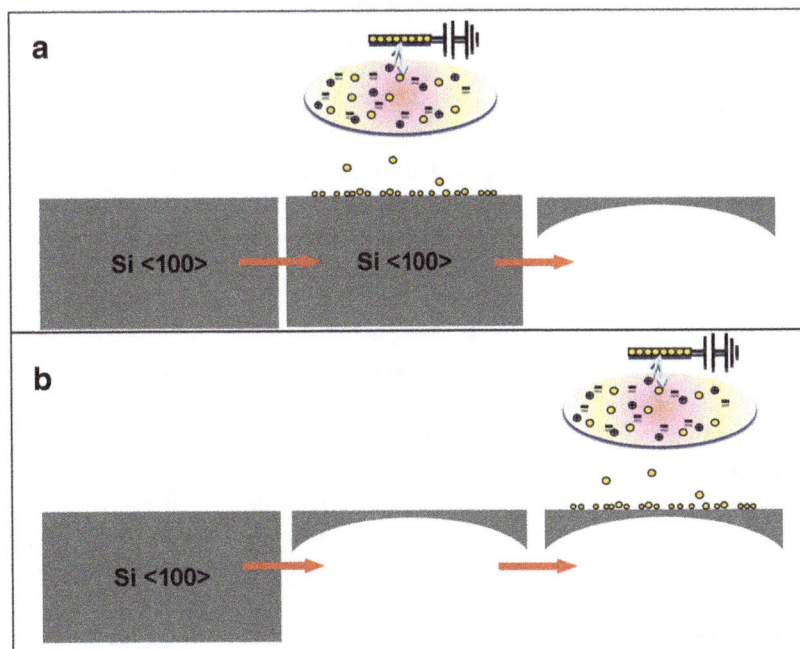

Fig. 1 Schematics depicting the two procedures followed for the synthesis of the samples. **a** Si substrates have been HF etched and subjected to the gold sputtering deposition, and then to the TEM thinning procedure. **b** Si substrates have been thinned for the TEM analysis before the HF and gold sputtering deposition

Fig 2 a Bright field TEM in planar view of the sample obtained by using the synthesis procedure indicated in Fig. 1a, i.e., first, the gold sputtering on Si (100) and then specimen thinning. **b** Bright field TEM micrograph in planar view of the sample prepared by using the synthesis procedure indicated in Fig. 1b, i.e., first TEM specimen preparation and then HF etch plus gold deposition. **c** Diffraction pattern of the sample imaged in **a**, **c** diffraction pattern of the sample imaged in **b**

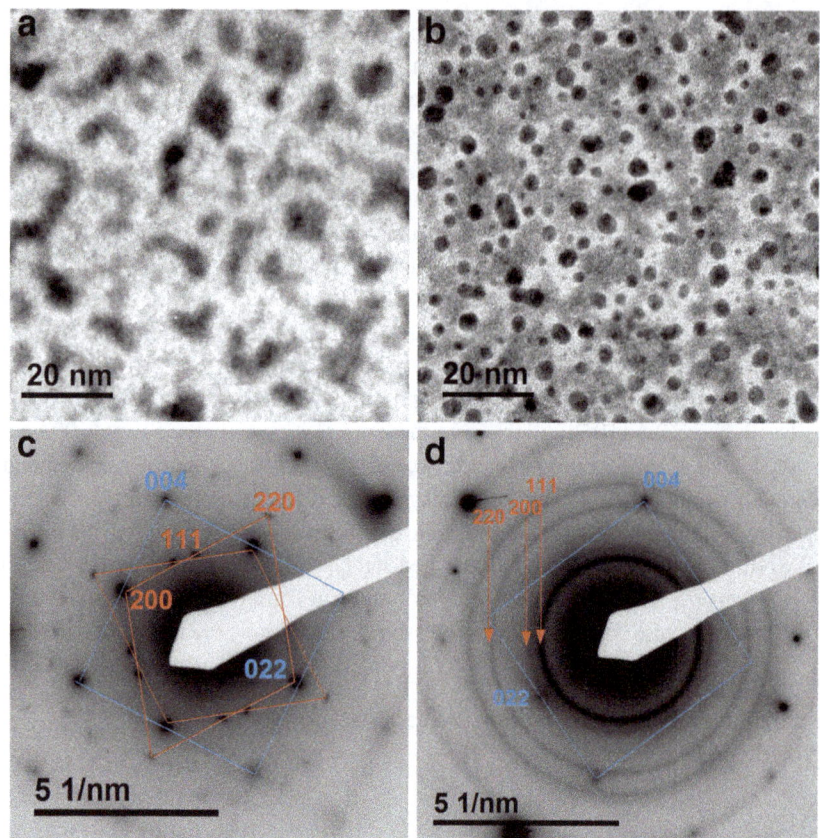

indicating that here gold has intermixed with silicon. Since the characterization is ex situ, it is not possible to establish whether the interaction has proceeded during the deposition because of the energetic plasma, or right after the deposition because of the few hours of room temperature exposure. However, it is reasonable to think that the plasma has contributed to it, as in case of sputtered continuous films [15], although with differences due to the discontinuity of the Au layer which will be discussed in the following. We can conclude that the process performed during the TEM preparation in the case of Fig. 1a modified the original Au morphological and structural characteristics. For this reason, all the results reported in the following have been obtained only by using the procedure in Fig. 1b, i.e., TEM specimen preparation before than the HF plus gold deposition.

Figure 3 shows the TEM micrographs in bright field conditions taken during the time evolution of the sample prepared as in Fig. 1b and analyzed a few hours after the deposition (Fig. 3a), after 14 days (Fig. 3b), 43 days (Fig. 3c), and 99 days (Fig. 3d). The first micrograph shows as in Fig. 2c two phases of gold: Au nanodots, the dark gray circular areas indicated by the red arrows, and amorphous regions of intermixed Au/Si, the large cloudy gray regions highlighted with the orange arrows. The interaction between Si and Au is known in literature for thin continuous films annealed at $T < 150$ °C [14, 15] and will be discussed in the following by comparing the previous results to

the new ones found in this work at room temperature [13, 21], while the agglomeration of gold in nanodots is presumably due to some local residual or regrown Si-O bonds or C atoms that prevent the Si/Au intermixing [21]. From a preliminary observation of the four micrographs in Fig. 3, it is possible to understand that the evolution of the gold nanodots does not show drastic changes, while the amorphous Au/Si agglomerates, corresponding to the gray areas, evidence a morphological modification because they shrink in density and size leaving room to Si uncovered areas. Furthermore, after 43 days of aging, some of the cloudy areas appear darker than in the as deposited and present a contrast more similar to the one of the Au dots. In some cases, these areas (like the ones indicated by the green arrows) exhibit the fringes typical of a crystalline phase lying over a crystalline substrate, indicating that these nanostructures after aging are crystalline. This result demonstrates that to ensure the accuracy of the results, besides than properly preparing the sample for analysis, it is crucial to analyze the samples immediately after the gold deposition. The TEM micrographs have been analyzed by a digital elaboration to measure the size of the dots and of the interacted regions, by averaging on at least 100 objects for each condition. The next results refer to the characterization performed on the nanodots first, and then to the amorphous agglomerates. Figure 4 shows the size distribution of the Au nanodots as a function of their radius for the several days of observation, and Table 1 reports the measured data

Fig. 3 TEM micrographs in bright field conditions reporting the evolution of the sample during the aging in air at room temperature after a few hours (**a**), 330 h (**b**), 1×10^3 h (**c**), and 2.4×10^3 (**d**) of storage. The *colored arrows* indicate the different materials: Au dots (*red arrows*) and Au/Si cloudy gray zones (*orange arrows*). The *green arrows* indicate the fringes inside some of the aligned Au nanostructures, formed after segregation

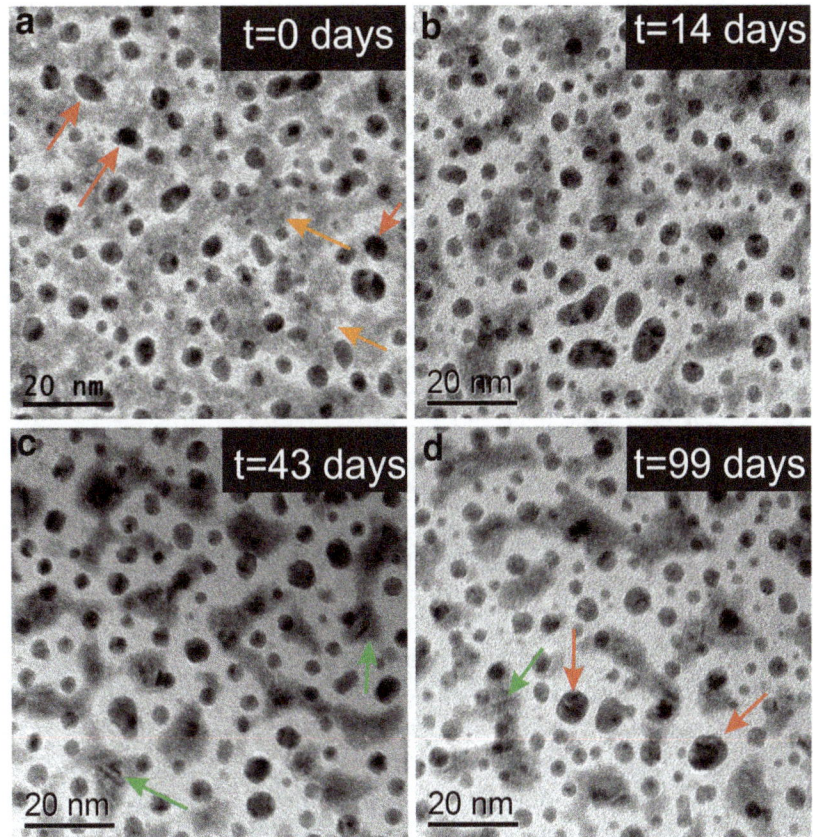

about the average nanodot radius, defined as explained in the "Experimental" section by approximating the shape planar projection of the dots to circles, and their surface density.

From the graph and the data reported in Table 1, we can say that the size distribution of the Au nanodots as a function of the radius presents a shape that is maintained all over the time with small differences in the peak value and in the centroid.

The quantitative analysis of the data then confirms the previous observations, i.e., the nanodots morphological characteristics do not show appreciable evolution over time, supporting the hypothesis that at the interface, the presence of residual or regrown SiO_x, or even carbon atoms, have inhibited the intermixing between Au and Si.

The previous data refer to the characteristics of the agglomerated Au nanodots. The regions where the gold is mixed with Si have been analyzed too. Since their shape cannot be approximated as circular, the parameter chosen to investigate their properties has been their area. The percentage of area covered with Au/Si intermixed nanostructures, Θ_g, has been obtained from the difference of two quantities measured experimentally: the coverage of the substrate with agglomerated gold nanodots, Θ_d, and the percentage of uncovered substrate Θ_s (the white areas present in the TEM micrographs of Fig. 3).

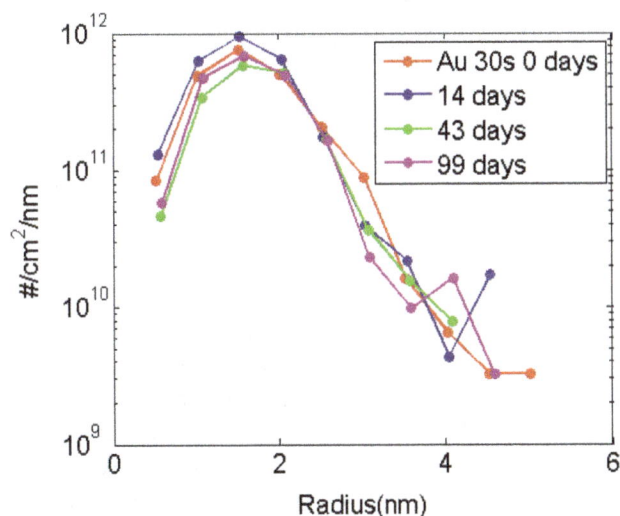

Fig 4 Gold dot radius distribution for the as deposited samples (*red*) and for the sample stored for 330 h (*blue*), 1×10^3 h (*green*), and 2.4×10^3 h (*magenta*) in air

Table 1 Average radius and dot surface density as a function of the evolution time

Evolution time (days)	Dot radius r_d (nm)	Dot density (#/cm^2)
0	1.7±0.6	2.2×10^{12}
14	1.6±0.6	2.6×10^{12}
43	1.8±0.6	1.7×10^{12}
99	1.7±0.6	1.9×10^{12}

We calculated the total Au coverage, Θ_G, as the difference between 1 and the fraction of uncovered Si substrate Θ_s. After this, Θ_g has been calculated as $\Theta_G - \Theta_d$. Figure 5 reports the results on Θ_d (light blue square symbols), Θ_s (diamond red symbols), Θ_G (green circles), and Θ_g the region of interacted Au/Si(magenta triangles).

The first consideration that can be inferred from Fig. 5 is that Θ_d is approximately constant, as anticipated by the results of Fig. 4. Instead, if we look at the amount of gold that has intermixed with Si, Θ_g, it decreases rapidly in the first 15 days and then tends to saturate. The fact that the quantity of Au interacted with Si decreases after aging will be discussed later in the paper, in the comments to Fig. 7.

The observation of the samples through the weeks after the deposition has also highlighted certain aspects of the crystallographic orientations of the aged material. The Fig. 6a, b, c reports the diffraction analysis for the 330 h (a), 1×10^3 h (b), and 2.4×10^3 h (c) aged samples. The diffraction patterns for all the samples have been taken off axis to avoid that the spots relative to the Si substrate, much more intense than the Au diffraction rings and spots, would saturate the image intensity and obscure the Au diffraction features. Figure 6a shows a sample after 14 days of aging. The sample area where the diffraction pattern is taken is about 500×500 nm. In this area, the sample resulted to be bent, and this fact has created different diffraction spots with respect the expected ones.

A blue square is superimposed to all the diffraction patterns as a guide to the eyes to identify Si spots. The concentric circles, present in Fig. 6a, are formed by the presence of the randomly oriented Au nanodots that behave as powders. In the diffraction pattern relative to the 43rd (6b) and 99th (6c) days, the weak

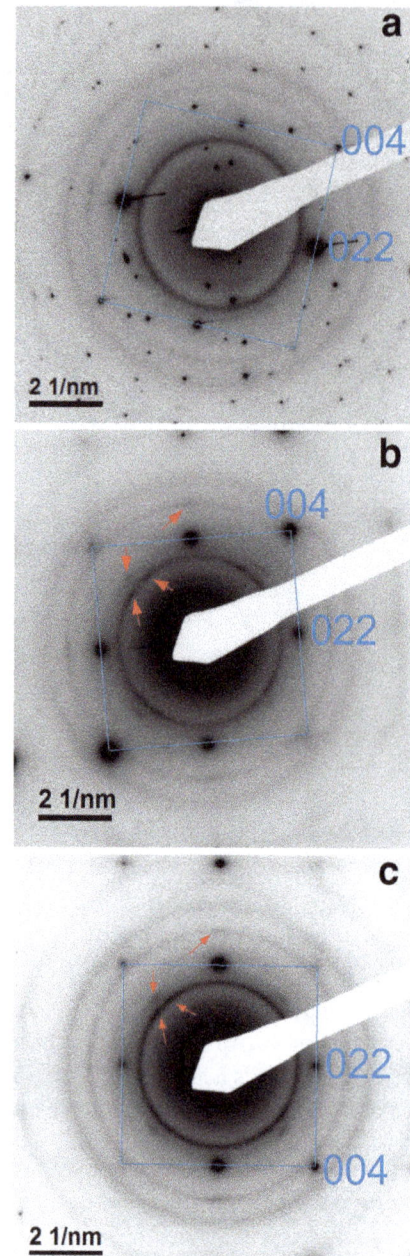

Fig. 6 Diffraction patterns of the samples aged for 14 (**a**), 43 (**b**), and 99 days (**c**). In the diffraction pattern relative to the 43rd and 99th day, the weak spots (identified by the *red arrows*) correspond to the orientation relationship Si (Au(110)‖Si(100) with Au[002]‖Si[022] and Au[002]‖Si[02−2])

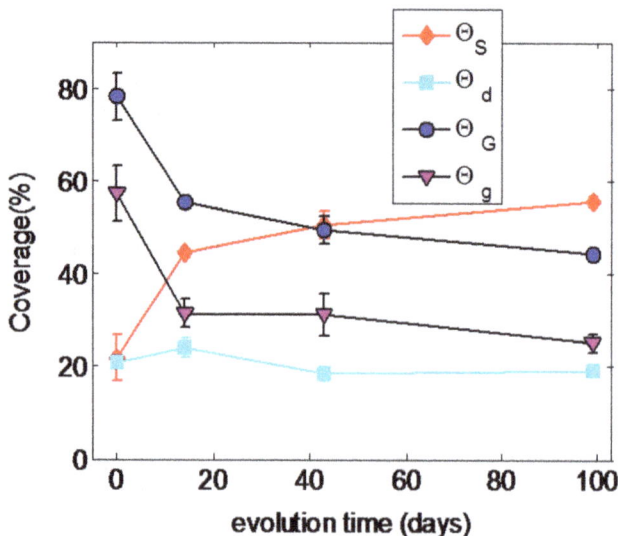

Fig. 5 Measured data on the percentage of uncovered substrate Θ_s (*diamond red symbols*) and on the coverage of gold nanodots Θ_d (*cyan square symbols*) as a function of evolution time; calculated values of coverage as a function of evolution time for the total Au, Θ_G (*blue circles*), and interacted gold, Θ_g (*magenta triangles*)

spots (identified by the red arrows) correspond to two Au patterns rotated by 90° relative to each other with the same zone axis [110] on Si [100]. This is a similar result to the alignment previously observed in the case of the sample prepared by sputtering Au before the TEM thinning procedure, presented in Fig. 2c. The orientation relationship is Au(110)‖Si(100) with Au[002]‖Si[022] and Au[002]‖Si[02−2].

The experimental results coming from the diffraction patterns coupled to those from the bright field analysis indicate an evolution of the sample toward the crystalline phase, the presence of the ring in every diffraction figure is due to the nanodots that are randomly oriented, and the interacted alloyed phase that is amorphous in the as deposited sample presents a phase transition demonstrated by the presence of weak spots that indicated the alignment. After 99 days of aging, experimental data shows the evidence of a transition of the material. The gold atoms from the alloy, tend to segregate in subsurface Au clusters similar to the "inert" ones, but are preferentially oriented with silicon. Regions of intermixed alloys are however still present in the sample.

In summary, the morphological and structural data have thus shown that gold deposited by sputtering on clean Si shows two phases: gold agglomerated as nanodots of a few nanometers in size and amorphous Au/Si agglomerates. The latter evolves at room temperature and tends to form smaller and fewer crystalline regions of pure gold aligned with some crystallographic orientations of the silicon substrate.

An explanation of the results obtained in this study relatively to the Si/Au intermixing is provided by a work of J. K. Bal and S. Hazra [21]. They studied the room temperature diffusion of continuous thin layers of gold on Si (001) substrates and have observed that in samples whose surface was initially treated with HF to remove the native oxide, as in our case, gold interacts with silicon diffusing through the surface. The amount of diffusing gold saturated after about 2 weeks from the date of deposition due to the regrowth of an oxide layer, which slows or prevents the intermixing Au-Si. The saturation time and the decrease in the interaction kinetics are very similar to our case. The literature data refer however to the characterization of continuous gold layers diffused in Si and does not present data on the morphological and structural characteristics of the nanostructures.

Relatively to the Au-Si interaction, it is known in literature that the intermixing between Au and Si even at low temperatures, produces a gold-rich (70%) amorphous surface alloy [22, 23]. Moreover, Kim al. [24] proposed a structural model, indicating the arrangement of atoms in the layers as they are deposited. The proposed model shows that at the interface between the deposited material and the substrate, the system Au/Si tends to reach the concentration of a 50-50 of compound Au-Si. Once reached this level, the system tends to preserve this condition, so it behaves as a "stable" condition.

In the as deposited sample, we expect to find 1 ML at the interface with Si with a concentration of 50 %; in the following discussion, we neglect this contribution. We start from the total amount of deposited gold and separate the contribution of the crystalline surface dots and the amorphous subsurface intermixed alloys. The amount of gold in the clusters is directly calculated from the TEM data, from the measured radius of each dot, by considering that gold dots have a circular shape, and from the *equivalent* gold thickness, by using the standard gold bulk density (because the dots present random crystal

orientations). The amount of gold which is contained in the amorphous Si-Au alloy is calculated by subtracting the quantity of gold in the crystalline dots from the total amount of gold. By taking into account that the amorphous zones, in the as deposited case, present a concentration of 70 % Au rich (by considering the literature results from AES analysis [22]), we calculate the equivalent thickness of the amorphous Au-Si alloy and report the results in Fig. 7. An analogous procedure has been carried for the 99th day of aged sample, but in this case, considering that the sample is made of crystalline pure gold, we considered a 100 % Au concentration. For the intermediate samples, the 14th and 43rd days of aging, a concentration from 70 to 100 % has been taken into account. In Fig. 8, we present the data relative to the calculated thickness for each case. The green square represent the thickness calculated by using 70 % for gold concentration, and the blue square represent the data calculated by using the 100 % concentration for the gold dots. The rectangles for the 15th and 43rd days indicate the possible thickness of the nanophases if the concentration of gold in the alloy goes from 70 % (green shade) to 100 % (blue shade). In the graph, the amount of gold in form of nanodots is not plotted because it does not change over time.

The peculiarity of the analysis performed in the present work relies on the capability to quantify only the data relative to the Au-Si intermixed compound, distinguishing from the inert surfacial nanodots by means of the TEM analysis in planar view which allows to quantify the effective area covered by the gold and alloyed islands. Moreover, thanks to the

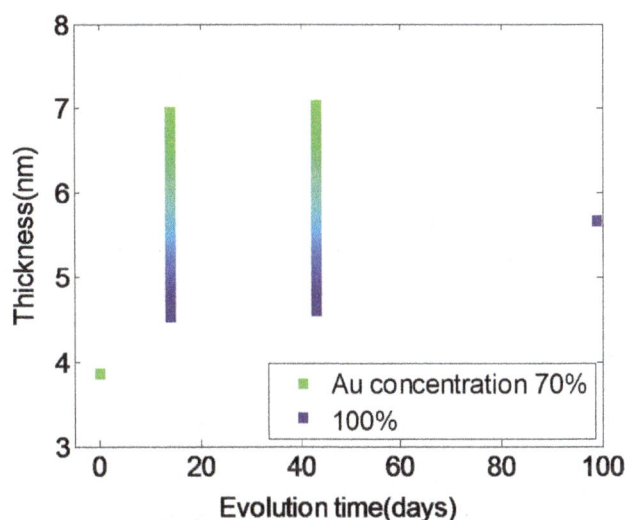

Fig. 7 Calculated equivalent thickness of the amorphous Au/Si alloys, during the aging (from the as deposited case to the aged samples) and the transition to the crystalline Au dots. The *green square* represent the thickness calculated by using 70 % for gold concentration and the *blue square* represent the data calculated by using the 100 % concentration for the *gold dots*. The *rectangles* for the 15th and 43rd days indicate the possible thickness of the nanophases if the concentration of gold in the alloy goes from 70 % (*green shade*) to 100 % (*blue shade*). In the graph, the amount of gold in form of nanodots is not reported because it does not change over time

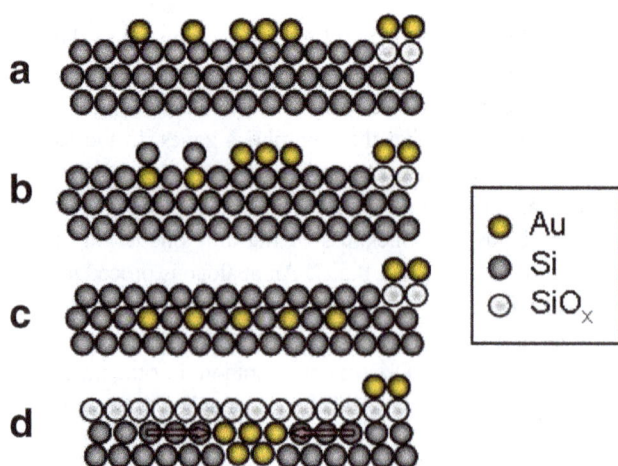

Fig. 8 Schematic summarizing the gold nanocluster characteristics during deposition (**a** to **c**) and aging (**d**). **a** Gold deposited on clean Si. **b** Au-Si interdiffusion all over the surface except where there is residual or regrown SiO$_x$. **c** The intermixing creates a surface Si layer and subsurface amorphous agglomerates of Au-Si, forming a stable stoichiometry of 70 %. **d** After the deposition and during the weeks later, Au starts to diffuse segregating in crystalline dots aligned with some crystallographic orientations of Si and occupying smaller and less dense areas

special specimen preparation, Au does not undergo to thermal treatment, except the 20–23 °C temperature exposure, and does not modify its original morphological characteristics. It should be noted that the data relative to the aged samples in Fig. 7, takes only into account the areal projection of the gold alloys and does not take into account their thickness, so they have to be understood as an estimation. From our data, we can then conclude that the Au/Si interdiffusion in the as deposited samples takes place also in case of nanoislands, when in contact with Si producing amorphous alloys of about 70:30 Au:Si composition and that after aging, it tends to the 100 % Au composition realigned in both <100> and <200> orientations. From Figs. 5 and 7, it can also be concluded that the coverage of Au interacted with Si decreases after aging, and this is correlated to the relatively slow spinodal decomposition kinetics occurring at RT (which is well below the eutectic temperature T_e=363 °C). Considering the equilibrium phases diagram and the Au-Si system, this process should cause after a certain time the complete separation of the mixed AuSi region in the two elemental phases which nucleate (in our case only pure Au nucleates) and grow at expenses of the intermixed region. However, the shrinking of the amorphous Au-Si nanostructures in size and density can be followed in our aging experiment since the mobility of the phase boundary is low due to the small temperature and the surface proximity.

Figure 8 is a schematic summarizing the results found in this work, supported by literature findings. First, gold is deposited on clean Si, and the gold atoms are randomly placed over the surface (a). Gold starts mixing with Si by interdiffusion. Due to some residual or regrown local SiO$_x$ the gold deposited on the oxide does not interact with Si and form

nuclei for stable crystalline dots (b). The intermixing creates a surface Si layer and subsurface amorphous islands of mixed Au-Si, which tend to form a stable stoichiometry of 70 % [22] (c). Figure 8a, c refer to processes happening during the growth. After the sample is taken out of the chamber, gold starts to diffuse segregating in crystalline islands aligned with some crystallographic orientations of Si and occupying smaller and less dense areas through a process similar to ripening. The exchange of gold atoms between the crystalline dots and the subsurface agglomerates is inhibited by the presence of contaminants (d).

If we compare the sample characteristics after 99 days of aging, Fig. 3d, and the sample prepared by using the standard TEM preparation, Fig. 2a, we see that they are similar, i.e., the areas covered by the interacted Au/Si phase are similar in size and shape and both present crystalline Au segregated regions. The only difference is that the gold clusters in the second case present a smaller density and are placed over the interacted regions. This suggests that the 150 °C annealing or the ion milling, can have promoted the "interaction" between the agglomerated clusters and the subsurface mixed phases, meaning that the barrier in this case has been removed or it was absent. However, the result let to understand that TEM, with the proper specimen preparation, is essential to characterize small amounts of alloyed Au/Si islands, in the early film growth even if they are still amorphous and before the elementary surface processes, such as surface diffusion, chemical reaction, or phase transition occur. The presence of islands growth of gold and gold alloys, and their evolution helps to explain also the controversy present in literature about the composition of the top part of the intermixed phase. The data indeed depend (1) on the initial gold coverage: for low coverage the Si signal coming from the matrix prevails, while for high coverage the signal of the pure and alloyed Au is present; (2) on the aging time: in the "fresh" case, both amorphous alloyed and crystalline pure Au phases (with random orientations) can be found; after aging, predominantly crystalline phases (either random or aligned) of pure gold immersed in a silicon matrix are contributing, together with some residual amorphous phases.

Conclusions

In this manuscript, a study on the morphological and structural characteristics of gold nanodots deposited by sputtering on Si <100> substrates and their time evolution during storage at room temperature is reported. This material is interesting for a number of applications such as light harvesting through plasmonic effects, metallic contacts in standard miniaturized devices, and catalysis of Si-NWs. As it is known in literature, the performance of all these materials in the final devices depend on the Au/Si system morphology and characteristics. The time evolution at room temperature of gold sputtered samples is monitored in this paper by TEM analysis, from a few hours up to 2.4×10^3 h after deposition. The variation of density and size of gold nanodots are obtained averaging on a large statistics.

The characterization shows the presence of two phases for Au nanostructures after deposition, i.e., (1) gold nanodots with diameter of few nanometers that do not have interaction with substrate and (2) amorphous nanophases of intermixed Au/Si. The structural characteristics of gold nanodots do not show a sensitive variation with time, attributed to the fact that locally, the presence contaminants slows down or inhibits the interdiffusion with Si. Au/Si amorphous agglomerates evolve, and their coverage rapidly decreases in the first 15 days after the deposition. Diffraction patterns of the interacted regions show that during the evolution, these areas tend to segregate forming pure gold crystalline regions aligned with some crystallographic orientations of Si (Au(110)||Si(100) with Au[002]||Si[022] and Au[002]||Si[02–2]).

We can conclude that the nanostructured Au-Si system evolves rapidly with time and this demonstrates the importance of analyzing the samples immediately after their deposition and without any modification induced by low thermal budget processes.

Acknowledgments The authors wish to thank Markus Italia, Salvatore Di Franco, and Rosa Ruggeri (CNR-IMM) for their valuable work and help during the samples preparation and analysis, and F. Ruffino (University of Catania) for RBS analysis.

References

1. Atwater HA, Polman A (2010) Plasmonics for improved photovoltaic devices. Nat Mater 9:205–213.
2. Catchpole KR, Polman A (2008) Plasmonic solar cells. Opt Express 6(26):21793
3. Catchpole KR, Polman (2008) A design principles for particle plasmon enhanced solar cells. Appl Phys Lett 93:191113.
4. Sivakov V, Andrä G, Himcinschi C, Gösele U, Zahn DRT, Christiansen S (2006) Growth peculiarities during vapor–liquid–solid growth of silicon nanowhiskers by electron-beam evaporation. Appl Phys A 85:311–315.
5. Irrera A, Pecora EF, Priolo F (2009) Control of growth mechanisms and orientation in epitaxial Si nanowires grown by electron beam evaporation. Nanotechnology 20:135601.
6. Garozzo C, Puglisi RA, Bongiorno C, Scalese S, Rimini E, Lombardo S (2011) Selective diffusion of gold nanodots on nanopatterned substrates realized by self-assembly of diblock copolymers. J Mater Res 26:240–246.
7. Garozzo C, La Magna A, Mannino G, Privitera V, Scalese S, Sberna PM, Simone F, Puglisi RA (2013) Competition between uncatalyzed and catalyzed growth during the plasma synthesis of Si nanowires and its role on their optical properties. J Appl Phys 113:214313.
8. Sivakov VA, Brönstrup G, Pecz B, Berger A, Radnoczi GZ, Krause M, Christiansen SH (2010) Realization of vertical and zigzag single crystalline silicon nanowire architectures. J Phys Chem C 114:3798–3803.
9. Schmidt V, Senz S, Gösele U (2005) Diameter-dependent growth direction of epitaxial silicon nanowires. Nano Lett 5(5):931–935.
10. Huang Z, Shimizu T, Senz S, Zhang ZX, Lee W, Geyer N, Gösele U (2009) Ordered arrays of vertically aligned [110] silicon nanowires by suppressing the crystallographically preferred <100> etching directions. Nano Lett 9:2519.
11. Chang S-W, Chuang VP, Boles ST, Thompson CV (2010) Metal-catalyzed etching of vertically aligned polysilicon and amorphous silicon nanowire arrays by etching direction confinement. Adv Funct Mater 20(24):4364–4370.
12. Milazzo RG, D'Arrigo G, Spinella C, Grimaldi MG, Rimini E (2012) Ag-assisted chemical etching of (100) and (111) n-type silicon substrates by varying the amount of deposited metal. J Electrochem Soc 159(9):D521–D525.
13. Akhtari-Zavareh A, Li W, Maroun F, Allongue P, Kavanagh KL (2013) Improved chemical and electrical stability of gold silicon contacts via epitaxial electrodeposition. J Appl Phys 113:063708.
14. Hiraki A, Lugujjo E, Mayer JW (1972) Formation of silicon oxide over gold layers on silicon substrates. J Appl Phys 43:3643.
15. Okuno K, Ito T, Iwami M, Hiraki A (1980) Presence of critical Au-film thickness for room temperature interfacial reaction between Au(film) and Si(crystal substrate). Solid State Commun 34:493–497.
16. Andersson TG (1982) The initial growth of vapour deposited gold films. Gold Bull 15(1):7–18.
17. Hiraki A, Kim SC, Imura T, Iwami M (1979) Si(LMM) Auger electron emission from Si alloys by keV Ar$^+$ ion bombardment, new effect and application. Jpn J Appl Phys 18:1767–1772.
18. Braicovich L, Garner CM, Skeath PR, Su CY, Chye PW, Lindau I, Spicer WE (1979) Photoemission studies of the silicon-gold interface. Phys Rev B 20:5131–5141.
19. Chang CA, Ottaviani G (1984) Outdiffusion of Si through gold films: the effects of Si orientation, gold deposition techniques and rates, and annealing ambients. Appl Phys Lett 44:901.
20. Puglisi RA, Nicotra G, Lombardo S, Spinella C, Ammendola G, Gerardi C (2005) Partial self-ordering observed in silicon nanoclusters deposited on silicon oxide substrates by chemical vapor deposition. Phys. Rev. B 71 (12)
21. Bal JK, Hazra S (2007) Interfacial role in room-temperature diffusion of Au into Si substrates. Phys Rev B 75:20541.
22. Hiraki A, Iwami M (1974) Electronic structure of thin gold film deposited on silicon substrate studied by Auger electron and X-ray photoelectron spectroscopies. Jpn J Appl Phys 2(Supplement 2–2):749–752
23. Hiraki A, Shimizu A, Iwami M, Narusawa T, Komiya S (1975) Metallic state of Si in Si noble metal vapor quenched alloys studied by Auger electron spectroscopy. Appl Phys Lett 26:57.
24. Kim JH, Yang G, Yang S, Weiss AH (2001) Study of the growth and stability of ultra-thin films of Au deposited on Si(1 0 0) and Si(1 1 1). Surf Sci 475:37–46.

24

Generation of gold nanoparticles according to procedures described in the eighteenth century

Álvaro Mayoral · Javier Agúndez ·
Ignacio Miguel Pascual-Valderrama · Joaquín Pérez-Pariente

Abstract Gold nanoparticles have received much attention in recent years due to their unique size-dependent properties, as they find useful applications in materials science [Mayoral et al. (Nanoscale 2:335–342, 2010)], catalysis [Schwerdtfeger (Angew Chem Int Ed 42:1892–1895, 2003)] [Hashmi and Hutchings (Angew Chem Int Ed 45:7896–7936, 2006)] and biology [Sperling et al. (Chem Soc Rev 37:1896–1908, 2008)]. The preparation of such nanoparticles benefits from modern chemical knowledge, and a large variety of several procedures have been developed aiming at controlling the size and shape of these metal nanoparticles. Here, we show that two eighteenth-century recipes (Online Resource 1) used at that time to prepare drinkable solutions of gold, used as drugs, actually generate gold nanoparticles, clusters and even monoatomic species of gold. These simple methods involve the dissolution of gold in a solution of ammonium chloride in nitric acid (*aqua regia*) and the mixing of the resulting solution with rosemary or cinnamon essential oils. The complex mixture of compounds resulting from the fast reaction between *aqua regia* and the essential oils behave simultaneously as reductants and stabilisers of the nascent gold particles. These results not only prove that historical speculations on the presence of finely divided gold particles floating in these solutions were basically correct but they could also serve as a source of inspiration for new experimental approaches procuring the generation of stable sub-nanometer gold nanoparticles.

Keywords Gold nanoparticles · STEM · Eighteenth-century recipes · Essential oil

There is a general agreement that the first scientific approach to the nature of what are now known as solutions of colloidal gold can be traced back to the works of the British scientist Michael Faraday. In a seminal paper published in 1857 [1, 2], he clearly concluded that the ruby or purple colours exhibited by solutions containing gold were due to the presence of very small particles of this metal, so small that they were invisible to the human eye, yet they could be easily detected by their effect in dispersing incident light.

As important as this publication is as pioneering the foundation of modern colloid chemistry, Faraday was not the first in preparing such ruby-coloured solutions neither in suggesting the presence in them of gold in such an extremely divided state.

There is a long tradition behind the preparation of colloidal gold, which can be briefly summarised as follows. Historically, two main research lines were developed aiming at two different purposes. One of them was directed toward the preparation of colloidal gold to be used for colouring the transparent glass into red colour, resulting in the much sought, valuable and beautiful ruby glass [3]. The second approach belongs to the field of medicine, where the goal was to prepare a gold drinkable solution which behaved as a powerful drug able to cure even the most severe ailments, the so-called potable gold, known in old Latin texts as *aurum potabile* [4, 5]. Although its origin can be traced back to the Middle Ages,

Á. Mayoral
Laboratorio de Microscopías Avanzadas,
Instituto de Nanociencia de Aragón, Universidad de Zaragoza,
Mariano Esquillor, Edificio I+D, 50018 Zaragoza, Spain

J. Agúndez · J. Pérez-Pariente (✉)
Instituto de Catálisis y Petroleoquímica,
Marie Curie, 2, Cantoblanco 28049, Madrid, Spain
e-mail: jperez@icp.csic.es

I. M. Pascual-Valderrama
I. E. S. Lope de Vega, San Bernardo, 70, Madrid, Spain

the preparation of potable gold was still a subject of interest in the eighteenth century.

The editor and commentator of the 1757 edition of the French chemist and pharmacist Nicholas Lemery's *Course of Chemistry*, M. Baron, included in this edition a recipe that described the preparation of the so-called potable gold of Mademoiselle Grimaldi [6]. Few years later, the French chemist Pierre-Joseph Macquer commented extensively on the recipe in his *Dictionary of Chemistry* (1766) [7]. He argued that the red solution resulting from the Grimaldi's procedure contains elemental gold in a state of extreme division, and the particles were "floating in the oily fluid" [8].

Would it be possible that the Grimaldi's recipe actually generates colloidal gold, i.e., gold nanoparticles and/or clusters? Could the conclusion of Macquer be experimentally corroborated, anticipating in this way for one century the Faradays' view? The Grimaldi's procedure is described with enough detail as to be suitable for replication in a modern laboratory, and moreover, as we will see, it involves the use of natural products, plant-derived organic substances, which would make it particularly appealing in the view of decreasing the input of synthetic molecules, which in many cases come from more complicated reactions involving different chemicals and producing a large amount of unwanted side products, in the growing field of eco-friendly nanotechnology.

The procedure described in the Lemery's treatise starts by dissolving gold in *aqua regia*. Then, the resulting golden-yellow solution is mixed with rosemary essential oil, and the mixture is let to stand for a while, in such a way that two layers are formed. The aqueous lower layer loses rapidly its colour, while the upper one is coloured in yellow. From this observation, the author of the recipe concluded that the gold was being taken up from the aqueous phase to the supernatant rosemary oil phase. Then, a portion of this oily layer is dissolved in alcohol (ethanol), which results in a beautiful red solution, which is the Grimaldi's potable gold, the solution that Macquer supposed shall contain gold in a state of extreme division.

We have replicated this procedure by using the same proportion of the recommend reagents, assuming that the *aqua regia* was prepared by mixing concentrated nitric acid and ammonium chloride (see "Experimental" section). By following this procedure, a dark-brown supernatant solution is immediately obtained after mixing the rosemary oil and the gold solution placed in a separating funnel (Online Resource 2). Twenty-four hours later, golden particles are observed in the interphase between the two layers, the upper oil and the lower aqueous phases, while the yellow colour of the aqueous phase is much less intense (Online Resource 3). This behaviour suggests that the gold has been driven from the aqueous to the oily phase, but at least part of it is reduced again to the metal state and concentrates just at the interphase.

The blackening of the oily phase is not described in the 1757 recipe. We then decided to let the mixture stand still to observe the evolution of the system. The oily layer evolved with time; and after 15 days, its colour was dark red, while that of the aqueous lower layer was pale yellow. At this point, we took an aliquot of the oily red phase, which was readily dissolved in absolute ethanol, while the aqueous phase was also sampled to analyse the presence of gold in there. Chemical analysis revealed that more than 99 % of the gold initially present in the acid aqueous phase had been removed, which was in agreement with the observation described in the old recipe. Moreover, the transparent ethanol solution exhibited an orange-red colour with a concentration of gold of 220 mg/L. (Online Resource 4). The replication of the Grimaldi's recipe reported here is in this way more complete and different than what Vanino reported in 1906, where he used rosemary oil as a reducing agent of a diluted solution of gold chloride in water without the intervention of *aqua regia* [9].

Traditionally, the analysis and therefore the observation of the nanoparticles presumably formed was an extremely complicated task due to the low resolution of the available transmission electron microscopes. Nowadays, with the implementation of the spherical aberration (C_s) correctors [10, 11], sub-angstrom resolution can be commonly attainable allowing imaging small clusters or even single atoms [12]. For the current analysis, we used an aberration-corrected scanning transmission electron microscope (STEM), which is the most appropriate technique for imaging heavy elements as gold. A drop of the as-synthesised ethanol solution was placed onto a holey carbon copper microgrid, which was subsequently let to dry. Further details can be found in the "Experimental" section.

Two main kinds of gold species were observed in the solution. Gold nanoparticles of an average size of 5 nm, being the smallest one of 2 nm and the largest 19 nm, but 90 % fall in the range of 2 to 8 nm, see Fig. 1. The most typical type of nanoparticles formed (3 to 4 nm) mainly adopted decahedral

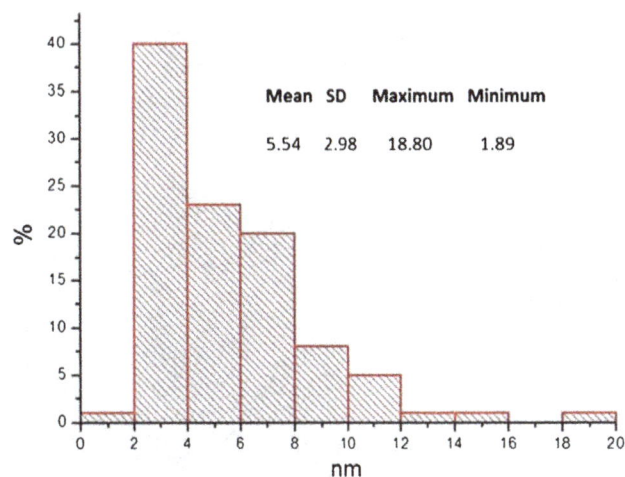

Fig. 1 Histogram of gold nanoparticle size distribution

Fig. 2 Cs-corrected STEM-HAADF images of three different types of Au nanoparticles **a** decahedra, **b** icosahedra and **c** *fcc* symmetry with the correspondent FFT diffractograms at the bottom

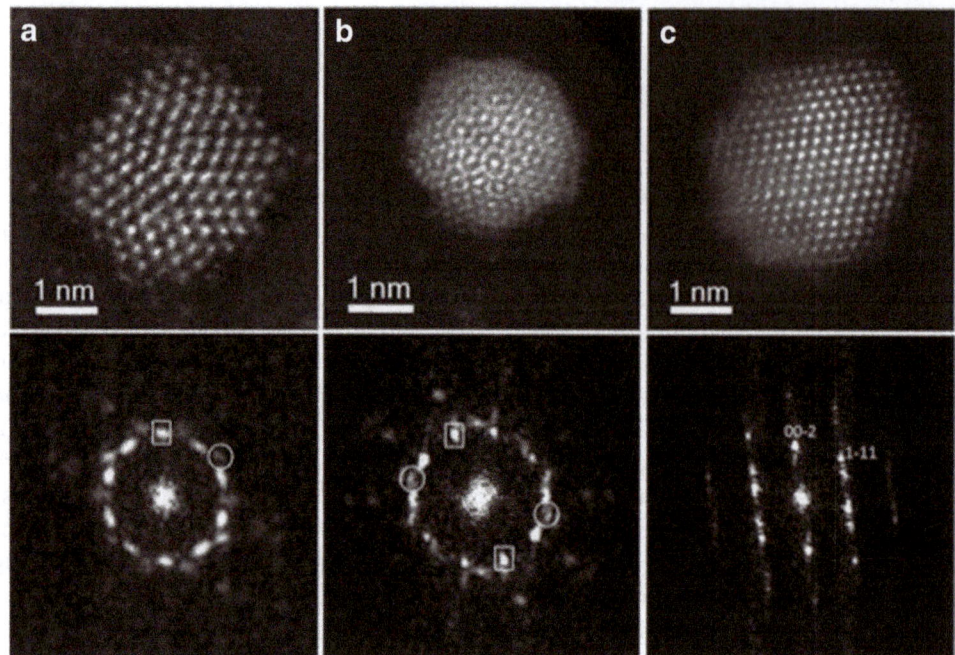

or icosahedral morphology, while *fcc*-twinned nanoparticles [13] are less abundant. Figure 2a displays a 2.7 nm Au decahedron on its 5-fold axes displaying two main crystallographic distances of 2.37 and 2.07 Å corresponding to the (111) and (200) planes, respectively. The fast Fourier transform (FFT) corroborating the 5-fold symmetry is shown at the bottom. Figure 2b corresponds to a 3.45 nm Au icosahedron also displaying its 5-fold symmetry axes, both FFTs exhibits the (111) and (200) reflections, marked with rectangles and circles, respectively. Figure 2c corresponds to a twinned *fcc* 4.2 nm nanoparticle along the [110] orientation. The second clearly differenced type of gold species observed correspond to small disordered clusters of about 1–1.5 nm, which also appeared together with isolated gold atoms (Fig. 3).

The recipe reported in the Lemery's treatise is not the only one described in academic texts in the eighteenth century on the preparation of gold solutions obtained by means of essential oils. The German physician F. Hoffmann reported in 1722 [14] the preparation of a "gold tincture" by mixing a solution of gold in *aqua regia* with a mixture of cinnamon essential oil

and ethanol under gentle heating. A dark-brown, resin-like product is obtained, from which a supernatant solution can be separated (Online Resource 5). The examination of this product by TEM reveals the presence of gold nanoparticles that reached sizes up to 20 nm. Different structural conformations of gold could be observed, which varied from multi-twinned crystals (Fig.4a), single crystal *fcc* nanoparticles, icosahedra (Fig.4b, c) and small clusters of few atoms surrounded by isolated gold atomic species (Fig. 4d).

These two procedures render quite similar results and confirm the Macquer's hypothesis regarding the presence of extremely divided gold.

It is noticeable the presence in these solutions of very small clusters of gold and even isolated atoms of this metal, which points out to a remarkable role of the oily phase resulting from the fast chemical reaction between the *aqua regia* and the essential oils in preventing the overgrowth of the primary gold particles into larger aggregates. In this way, the oily phase behaves at the same time as a reducing and a very efficient capping agent [15–17], leading to a simple method to prepare

Fig. 3 C$_s$-corrected STEM-HAADF images of **a** several Au clusters, **b** a magnified image of a 1-nm Au cluster and **c** Au cluster together with several isolated Au atoms

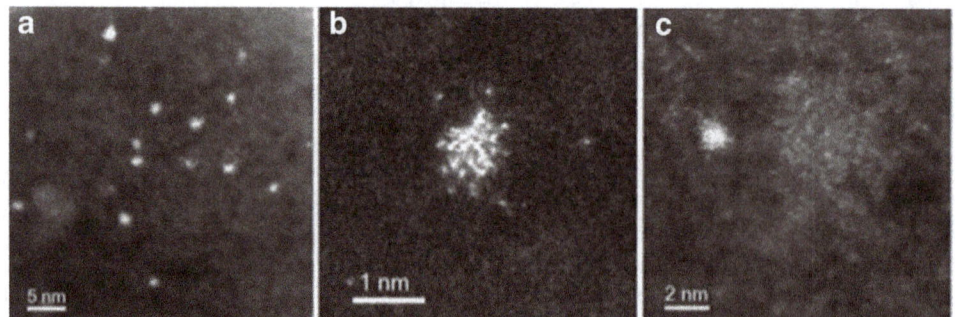

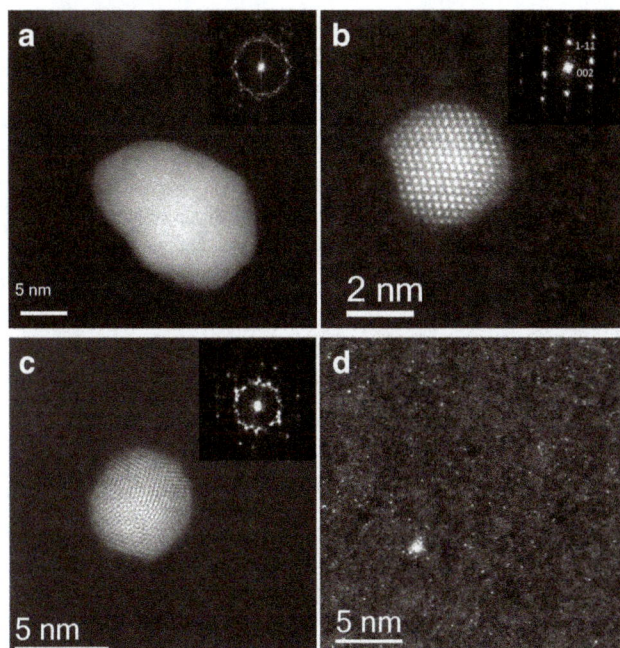

Fig. 4 C_s-corrected STEM-HAADF images of **a** multi-twinned crystal, **b–c** single crystal *fcc* nanoparticles and icosahedra and **d** small Au clusters of few atoms surrounded by isolated gold atomic species

gold nanoparticles constituted by a very small number of atoms and even single-atom species. This effect of the essential oils shall be related to their specific chemical composition and the way their individual chemical components react with the *aqua regia* to yield complex mixtures of oxidation compounds. In this regard, it is worth mentioning that the rosemary essential oil used in this work contains a mixture of terpenic hydrocarbons dominated by alpha-pinene and camphene and the oxygen-containing terpenes 1,8-cineole and camphor, while the major component of the cinnamon essential oil is eugenol (see "Experimental" section).

This aspect would be worth of further investigation, for it could inspire new experimental approaches toward the generation of stable sub-nanometer gold nanoparticles, which have been reported to be extremely active in the oxidation of organic substrates with oxygen [18].

Experimental section

Replication of the Hoffmann's recipe

A gold nugget (0.154 g) was dissolved in 18.5 g of *aqua regia* (NH_4Cl/HNO_3; 1:4 by mass; HNO_3, 65 wt%). The mixture was stirred in a 50-mL glass beaker placed in a sand bath at 55°C. A solution of essential cinnamon oil in ethanol (1:3 by mass) was added (essential oil solution/Au solution 3:1 by mass). Some bubbles were immediately produced causing violent jets of liquid shooting the walls of the beaker, while the brown colour of the mixture becomes darker. In a few

minutes, it evolved to a black viscose substance, as described in the original recipe, although careful examination revealed that it was in fact brown-reddish. The solution was cooled down to room temperature, and ethanol was added to dissolve a portion of the dense substance (2:1 by weight). The mixture was separated by decantation, and the supernatant liquid was observed under the transmission electron microscope to search for gold nanoparticles in it.

Replication of the Lemery's recipe

A gold nugget (0.470 g) was dissolved in 15.0 g of aqua regia (NH_4Cl/HNO_3; 1:4 by mass; HNO_3, 65 wt%). The resulting golden solution was placed in a separating funnel, to which 7.5 g of rosemary essential oil were added. The oil became immediately dark brown by mixing. The system is let to stand for 15 days; and after that, the oily phase gets dark red. Then, a portion of the resulting red supernatant solution is taken and mixed with ethanol (1:5, *w/w*). This solution is examined by TEM according to the method described in the paper.

TEM images

Electron microscopy was performed in a spherical aberration (C_s)-corrected transmission electron microscopy, Titan X-FEG FEI 60–300 kV (operated at 300 kV) equipped with a monochromator and a Gatan 2 k×2 k CCD camera. In addition to the STEM unit, the microscope is also fitted with a CEOS C_s corrector for the electron probe capable of achieving a resolution of 0.8 Å. For analytical measurements, the microscope is equipped with a Gatan Energy Filter Tridiem 866 ERS and an EDS detector.

GC-MS analysis of essential oils

The essential oils (rosemary and cinnamon) were analysed by gas chromatography-mass spectrometry (GC-MS) employing a gas chromatograph (Agilent 6890) coupled with a mass spectrometer (Agilent 5973 N) using a capillary column made of methylpolysiloxane (30 m×0.25 mm×0.25 μm), heating from 70 to 290 °C at 6°C/min. The composition (wt%) of the rosemary oil was 24.9 % 1,8-cineole, 21.9 % *alpha*-pinene, 20.91 % camphor, 9.06 % camphene, 3.81 % borneol, 3.34 % verbenone, 2.59 % myrcene, 2.41 % *beta*-pinene, 2.01 % caryophyllene, 2.00 % *p*-cymene, 1.16 % *alpha*-humulene, 0.98 % bornyl acetate, 0.94 % *gamma*-terpinene, 0.63 % 4-terpineol, 0.63 % *alpha*-terpineol, 0.29 % *alpha*-terpinolene, and 0.28 % fenchone. The composition (wt%) of the cinnamon oil was 87.96 % eugenol, 3.14 % caryophyllene, 1.13 % *alpha*-pinene, 0.76 % *alpha*-cubebene, 0.35 % a-humulene, 0.26 % *beta*-pinene, 0.24 % linalool, 0.20 % *p*-cymene, 0.13 % methylphenyl benzoate, and 0.12 % safrol.

Acknowledgements The research leading to these results has received funding from the European Union Seventh Framework Programme under Grant Agreement 312483-ESTEEM2 (Integrated Infrastructure Initiative-I3). The Zentralbibliothek Zürich (service of Handschriften, Musik und Alte Drucke) is acknowledged for giving us access to the first edition of the book Observationum Physico-chymicarum selectiorum libri III, Haal, 1722, by F. Hoffman, and Dr. Ana B. Pinar is acknowledged for obtaining copies of the Observatio XIII and Observatio XXI of this book. We are also grateful to Dr. Hereward Tilton (Department of History, University of Exeter) for providing us with copies of some of the references. Prof. V. Fernández Herrero is acknowledged for his assistance in translating into English some of the cited references that are written in German. We are grateful to Dr. A. Cristina Soria Monzón and Prof. Jesús Sanz (Institute of Organic Chemistry, CSIC) for the GC-MS analyses of the essential oils. The help of Prof. Ermias Dagne (Dept of Chemistry, Addis Ababa University) in preliminary analyses of the oils is also acknowledged. Dr. L. Gómez-Hortigüela is acknowledged for the pictures of the rosemary oil experiment included in the extended data.

References

1. Faraday M (1857) The Bakerian Lecture: experimental relations of gold (and other metals) to light. Philos Trans R Soc Lond 147:145–181
2. Tweney RD (2006) Discovering discovery: how Faraday found the first metallic colloid. Perspect Sci 14:97–121
3. Hunt LB (1976) The true story of Purple of Casius. Gold Bull 9:134–139
4. Hauser EA (1952) Aurum potabile. J Chem Educ 29:456–458
5. Highby GJ (1982) Gold in medicine. Gold Bull 15:130–140
6. Lemery N (1757) Cours de chymie, Paris.
7. Macquer PJ (1766) Dictionnaire de chymie. Paris, pp 174–177
8. Svedberg T (1921) The formation of colloids. J. and A. Churchill, London, p 74
9. Vanino L, Hartl F (1906) Ueber die Bildung colloïdaler Goldlösungen mittels ätherischer Oele. Chem Ber 39:1696–1700
10. Krivanek OL, Delby N, Lupini AR (1999) Proceedings of the International Workshop towards Atomic Resolution Analysis—Port Ludlow, Washington, USA, 6–11 September 1998—Part 1: Techniques and instrumentation—Foreword. Ultramicroscopy 78:1–11
11. Haider M, Uhlemann S, Zach J (2000) Upper limits for the residual aberrations of a high-resolution aberration-corrected STEM. Ultramicroscopy 81:163–175
12. Mayoral A, Blom DA, Mariscal MM, Guiterrez-Wing C, Aspiazu J, Jose-Yacaman M (2010) Gold clusters showing pentagonal atomic arrays revealed by aberration-corrected scanning transmission electron microscopy. Chem Commun 46:8758–8760
13. Mayoral A, Barron H, Estrada-Salas R, Vazquez-Duran A, Jose-Yacaman M (2010) Nanoparticle stability from the nano to the meso interval. Nanoscale 2:335–342
14. Hoffmann F (1722) Observationum physico-chymicarum selectiorum. Hale, pp 375
15. Dumur F, Guerlin A, Dumas E, Bertin D, Gigmes D, Mayer CR (2011) Controlled spontaneous generation of gold nanoparticles asited by dual reducing and capping agents. Gold Bull 44:119–137
16. Brust M, Walker M, Bethel D, Schiffrin DJ, Whyman R (1994) Synthesis of thiol-derivatised gold nanoparticles in a two-phase liquid-liquid system. Chem Commun 801–802
17. Liz-Marzán LM (2013) Gold-nanoparticles research before and after Brust-Schiffrin method. Chem Commun 49:16–18
18. Corma A, Concepcion P, Boronat M, Sabater MJ, Navas J, Yacaman MJ, Larios E, Posadas A, Lopez-Quintela MA, Buceta D, Mendoza E, Guilera G, Mayoral A (2013) Exceptional oxidation activity with size-controlled supported gold clusters of low atomicity. Nat Chem 5:775–781

Permissions

List of Contributors

Sónia A. C. Carabineiro
Laboratory of Catalysis and Materials, Department of Chemical Engineering, Faculty of Engineering, University of Porto, Rua Dr. Roberto Frias, s/n, 4200-465 Porto, Portugal

Martin Makosch
Institute for Chemical and Bioengineering, ETH Zurich, Wolfgang-Pauli Strasse, 8093 Zurich, Switzerland

Václav Bumbálek
Institute for Chemical and Bioengineering, ETH Zurich, Wolfgang-Pauli Strasse, 8093 Zurich, Switzerland

Jacinto Sá
Paul Scherrer Institute (PSI), Villigen, Switzerland

Jeroen A. van Bokhoven
Institute for Chemical and Bioengineering, ETH Zurich, Wolfgang-Pauli Strasse, 8093 Zurich, Switzerland

Paolo Battaini
8853 Spa, Pero, Milan, Italy

Edoardo Bemporad
Mechanical and Industrial Engineering Department, University of Rome "Roma Tre", Rome, Italy

Daniele De Felicis
Mechanical and Industrial Engineering Department, University of Rome "Roma Tre", Rome, Italy

Miguel Peixoto de Almeida
Laboratory of Catalysis and Materials (LCM), Associate Laboratory LSRE/LCM, Faculdade de Engenharia, Universidade do Porto, Rua Dr. Roberto Frias, 4200-465 Porto, Portugal

Sónia A. C. Carabineiro
Laboratory of Catalysis and Materials (LCM), Associate Laboratory LSRE/LCM, Faculdade de Engenharia, Universidade do Porto, Rua Dr. Roberto Frias, 4200-465 Porto, Portugal

Meng Shan
Key Laboratory of Analytical Chemistry for Life Science of Shaanxi Province, School of Chemistry and Chemical Engineering, Shaanxi Normal University, Xi'an 710062, People's Republic of China

Min Li
Key Laboratory of Analytical Chemistry for Life Science of Shaanxi Province, School of Chemistry and Chemical Engineering, Shaanxi Normal University, Xi'an 710062, People's Republic of China

Xiaoying Qiu
Key Laboratory of Analytical Chemistry for Life Science of Shaanxi Province, School of Chemistry and Chemical Engineering, Shaanxi Normal University, Xi'an 710062, People's Republic of China

Honglan Qi
Key Laboratory of Analytical Chemistry for Life Science of Shaanxi Province, School of Chemistry and Chemical Engineering, Shaanxi Normal University, Xi'an 710062, People's Republic of China

Qiang Gao
Key Laboratory of Analytical Chemistry for Life Science of Shaanxi Province, School of Chemistry and Chemical Engineering, Shaanxi Normal University, Xi'an 710062, People's Republic of China

Chengxiao Zhang
Key Laboratory of Analytical Chemistry for Life Science of Shaanxi Province, School of Chemistry and Chemical Engineering, Shaanxi Normal University, Xi'an 710062, People's Republic of China

Jian Zhu
The Key Laboratory of Biomedical Information Engineering of Ministry of Education, School of Life Science and Technology, Xi'an Jiaotong University, Xi'an, 710049, People's Republic of China

Fan Zhang
The Key Laboratory of Biomedical Information Engineering of Ministry of Education, School of Life Science and Technology, Xi'an Jiaotong University, Xi'an, 710049, People's Republic of China

Jian-Jun Li
The Key Laboratory of Biomedical Information Engineering of Ministry of Education, School of Life Science and Technology, Xi'an Jiaotong University, Xi'an, 710049, People's Republic of China

Jun-Wu Zhao
The Key Laboratory of Biomedical Information Engineering of Ministry of Education, School of Life Science and Technology, Xi'an Jiaotong University, Xi'an, 710049, People's Republic of China

Kun Luo
Key Laboratory of New Processing Technology for Nonferrous Metals and Materials, Ministry of Education, College of Materials Science and Engineering, Guilin University of Technology, Guilin 541004, China

Haiming Wang
Key Laboratory of New Processing Technology for Nonferrous Metals and Materials, Ministry of Education, College of Materials Science and Engineering, Guilin University of Technology, Guilin 541004, China

Xiaogang Li
Key Laboratory of New Processing Technology for Nonferrous Metals and Materials, Ministry of Education, College of Materials Science and Engineering, Guilin University of Technology, Guilin 541004, China

M. del P. Rodríguez-Torres
Grupo de Espectroscopia de Materiales Avanzados yNanoestructurados (GEMANA), Centro de Investigaciones en Optica A. C., León Gto. 37150, Mexico

Luis Armando Díaz-Torres
Grupo de Espectroscopia de Materiales Avanzados yNanoestructurados (GEMANA), Centro de Investigaciones en Optica A. C., León Gto. 37150, Mexico

Pedro Salas
Centro de Física Aplicada y Tecnología Avanzada, Universidad Nacional Autónoma de México, A.P. 1-1010, Querétaro 76000, Mexico

Claramaría Rodríguez-González
Centro de Física Aplicada y Tecnología Avanzada, Universidad Nacional Autónoma de México, A.P. 1-1010, Querétaro 76000, Mexico

Martin Olmos-López
Grupo de Espectroscopia de Materiales Avanzados yNanoestructurados (GEMANA), Centro de Investigaciones en Optica A. C., León Gto. 37150, Mexico

Pierre-Jean Debouttière
CNRS, LCC (Laboratoire de Chimie de Coordination), 205 route de Narbonne, BP 44099, 31077 Toulouse Cedex 4, France
UPS, INPT, Université de Toulouse, 31077 Toulouse Cedex 4, France
RTRA "Sciences et Technologies pour l'Aéronautique et l'Espace", 31030 Toulouse, France

Yannick Coppel
CNRS, LCC (Laboratoire de Chimie de Coordination), 205 route de Narbonne, BP 44099, 31077 Toulouse Cedex 4, France
UPS, INPT, Université de Toulouse, 31077 Toulouse Cedex 4, France

Philippe Behra
INPT, INRA, Université de Toulouse, UMR 1010, ENSIACET, 4 allée Emile Monso, 31030 Toulouse CEDEX 4, France
LCA (Laboratoire de Chimie Agro-industrielle), INRA, 31030 Toulouse, France

Bruno Chaudret
INSA, CNRS, UPS, LPCNO, Université de Toulouse, 31077 Toulouse, France

Katia Fajerwerg
CNRS, LCC (Laboratoire de Chimie de Coordination), 205 route de Narbonne, BP 44099, 31077 Toulouse Cedex 4, France
UPS, INPT, Université de Toulouse, 31077 Toulouse Cedex 4, France

Laure Bertry
Laboratoire de Chimie de la Matière Condensée, UPMC, CNRS, Collège de France, UMR 7574, 11 place Marcelin Berthelot, 75005 Paris, France

Olivier Durupthy
Laboratoire de Chimie de la Matière Condensée, UPMC, CNRS, Collège de France, UMR 7574, 11 place Marcelin Berthelot, 75005 Paris, France

Patrick Aschehoug
Laboratoire de Chimie de la Matière Condensée, UPMC, CNRS, Chimie Paristech, UMR 7574, 11 rue Pierre et Maris Curie, 75005 Paris, France

Bruno Viana
Laboratoire de Chimie de la Matière Condensée, UPMC, CNRS, Chimie Paristech, UMR 7574, 11 rue Pierre et Maris Curie, 75005 Paris, France

Corinne Chanéac
Laboratoire de Chimie de la Matière Condensée, UPMC, CNRS, Collège de France, UMR 7574, 11 place Marcelin Berthelot, 75005 Paris, France

Xiaoyun Qin
Chemical Synthesis and Pollution Control Key Laboratory of Sichuan Province, School of Chemistry and Chemical Industry, China West Normal University, Nanchong 637002 Sichuan, China

Qingzhen Li
Chemical Synthesis and Pollution Control Key Laboratory of Sichuan Province, School of Chemistry and Chemical Industry, China West Normal University, Nanchong 637002 Sichuan, China

Abdullah M. Asiri
Chemistry Department, Faculty of Science, King Abdulaziz University, Jeddah 21589, Saudi Arabia

Center of Excellence for Advanced Materials Research, King Abdulaziz University, Jeddah 21589, Saudi Arabia

Abdulrahman O. Al-Youbi
Chemistry Department, Faculty of Science, King Abdulaziz University, Jeddah 21589, Saudi Arabia
Center of Excellence for Advanced Materials Research, King Abdulaziz University, Jeddah 21589, Saudi Arabia

Xuping Sun
Chemical Synthesis and Pollution Control Key Laboratory of Sichuan Province, School of Chemistry and Chemical Industry, China West Normal University, Nanchong 637002 Sichuan, China
Chemistry Department, Faculty of Science, King Abdulaziz University, Jeddah 21589, Saudi Arabia
Center of Excellence for Advanced Materials Research, King Abdulaziz University, Jeddah 21589, Saudi Arabia

Marco Demurtas
Interdisciplinary Biomedical Research Centre, School of Science and Technology, Nottingham Trent University, Clifton Lane, Nottingham NG11 8NS, UK

Carole C. Perry
Interdisciplinary Biomedical Research Centre, School of Science and Technology, Nottingham Trent University, Clifton Lane, Nottingham NG11 8NS, UK

Jie Wang
State Key Laboratory of Fine Chemicals, School of Chemical Engineering, Dalian University of Technology, Dalian 116024, People's Republic of China

Zhen-Hao Hu
State Key Laboratory of Fine Chemicals, School of Chemical Engineering, Dalian University of Technology, Dalian 116024, People's Republic of China

Yu-Xin Miao
State Key Laboratory of Fine Chemicals, School of Chemical Engineering, Dalian University of Technology, Dalian 116024, People's Republic of China

Wen-Cui Li
State Key Laboratory of Fine Chemicals, School of Chemical Engineering, Dalian University of Technology, Dalian 116024, People's Republic of China

Olivier Pluchery
Institut des NanoSciences de Paris, UMR 7588 CNRS, Université Pierre etMarie Curie-UPMC, 4 place Jussieu, 75252 Paris Cedex 05, France

Hynd Remita
Laboratoire de Chimie Physique, UMR 8000 CNRS, Université Paris-Sud, 91405 Orsay Cedex, France

Delphine Schaming
ITODYS, UMR 7086 CNRS, Sorbonne Paris Cité, Université Paris Diderot, 15 rue Jean-Antoine de Baïf, 75205 Paris Cedex 13, France

Md. Abdul Aziz
Center of Research Excellence in Nanotechnology, King Fahd University of Petroleum and Minerals, Dhahran 31261, Saudi Arabia

Jong-Pil Kim
Department of Material Chemistry, Graduate School of Engineering, Kyoto University, Nishikyo-ku, Kyoto 615-8520, Japan
Surface Properties Research Team, Korea Basic Science Institute Busan Center, Busan 609-735, South Korea

Munetaka Oyama
Department of Material Chemistry, Graduate School of Engineering, Kyoto University, Nishikyo-ku, Kyoto 615-8520, Japan

Tung-Tso Ho
Institute of Polymer Science and Engineering, National Taiwan University, No. 1, Sec. 4 Roosevelt Road, Taipei 10617 Taiwan, Republic of China
Institute of Biomedical Engineering, National Chung Hsing University, Taichung, Taiwan, Republic of China

Yu-Chun Lin
Department of Surgery, Fong-Yuan Hospital Department of Health Executive Yuan, Taichung, Taiwan, Republic of China

Shan-hui Hsu
Institute of Polymer Science and Engineering, National Taiwan University, No. 1, Sec. 4 Roosevelt Road, Taipei 10617 Taiwan, Republic of China

Xuhui Liu
Key Laboratory of Applied Surface and Colloid Chemistry, Ministry of Education, School of Chemistry and Chemical Engineering, Shaanxi Normal University, Xi'an 710062, China

Rui Zhang
Key Laboratory of Applied Surface and Colloid Chemistry, Ministry of Education, School of Chemistry and Chemical Engineering, Shaanxi Normal University, Xi'an 710062, China

Xiaqing Yuan
Key Laboratory of Applied Surface and Colloid Chemistry, Ministry of Education, School of Chemistry and Chemical Engineering, Shaanxi Normal University, Xi'an 710062, China

Lu Liu
Key Laboratory of Applied Surface and Colloid Chemistry, Ministry of Education, School of Chemistry and Chemical Engineering, Shaanxi Normal University, Xi'an 710062, China

Yiying Zhou
Key Laboratory of Applied Surface and Colloid Chemistry, Ministry of Education, School of Chemistry and Chemical Engineering, Shaanxi Normal University, Xi'an 710062, China

Qiang Gao
Key Laboratory of Applied Surface and Colloid Chemistry, Ministry of Education, School of Chemistry and Chemical Engineering, Shaanxi Normal University, Xi'an 710062, China

Blake J. Plowman
School of Applied Sciences, RMIT University, GPO Box 2476V, Melbourne, VIC 3001, Australia

Nathan Thompson
School of Applied Sciences, RMIT University, GPO Box 2476V, Melbourne, VIC 3001, Australia

Anthony P. O'Mullane
School of Applied Sciences, RMIT University, GPO Box 2476V, Melbourne, VIC 3001, Australia
School of Chemistry, Physics and Mechanical Engineering, Queensland University of Technology, GPO Box 2434, Brisbane, QLD 4001, Australia

Yu-Xin Miao
State Key Laboratory of Fine Chemicals, School of Chemical Engineering, Dalian University of Technology, Dalian 116024, People's Republic of China

Lei Shi
State Key Laboratory of Fine Chemicals, School of Chemical Engineering, Dalian University of Technology, Dalian 116024, People's Republic of China

Li-Na Cai
State Key Laboratory of Fine Chemicals, School of Chemical Engineering, Dalian University of Technology, Dalian 116024, People's Republic of China

Wen-Cui Li
State Key Laboratory of Fine Chemicals, School of Chemical Engineering, Dalian University of Technology, Dalian 116024, People's Republic of China

C. L. Gan
Spansion (Penang) Sdn Bhd, Phase II Free Industrial Zone, Penang 11900 Bayan Lepas, Malaysia

F. C. Classe
Spansion (Penang) Sdn Bhd, Phase II Free Industrial Zone, Penang 11900 Bayan Lepas, Malaysia

B. L. Chan
Spansion (Penang) Sdn Bhd, Phase II Free Industrial Zone, Penang 11900 Bayan Lepas, Malaysia

U. Hashim
Institute of Nanoelectronic Engineering (INEE), Universiti Malaysia Perlis, Perlis 01000, Kangar, Malaysia

Rebeka Rudolf
Faculty of Mechanical Engineering, University of Maribor, 2000 Maribor, Slovenia
Zlatarna Celje d.d., Celje, Slovenia

Sergej Tomić
Medical Faculty of the Military Medical Academy, University of Defence, 11000 Belgrade, Serbia

Ivan Anžel
Faculty of Mechanical Engineering, University of Maribor, 2000 Maribor, Slovenia

Tjaša Zupančič Hartner
Ortotip d.o.o., Murska Sobota, Slovenia

Miodrag Čolić
Medical Faculty of the Military Medical Academy, University of Defence, 11000 Belgrade, Serbia
Medical Faculty, University of Niš, 18000 Niš, Serbia

Mariana Erasmus
Department of Microbial, Biochemical and Food Biotechnology, Faculty of Natural and Agricultural Sciences, University of the Free State, Bloemfontein 9300, South Africa

Errol Duncan Cason
Department of Microbial, Biochemical and Food Biotechnology, Faculty of Natural and Agricultural Sciences, University of the Free State, Bloemfontein 9300, South Africa

Jacqueline van Marwijk
Department of Microbial, Biochemical and Food Biotechnology, Faculty of Natural and Agricultural Sciences, University of the Free State, Bloemfontein 9300, South Africa

Elsabé Botes
Department of Microbial, Biochemical and Food Biotechnology, Faculty of Natural and Agricultural Sciences, University of the Free State, Bloemfontein 9300, South Africa

Mariekie Gericke
Biotechnology Division, Mintek, Private Bag X3015, Randburg 2125, South Africa

Esta van Heerden
Department of Microbial, Biochemical and Food Biotechnology, Faculty of Natural and Agricultural Sciences, University of the Free State, Bloemfontein 9300, South Africa

C. Garozzo
Consiglio Nazionale delle Ricerche, Istituto per laMicroelettronica e Microsistemi, Strada Ottava 5 Zona Industriale, 95121 Catania, Italy

A. Filetti
Consiglio Nazionale delle Ricerche, Istituto per laMicroelettronica e Microsistemi, Strada Ottava 5 Zona Industriale, 95121 Catania, Italy

C. Bongiorno
Consiglio Nazionale delle Ricerche, Istituto per laMicroelettronica e Microsistemi, Strada Ottava 5 Zona Industriale, 95121 Catania, Italy

A. La Magna
Consiglio Nazionale delle Ricerche, Istituto per laMicroelettronica e Microsistemi, Strada Ottava 5 Zona Industriale, 95121 Catania, Italy

F. Simone
Dipartimento di Fisica e Astronomia, Università di Catania, Via Santa Sofia 64, 95123 Catania, Italy

R. A. Puglisi
Consiglio Nazionale delle Ricerche, Istituto per laMicroelettronica e Microsistemi, Strada Ottava 5 Zona Industriale, 95121 Catania, Italy

Álvaro Mayoral
Laboratorio de Microscopías Avanzadas, Instituto de Nanociencia de Aragón, Universidad de Zaragoza, Mariano Esquillor, Edificio I+D, 50018 Zaragoza, Spain

Javier Agúndez
Instituto de Catálisis y Petroleoquímica, Marie Curie, 2, Cantoblanco 28049, Madrid, Spain

Ignacio Miguel Pascual-Valderrama
I. E. S. Lope de Vega, San Bernardo, 70, Madrid, Spain

Joaquín Pérez-Pariente
Instituto de Catálisis y Petroleoquímica, Marie Curie, 2, Cantoblanco 28049, Madrid, Spain